Charge and Field Effects in Biosystems-3

Charge and Field Effects in Biosystems—3

Edited by:
Milton J. Allen
Stephen F. Cleary
Arthur E. Sowers
Donald D. Shillady

1992

Birkhäuser
Boston • Basel • Berlin

Milton J. Allen
Department of Chemistry and Biology
Virginia Commonwealth University
Richmond, VA 23284

Stephen F. Cleary
Department of Physiology and
    Biophysics
Virginia Commonwealth University
Richmond, VA 23284

Arthur E. Sowers
Department of Biophysics, School of
    Medicine
University of Maryland
Baltimore, MD 21201

Donald D. Shillady
Department of Chemistry
Virginia Commonwealth University
Richmond, VA 23284

Library of Congress Cataloging-in-Publication Data

Charge and field effects in biosystems-3 / edited by Milton J. Allen,
    Stephen F. Cleary, Arthur E. Sowers, Donald D. Shillady
        p.   cm. -- Based on the Third International Symposium on Charge and Field Effects
    in Biosystems, July 21-27, 1991, Virginia Commonwealth University
    Includes bibliograpical references.
    ISBN 0-8176-3564-5 (H : alk. paper).-- ISBN 3-7643-3564-5 (H : alk. paper)
    1. Bioelectrical Chemistry--Congresses. 2. Electric fields--Physiological effects--Congresses.
    3. Electric charge and distribution--Physiological effects--Congresses.   I. Allen, M.J.
    (Milton Joel), 1918-   . II. International Symposium on Charge and Field Effects in
    Biosystems (3rd : 1991 : Virginia Commonwealth University)
        [DNLM: 1. Biological Transport--congresses.   2. Cell Membrane-Physiology--congresses.
    3. Electrochemistry--congresses.   4. Energy Transfer--physiology--congresses.
    5. Ions--congresses   QD 551 C472 1991]
QP341.C42 1992
574.19'127--dc20
DNLM/DLC
for Library of Congress   91-33451
                                                                                 CIP

Printed on acid-free paper.

ISBN 0-8176-3564-5
ISBN 3-7643-3564-5

Camera-ready copy prepared by the editors.
Printed and bound by Quinn-Woodbine, Woodbine, New Jersey
Printed in U.S.A.

9 8 7 6 5 4 3 2 1

Preface

We have again brought together for the Third International Symposium on Charge and Field Effects in Biosystems (July 21–27, 1991), a group of scientists whose interests reside in the fields of bioelectrochemistry, bioenergetics, and bioelectric phenomena.

Like the previous symposia at the University of Nottingham (1983) and Virginia Commonwealth University (1989) the topics discussed were related to bioelectric phenomena, including solid state theoretical and experimental approaches to charge and energy transfer in biomolecular and cellular systems, ion and electron transport properties of biological and artifical membranes, the effects of electric fields on biological systems, photoinduced bioelectrochemical phenomena, and the applications of bioelectrochemical technology. The present conference also introduced procedures which may well serve to define the mechanisms of various bioelectrical phenomena, including electroporation for gene transfer and electro-fusion for hybridoma formation.

Favorable comments made during and after the Symposium indicated that a further conference should be held. Tentatively, plans are being considered for 1993 or 1994.

Milton J. Allen
Stephen F. Cleary
Arthur E. Sowers
Donald D. Shillady

Acknowledgments

The Editors wish to express their thanks to Rinnie O'Connor, Diane Ruff, Rae Gerber, and Jody Allen for their assistance in preparing the Symposium volume for publication. Our special thanks also to the reviewers who performed their tasks with enthusiastic promptness.

The Organizing Committe gratefully acknowledges the support and financial contributions made by Virginia Commonwealth University (College of Humanities and Sciences, Medical College of Virginia, and the Department of Chemistry), Philip Morris Research Center, and Oak Ridge Associated Universities.

Contents

**Effects of Electrochemical Processes and Electromagnetic Fields on
Biological Systems**

Photo-Induced Bioelectrochemical Processes

Applications of Bioelectrochemical Technology

Experimental Approaches to the Study of Charge and
Energy Transfer in Biomolecular and Intact Cellular Systems

FRESH APPROACHES AND NEW SURPRISES WITH IRON-SULFUR CLUSTERS

Fraser Armstrong and Julea Butt
Department of Chemistry,
University of California, Irvine, California, USA

Jacques Breton and Andrew J Thomson
School of Chemical Sciences,
University of East Anglia, Norwich, UK

Several major advances in our understanding of Fe-S clusters have occurred over the past decade. Among these are the following:

First, we are now aware that Fe-S clusters have important functions outside the realm of one-electron transfer (Beinert,1989). One particularly exciting development has been the discovery of Lewis acid catalytic activity in aconitase (Beinert, 1989; Beinert and Kennedy, 1989) and most recently, there is evidence to suggest that they may be directly involved in the genetic regulation of Fe biochemistry (Rouault et al 1991). Secondly, there is a growing awareness of structural diversity in the mode of coordination of clusters. Far from the old view of Fe-S clusters coordinated to the protein entirely through Fe-cysteine (RS^-) bonds, it is now known that they may be ligated by non-cysteine amino-acid residues, such as histidine (N) for Rieske-type centers (Gurbiel et al. 1985), or by exogenous ligands. The latter case is exemplified by aconitase in which one Fe subsite of the [4Fe-4S] core is not ligated by cysteine but by OH^- (Robbins and Stout, 1989; Werst et al., 1990). This coordinatively differentiated Fe also binds the organic substrates (Beinert and Kennedy, 1989). Thirdly, dynamic reorganization of cluster structures *within* certain proteins is now recognized. Specifically, it is now known that [3Fe-4S] and [4Fe-4S] clusters are interconvertible in a number of proteins (Moura et al., 1982; George et al., 1989; Conover et al., 1989; Beinert and Kennedy, 1989). Furthermore, it has been shown that metal ions other than Fe may be incorporated into Fe-S clusters by coordination to the [3Fe-4S] core (Moura et al., 1986; Surerus et al., 1987; Conover et al., 1990; Butt et al., 1991). As observed for the *in vitro* activation of aconitase, in which Fe^{2+} adds to a reduced $[3Fe-4S]^0$ cluster, such transformations may occur rapidly in response to a change in conditions, without significant disruption of the polypeptide chain (Robbins and Stout, 1989). Equilibrium between cluster types may serve as the basis for *sensing* Fe levels in eukaryotic cells. This suggestion stems from the recent discovery

(Rouault et al., 1991) that the sequence of aconitase is similar to that of the iron-regulated RNA-binding protein (IRE-BP) which regulates expression of ferritin and the transferrin receptor. The questions are raised: "What factors determine the ease with which such transformations can occur ? Is there a general basis for metal ion selectivity ?

Several factors make it difficult to study these reactions of Fe-S clusters directly, as they occur in the protein molecule. Reasons include: the broad, rather featureless UV-visible spectra of most Fe-S clusters and the consequential need to use low-temperature EPR, MCD or Moessbauer spectroscopy to determine if and to what extent reaction has occurred; substantial sample requirements of these techniques (which for proteins in scarce supply may be prohibitive); and the difficulty of manipulating solutions of sensitive proteins and metal ions under critical conditions of controlled low potential. With very labile clusters, any systemmatic patterns of behavior may fail to materialize.

In this article, we outline our application of voltammetry for solving some of these problems.

We have found that stable films of ferredoxins assemble spontaneously at edge-oriented pyrolytic graphite electrodes upon painting the surface with a mixture of the protein and an aminocyclitol. These reagents, examples of which include gentamycin or neomycin, are complex sugars and are commercially available as antibiotics [Rinehart and Shield, 1980]. Their co-adsorbate property derives from the arrangement of several $-NRH_2^+$ groups (R = H, alkyl) distributed across three or four rings. They are able to function as cross-linking agents, stabilizing interactions between negatively charged residues on protein molecules and oxides on the carbon surface. A cartoon illustrating an idealized configuration of the protein/co-adsorbate (Δ) film is shown in Figure 1.

FIGURE 1

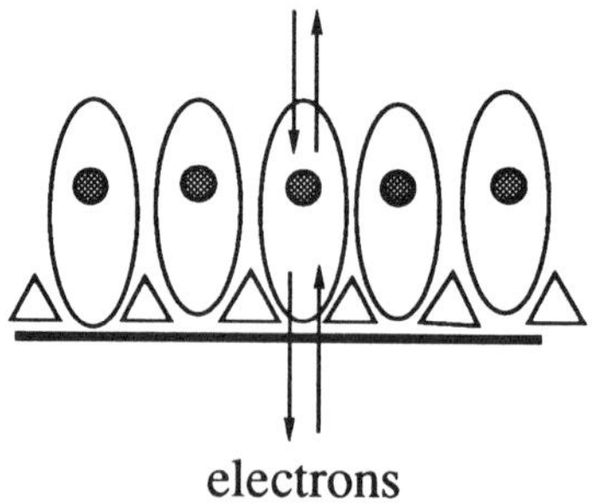

Such films, coverage equivalent to about one monolayer, remain intact throughout successive transfers of the electrode between protein-free electrolyte solutions. By varying the electrode potential, redox-state dependent activities of active sites can be initiated or terminated, and their status recorded in the time domain. Cyclic voltammetry generates signals that reflect redox reactions and coupled processes that are characteristic of the protein molecules in free solution. Unlike more traditional configurations for physical measurements of active site chemistry, surface voltammetry requires only a minuscule quantity of protein. Thus a large number of 'trailblazing' experiments can be performed with a small amount of material. Furthermore the technique permits quantitative, time-domain investigations of active-site reactions with sub-micromolar levels of ligands and metal ions in the contacting electrolyte. The discovery has been exploited to pursue and clarify some unusual, unexpected and (to-date) largely intractible, dynamic aspects of Fe-S clusters in proteins.

The main subject of our investigations has been Ferredoxin III from *Desulfovibrio africanus*. As isolated, this protein contains one [3Fe-4S] and one [4Fe-4S] cluster which are believed to occupy positions in the amino acid sequence as shown here in Figure 2 (Bovier-Lapierre et al., 1987; Armstrong et al., 1989)

FIGURE 2

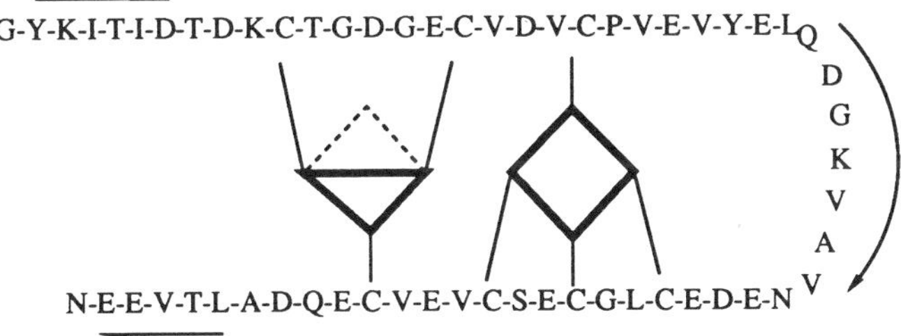

A voltammogram of a film of Fd III, formed with neomycin at a freshly polished PGE electrode and scanned in buffer (0.1 M NaCl, pH 7) is shown in Figure 3. Well defined signals are observed (Armstrong et al., 1989, Butt et al., 1991a). Couples A' and B' have been assigned to $[3Fe-4S]^{1+/0}$ (-140 mV) and $[4Fe-4S]^{2+/1+}$ (-390 mV) on the basis of the correspondence with couples observed by bulk solution voltammetry and characterisation of these states by EPR and MCD spectroscopy. However, additional reduction and re-oxidation waves (couple C') are observed at an average electrode potential of around -700 mV vs SHE. Since this value is somewhat lower than the thermodynamic limits of dithionite, the process would not be detectable by conventional chemical redox titrations. We have observed similar additional redox activity for other proteins that contain [3Fe-4S] clusters and we have noted that the low-potential couple vanishes if the [3Fe-4S] cluster is transformed by uptake of a fourth metal ion (see below). For Fd III, a comparison of the charge passed in the reoxidation wave with that passed for the $[3Fe-4S]^{1+/0}$ couple gives a stoiochiometry of 2:1, thus showing that two additional electrons can be added and removed reversibly from the protein. The reduction wave is more difficult to quantitate because it broadens and weakens as the pH is increased above pH 7. By measuring the average of peak positions for reduction and oxidation waves C', we have estimated the pH dependence of the effective reduction potential for the complex reaction. Results suggest that 2 -3 H^+ are taken up upon 2 e^- reduction, and released upon re-oxidation. Efforts are now aimed at spectroscopically characterising the highly reduced species in Fd III and other proteins.

FIGURE 3. Cyclic voltammogram of a film of 7Fe Fd III, scanned at 190 mV/sec in electrolyte composed of 0.1 M NaCl, pH 7, containing 2 mM neomycin and 10 mM EGTA.

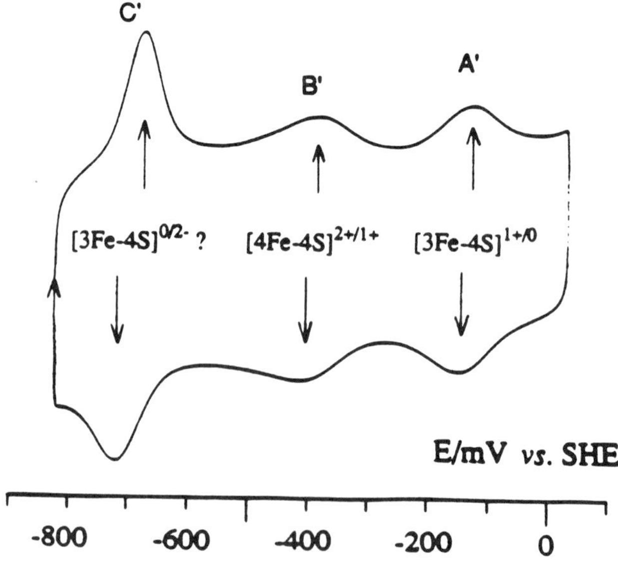

The reduced $[3Fe-4S]^0$ cluster of Fd III reacts with various metal ions M^{2+} to produce clusters of the type $[M3Fe-4S]^{2+/1+}$ (Butt et al., 1991a). Transformations are conveniently initiated and monitored with the film method. After transfer of a coated electrode (pre-scanned in buffer to remove unadsorbed protein molecules) to solutions containing the metal ions of interest, the reaction is initiated by reductive passage through couple A' (producing [3Fe-4S]^0). As shown in Figure 4, upon cycling, couples A' and C' disappear rapidly to be replaced by a new couple D'.

FIGURE 4. Cyclic voltammograms of films of 7Fe Fd III scanned in EGTA, Fe, Zn and Cd solutions (pH 7). Scan rate 190 mV/sec.

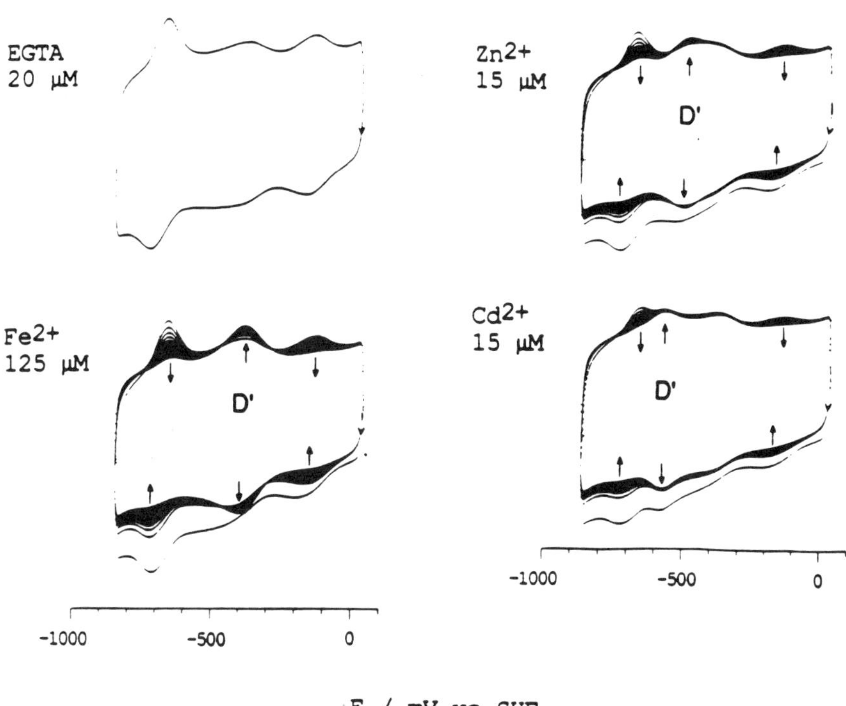

Reduction potentials of D' obtained after cycling in solutions containing Fe^{2+}, Zn^{2+}, or Cd^{2+} are very similar to values obtained by bulk-solution voltammetry on the products of addition of stoichiometric quantities of metal ion to electrochemically reduced ferredoxin solution. Examination of the latter by EPR and MCD spectroscopy shows the products to be the clusters $[4Fe-4S]^{2+/1+}$, $[Zn3Fe-4S]^{2+/1+}$ and $[Cd3Fe-4S]^{2+/1+}$.

The reversible cluster transformations and their associated equilibrium constants may be written as in Scheme 1

SCHEME 1

$$[M3Fe\text{-}4S]^{2+} \rightleftharpoons [3Fe\text{-}4S]^{0} + M^{2+}$$

$$[M3Fe\text{-}4S]^{2+} + e^{-} \rightleftharpoons [M3Fe\text{-}4S]^{1+}$$

$$K_d = \frac{\{[3Fe\text{-}4S]^{0}\}\ \{M^{2+}\}}{\{[4Fe\text{-}4S]^{2+}\}}$$

In order to measure and compare K_d values for various M, it is necessary to control the solution potential in such a way that the oxidation levels of the clusters are fixed as indicated in Scheme 1. It is also necessary to use a technique that is sensitive to changes induced by low levels of M^{2+}. With the adsorbed film technique, both of these requirements are readily dealt with. Ratios of cluster population [M3Fe-4S]/[3Fe-4S] are determined by measuring the relative attenuations of waves A' observed for rapidly scanned voltammograms upon contacting electrolytes containing M^{2+}, while the electrode potential has been poised at a value midway between the reduction potentials of the couples $[3Fe\text{-}4S]^{1+/0}$ and $[M3Fe\text{-}4S]^{2+/1+}$. Dissociation constants K_d are determined by plotting [M3Fe-4S]/[3Fe-4S] against $[M^{2+}]$. Values obtained are given in Table I.

TABLE 1

M	K_d / μM	$E^{o'}$ / mV vs SHE	
		film	bulk
Fe^{2+}	30 ± 15	-393	-400
Zn^{2+}	1.6 ± 1.0	-492	-480
Cd^{2+}	0.8 ± 0.5	-569	-580
Tl^{1+}	1.5 ± 1.0		

For $[Tl3Fe\text{-}4S]^{2+}$

Tl^{1+}	$34\,mM$	$+81$	

The order of decreasing affinity observed in these experiments i.e. Cd $\geq$ Zn $>>$ Fe is interesting in view of the apparently exclusive existence of M = Fe clusters [4Fe-4S] in Nature. Yet such a preference is indeed just as expected for a sulphur-rich site (Phillips and Williams, 1965). The non-cysteine ligand is either carboxylate (asp) or H_2O (OH^-). The transformed $[4Fe-4S]^{n+}$ cluster in Fd III is diamagnetic for the oxidized level (n = 2) and S = 3/2 for the reduced form (n = 1), the latter being a most unusual spin state for $[4Fe-4S]^{1+}$ (George et al., 1989, Conover et al., 1989). Both $[Zn3Fe-4S]^{n+}$ and $[Cd3Fe-4S]^{n+}$ clusters produced in Fd III are characterised by spin states S = 2 (n = 2) and S = 5/2 (n = 1) (Butt et al.,1991a). Reduced samples exhibit EPR spectra with g-values 4.46, 4.01 and 4.54, 4.14 respectively.

The [3Fe-4S] cluster of Fd III also reacts with Tl^+ ions (Butt et al., 1991b). The film voltammetry technique allows us to observe high-affinity interaction of Tl^+ with $[3Fe-4S]^0$ and a corresponding low-affinity interaction with $[3Fe-4S]^{1+}$. By contrast to the reactions observed with the divalent metal ions, equilibrium with Tl^+ is established very rapidly compared to the rate of voltammetric scanning. As the concentration of Tl^+ in the contacting electrolyte is increased, the apparent reduction potential of couple A' becomes more positive. Two K_d values are determined from a plot of observed reduction potential against log $[Tl^+]$. Substituting Tl^+ by Rb^+ or K^+ results in no change in the voltammetry. Formation of [Tl3Fe-4S] clusters is supported by the observation that addition of Tl^+ (final concentration 0.13 M) to a sample of oxidized Fd III results in replacement of the characteristic g = 2.01 signal by a rhombic spectrum having g-values 2.04, 1.99 and 1.95.

We have thus shown how a very economical voltammetric technique permits the study of reactions occurring at active sites of proteins, in this case a series of rapid transformation processes occurring at a [3Fe-4S] cluster due to interaction with trace metals in the electrolyte. Voltammetry provides a tool for surveying and analyzing these reactions and guides the investigator in generating bulk samples for spectroscopic characterization. Among the observations described here for ferredoxin III, the relatively high affinity of a $[3Fe-4S]^0$ cluster for metals other than Fe is interesting. Cadmium and Thallium are known poisons. By contrast, Zn is a biologically essential metal having an abundance in organisms that is comparable to that of Fe . It will be of interest to seek other examples of Zn incorporation in Fe-S clusters and determine whether species such as [Zn3Fe-4S] may exist naturally. Another observation is the ease and rapidity with which these metal ions enter the $[3Fe-4S]^0$ core. Further equilibrium, mechanistic and spectroscopic studies of these reactions are in progress.

ACKNOWLEDGEMENTS

This research was supported by the University of California and by an Exxon Education Foundation Award (to FAA), by a grant from the Molecular Recognition Initiative of the SERC (to AJT) and by a NATO Collaborative Research Grant (CRG 900302).

REFERENCES

Armstrong FA, George SJ, Cammack R, Hatchikian EC and Thomson AJ (1989a): Electrochemical and spectroscopic characterization of the 7Fe form of ferredoxin III from *Desulfovibrio africanus*. *Biochemical Journal* 264: 265 - 273.

Armstrong FA, Butt JN, George SJ, Hatchikian EC and Thomson AJ (1989b): Evidence for reversible multiple redox transformations of [3Fe-4S] clusters. *FEBS Letters* 259: 15-18.

Beinert H (1989): Recent developments in the field of iron-sulfur proteins. *FASEB J* 4: 2483 - 2491.

Beinert H and Kennedy MC (1989): Engineering of protein-bound iron-sulfur clusters. A tool for the study of protein and cluster chemistry and mechanism of iron-sulfur enzymes. *Eur. J. Biochem* 186: 5 - 15.

Bovier-Lapierre G, Bruschi M, Bonicel J and Hatchikian EC (1987): Amino-acid sequence of *Desulfovibrio africanus* ferredoxin III: a unique structural feature for accommodating iron-sulfur clusters. *Biochim. Biophys. Acta* 913: 20 - 26.

Butt JN, Armstrong FA, Breton J, George SJ, Thomson AJ and Hatchikian EC (1991a): Investigation of metal-ion uptake reactivites of [3Fe-4S] clusters in proteins: voltammetry of co-adsorbed ferredoxin-aminocyclitol films at graphite electrodes and spectroscopic identification of transformed clusters. *J. Amer. Chem. Soc* 113: 6663 - 6670.

Butt JN, Sucheta A, Armstrong FA, Breton J, Thomson AJ and Hatchikian EC (1991b): Binding of thallium (I) to a [3Fe-4S] cluster: Evidence for rapid and reversible formation of $[Tl3Fe-4S]^{2+}$ and $[Tl3Fe-4S]^{1+}$ centers in a protein. *J. Amer. Chem. Soc* 113: in the press.

Conover RC, Kowal AT, Fu W, Park J-B, Aono S, Adams MWW and Johnson MK (1990): Spectroscopic characterization of the novel iron-sulfur cluster in *Pyrococcus furiosus* ferredoxin. *J. Biol. Chem.* 265: 8533 - 8541.

George SJ, Armstrong FA, Hatchikian EC and Thomson AJ (1989): Electrochemical and spectroscopic characterization of the conversion of the 7Fe into the 8Fe form of ferredoxin III from *Desulfovibrio africanus*. *Biochemical Journal* 264: 275 - 284.

Gurbiel, RJ, Batie CJ, Sivaraja M, True AE, Fee JA, Hoffman BM and Ballou DP (1989): Electron-nuclear double resonance spectroscopy of ^{15}N-enriched phthalate dioxygenase from *Pseudomonas cepacia* proves that two histidines are coordinated to the [2Fe-2S] Rieske-type clusters. *Biochemistry* 28: 4861 - 4871.

Moura JJG, Moura I, Kent TA, Lipscomb JD, Huynh B-H, LeGall J, Xavier AV and Muenck E (1982): Interconversions of [3Fe-3S] and [4Fe-4S] clusters: Moessbauer and electron paramagnetic resonance studies of *Desulfovibrio gigas* ferredoxin II. *J. Biol. Chem* 257: 6259 - 6267.

Moura I, Moura JJG, Muenck E, Papaefthymiou V and LeGall J (1986): Evidence for the formation of a $CoFe_3S_4$ cluster in *Desulfovibrio gigas* ferredoxin II. *J. Amer. Chem. Soc* 108: 349 - 351.

Rouault TA, Stout CD, Kaptain S, Harford JB and Klausner RD (1991): Structural relationship between an iron-regulated RNA-binding protein (IRE-BP) and aconitase: functional implications. *Cell* 64: 881 - 883.

Rinehart KL and Shield LS (1980): Aminocyclitol antibiotics: an introduction. In: *Aminocyclitol Antibiotics*, Rinehart KL and Suami T, eds. ACS Symposium Ser. No. 125.

Robbins AH and Stout CD (1989): Structure of activated aconitase: Formation of the [4Fe-4S] cluster in the crystal. *Proc. Natl. Acad. Sci. USA* 86: 3639 - 3643.

Surerus KK, Muenck E, Moura I, Moura JJG and LeGall J (1987): Evidence for the formation of a $ZnFe_3S_4$ cluster in *Desulfovibrio africanus* ferredoxin II. *J. Amer. Chem. Soc* 109: 3805 - 3807.

Werst MM, Kennedy MC, Beinert H and Hoffman BM (1990): ^{17}O, ^{1}H, and ^{2}H electron nuclear double resonance characterization of solvent, substrate, and inhibitor binding to the $[4Fe-4S]^{+}$ cluster of aconitase. *Biochemistry* 29: 10527 - 10532.

ELECTROCHEMICAL CONTROL OF PROTEIN INTERACTIONS WITH SOLID SURFACES

Alexander N. Asanov
Institute of Chemical Physics
of the USSR Academy of Sciences, Moscow, 117977

Ludmila L. Larina
Institute of Chemical Physics
of the USSR Academy of Sciences

INTRODUCTION

The present chapter examines the effect of electro-chemical polarization of the SnO_2 surface on the albumin molecules behaviour at the solid/liquid interface.

The mechanism of protein interactions with surfaces of artificial materials as well as the ability to control these processes are of great interest for such areas as molecular electronics, biomimetics, biocompatible materials, biosensors, drug release systems, immunoassays, chromatography, etc. A variety of reactions in these systems involving cells, enzymes and others macromolecules occurs on the surface covered by irreversibly adsorbed proteins (Andrade, 1985; Norde et al., 1986). The response of the biological environment to the contact with the surface depends on the nature of adsorbed proteins, their surface concentrations, strength of adsorption, conformational changes or denaturation, and on the orientation of adsorbed molecules. There are evidences that protein adsorption and such response of the blood like thrombosis are strongly affected by electrochemical properties of the system (Norde et al., 1986; Sawyer et al., 1970). However, the electrochemical aspects of these processes as well as the

role of protein adsorption in the surface processes are
not well understood. The purpose of this work is to study
the effect of electrochemical polarization on the protein
interactions with the surface.

EXPERIMENTAL

A new spectroelectrochemical technique based on a
combination of the total internal reflection fluorescence
(TIRF) spectroscopy and the electrochemical system, allowing external control of the potential of the surface under
the study, was developed in this work.

TIRF technique uses an evanescently decaying wave of
the light to excite the fluorescence of molecules adsorbed
at the interface. In the present study this is SnO_2/water
interface. The intensity of evanescent wave in the liquid
phase (see inset in Fig. 1A) is characterized by equations:

$$E = E_0 \exp(-z/d_p) \qquad (1)$$

$$d_p = (\lambda_0/2\pi)(n_1^2 \sin^2\theta - n_2^2)^{-1/2} \qquad (2)$$

Here E_0 and E are the intensities of electric field of the
light at the interface (E_0) and at the depth z from the surface; d_p is penetration depth constant which depends on
refractive indexes n_1 and n_2, on the angle θ of the light
beam at the interface, and on λ_0 wavelength of light in
vacuum. In most practical cases the value of d_p is in the
range 100 - 300 nm.

The principles of TIRF spectroscopy have been presented elsewhere in details (Harrick, 1967; Beissinger and
Leonard, 1980; Hlady et al., 1985), and therefore we describe only special features of our experiment. Figure 1
shows the optical scheme of the TIRF flow cell (Fig. 1A)
and the electrochemical setup of the experiment (Fig. 1B).
The same TIRF electrochemical cell was used in two sets of
the experiment. Protein adsorption kinetics was registrated using Shimadzu RF-5000 fluorescence spectrophotometer.
TIRF cell on a special support was inserted into cuvette

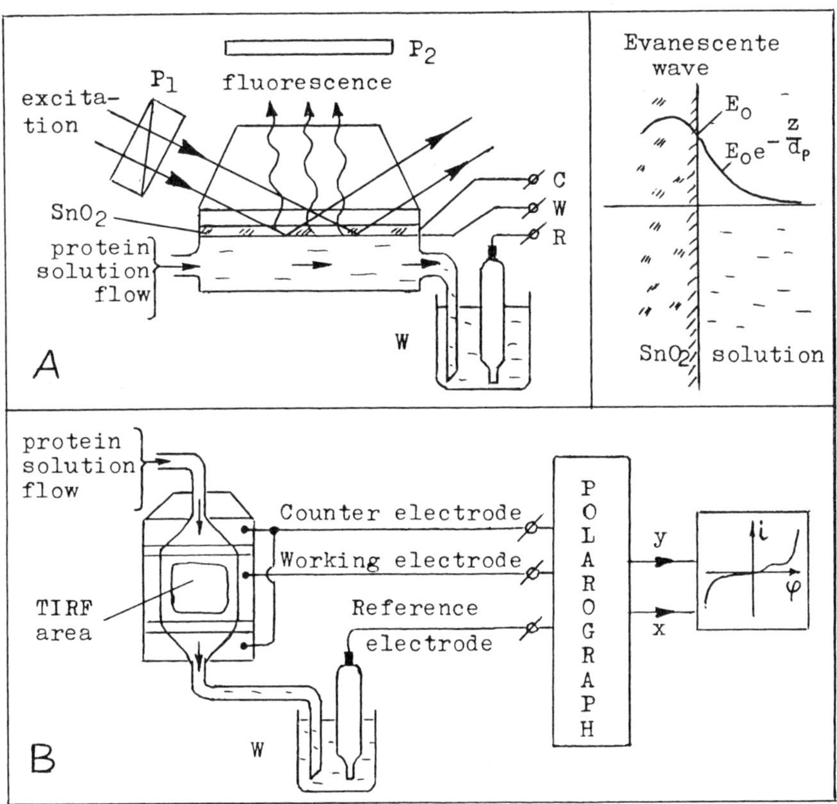

FIGURE 1. Optical scheme of TIRF flow cell (A) and
electrochemical setup of the experiment (B).

compartment of the RF-5000 spectrophotometer. Structural
changes of adsorbed protein were studied by the time-resolved fluorescence anisotropy. Special TIRF cell housing,
pulse laser excitation, and fluorescence registration by
fast photomultiplier and oscilloscope with memory were
used in this set of the experiment.

Absolute surface concentration of adsorbed protein
was not measured in this work. The amount of adsorbed protein was registrated in arbitrary units as the intensity
of fluorescence under indentical experimental conditions.
Experimental data and the results of calculations show
that the input into the fluorescence signal from the pro-

tein in the bulk of solution is negligible at protein concentrations lower than 1 mg/ml.

The system of tubing, stopcocks, and reservoirs supplies the TIRF cell with protein and pure buffer solutions. Due to small thickness of the flow compartment of TIRF cell (thickness b=0.015 cm, length l=3.0 cm, and width w = =1 cm) high values of wall shear rates γ:

$$\gamma = 6 \ V/w \ b^2 \ \cong \ 4.4 \cdot 10^3 \ sec^{-1} \tag{3}$$

is attained at reasonable protein solution flow rate V = $\pm$ 10 ml/min. According to (Beissinger and Leonard, 1980), the true kinetics of protein adsorption and desorption is observed in the time t after the flow was switched on:

$$t \cong 3 \ l^{2/3} \ \gamma^{-2/3} \ D^{-1/3} \ \cong 3 \ sec \tag{4}$$

In our case of human serum albumin adsorption, which diffusion coefficient $D = 6.1 \cdot 10^{-7} cm^2 sec^{-1}$, this lag time is about of three seconds.

Structural changes of adsorbed protein was studied in the present work by the technique of fluorescence anisotropy decay. The technique has been described elsewhere (Dobretsov, 1989). Recently, Fukumura and Hayashi (1990) applied the technique for TIRF studies of dynamic behaviour of adsorbed protein. Our experimental setup differs from that described by (Fukumura and Hayashi, 1990) in TIRF cell design, electrochemical scheme, TIRF cell housing, and in parameters of excitation pulses.

Nitrogen laser LGI-21 (USSR) was used to excite the fluorescence of pyrene sulfonyl label covalently bonded to albumin. The parameters of laser excitation were the following: wavelength 337 nm, pulse width ~5 ns, pulse energy 0.75 mJ, repetition rate 20 Hz. Exciting light passes through the Glan prism polariser P_1 (see Fig. 1A). The emmited fluorescence was measured by subnanosecond photomultiplier though film polarizer (P_2) and a combination of absorbing and interference filters with transmitance maximum at 400 nm, near the second peak (396 nm) of pyrene sulfonyl fluorescence.

SnO_2 film deposited on the quartz support was chosen as electrochemically polarizible surface because of the following properties: low attenuation of exciting light and emmiting fluorescence, high chemical stability, high overpotentials for water decomposition reactions, mechanical stability, and low electrical resistivity. SnO_2 film was sputtered on the quartz support from $SnCl_4$ vapour at the support temperature near $350^{\circ}C$. Films obtained were 0.5-0.6 μm thickness. At room temperature SnO_2 is a degenerate n-type semiconductor with conductivity 10^2 $Ohm^{-1} \cdot cm^{-1}$. The film on the support was divided by etching into three bands (see Fig. 1B). The band in the middle of the support (~ 1 cm width) was used as working electrode, while two side bands were connected with each other and used as the counter electrode. Two cables enter the cuvette compartment or TIRF cell housing giving connections with the working surface and the surface of counter electrode. The saturated calomel electrode (SCE) was used as reference electrode. This electrode dipped in the waste reservoir (W), which was connected with TIRF cell by 10 cm tubing. The electrochemical scheme was served by the polarograph PU-1 (USSR) with 3-electrode potentiostatic scheme and X-Y plotter.

Phosphate buffer saline (PBS) pH 7.35 was used as a solvent for protein solutions. PBS is consisted of 0.1 **M** NaCl, 0.086 M KH_2PO_4, and 0.041 M Na_2HPO_4. Human serum albumin (HSA) and fluorescein isothiocyanate (FITC) were used as recieved from Serva Chemical Company, and pyrene sulfonyl chloride (PyS) from Molecular Probes Company. FITC was covalently bonded to HSA following the procedure of Goldman (1968), and PyS according to the method described by Fukumura and Hayashi (1990). Dyes that did not react were removed by gel filtration on Sefadex G-15 previously equilibrated with PBS. The molar ratios of the dyes reacted with protein were 0.90 for HSA-FITC conjugate and 0.48 for HSA-PyS. FITC label was used in kinetic measurements, while PyS in anisotropy decay studies because of longer lifetime of PyS. The experiments were carried out at room temperature $21\pm1^{\circ}C$.

RESULTS AND DISCUSSION

Figure 2 shows the current-potential curve of the SnO_2 working electrode obtained when PBS pure solution flowed through the TIRF cell. Wave of the current starting at -0.4 V is decreased but not canceled by O_2 removal from PBS flow. After long cathodic polarization at -0.6÷ -0.8 Volt, the SnO_2 film becomes grey, obviously due to the oxide reduction to metallic Sn. One could conclude, that the superposition of two electrochemical reactions is responsible for this cathodic wave: i) oxygen reduction and ii) SnO_2 reduction to metallic Sn.

Sharp current rises at >1.3 V and at <-0.8 Volt are related to water molecules decomposition reactions: oxidation to O_2 and reduction to H_2, respectively. At these potentials trivial tearing off the adsorbed protein by evolving gases is observed. Because of severe changes of chemical and optical properties of the SnO_2 in the range -0.6 ÷ -0.8 V, these potentials were also avoided in this work.

In the range from +0.8 to -0.4 Volt, however, SnO_2 is quite ideal polarizible electrode. No net electrochemical reaction with charge transfer across the SnO_2/water interface is observed under our experimental conditions. Addition of HSA, HSA-FITC or HSA-PyS up to 10 mg/ml results in a decrease of the capacitance current in the range +0.8 ÷ -0.4 Volt. This fact means, that no electrochemical reaction with charge transfer to HSA or HSA-dye conjugates is observed in this potential range. The decrease of capacitance current could be related to a decrease of the capacitance of the double electric layer due to the protein adsorption at the SnO_2/water interface.

Although HSA is not involved in reactions with charge transfer across SnO_2/water interface, the behaviour of the protein was found to be strongly affected by the polarization of SnO_2 in the range +0.8 ÷ -0.4 Volt. The rate of HSA adsorption onto fresh SnO_2 surface, and the equilibrium surface concentration attained at the same HSA concentration in the solution, depend on the potential of SnO_2.

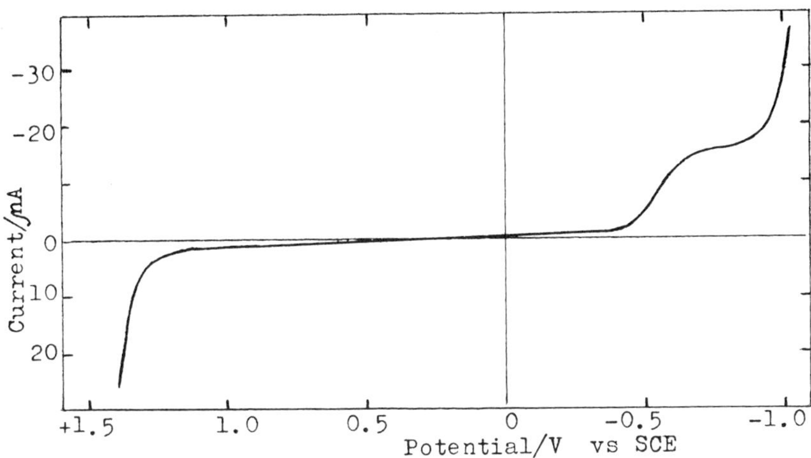

FIGURE 2. Current-potential curve of SnO_2 working electrode in the TIRF cell. PBS flow 10 ml/min, scanning rate 50 mV/min.

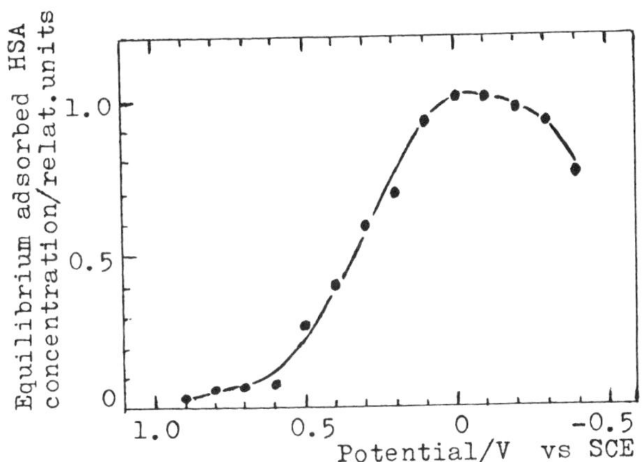

FIGURE 3. Equilibrium surface concentration of HSA adsorbed on SnO_2 from 0.03 mg/ml protein solution as a function of the SnO_2 potential.

Maximum adsorption is observed in the range 0.0 ÷ -0.3 V (see Fig. 3). Initial rate of adsorption onto fresh SnO_2 surface has the maximum value in the same range of potentials. Minimum adsorption and maximum values of desorption rates after fast changes of the potential are observed in the range +0.8 ÷ +0.5 V, as shown in Fig. 3 and can be seen from Figure 4.

Kinetics and isotherms of HSA adsorption at fixed SnO_2 potentials demonstrate features which are typical for energetically heterogeneous protein/surface interactions. The model describing protein adsorption was proposed earlier (Asanov et al., 1987). Isotherm and kinetics of adsorption in the case of wide rectangular distributions of activation energies of adsorption and energies of protein/ /surface binding are described by equations:

$$\Gamma(C) = K_1 RT \ln(b_{max}C) \qquad (5)$$

$$\Gamma(t) = K_2 RT \ln(k_{max}Ct) \qquad (6)$$

Here Γ is the surface concentration of adsorbed protein at equilibrium - $\Gamma(C)$ and in time t after the begining of the adsorption onto fresh surface - $\Gamma(t)$; C is the protein concentration in the bulk solution; K_1 and K_2 are constants; R - gas constant, T - temperature; b_{max} is maximum adsorption coefficient for the protein/surface system; k_{max} - maximum rate constant of adsorption; and t - time. The kinetic curves of HSA adsorption onto fresh SnO_2 surface, which are shown in Fig. 4, are well described by equation (6) with $k_{max} = 2.1$ ml/mg sec at SnO_2 potential -0.1 V, and $k_{max} = 0.11$ ml/mg sec - at +0.5 Volt.

After the surface concentration of HSA attains equilibrium value, the change of protein solution flow to the flow of pure PBS solution causes the desorption of only a limited amount of the adsorbed protein. About 70 percent of the protein molecules remain on the surface even after 5 hours of washing with PBS. This is so-called irreversibly adsorbed protein. Bohnert and Horbett (1986) provided evidence that the strength of this protein binding with the surface increases with time.

The system HSA/SnO_2 also demonstrates the tendency to the strengthening of the binding. The rate of the desorption and the amount of HSA desorbed in the PBS solution

were found to decrease with the time of the system being
in the equilibrium state.

The rate of desorption was found to be dramatically
depended on the manner of SnO_2 potential changes. If the
potential is switched from -0.1 V to +0.5 ⟶ +0.8 Volt,
the system responses to this polarization shock by fast
desorption of the major part of the protein (see Figure
4). Further potential increase by switching from +0.5 to
+0.8 ⟶ +1.0 Volt causes fast desorption of the remained
protein. The surface becomes practically free of the protein after shock polarization from -0.1 to +0.8 ⟶ +1.0 V.

However, slow scanning of the potential in the same
voltage range from -0.1 to +0.5 V results in desorption
of only about 20 percent of adsorbed HSA molecules. The
delay of giving the polarization shock has little effect
on the fast desorption. After the delay, the duration of
which is equal to the time of scanning, the total amount
of desorbed protein is less only for ≈ 0.06 of that after
the shock without delay; the difference of initial rates
of desorption is negligible. Thus, the above mentioned
strengthening of protein binding with the surface could
be responsible only for a small part of the difference
observed for fast and slow changes of potential.

One could supposes that the observed slow desorption
under slow potential scanning is accounted for by a transition of protein molecules into a new adsorbed state,
suitable for changed electrostatic situation   at the
interface. Indeed, lateral diffusion, rotation on the surface, conformational changes, and denaturation are known
to be important features of protein behaviour at polymeric and inorganic surfaces (Bohnert and Horbett, 1986;
Burghardt and Axelrod, 1981).

In order to study the changes in the state of adsorbed protein, fluorescence anisotropy decay measurements
were carried out with HSA-PyS conjugate at different SnO_2
potentials. Two sets of measurements were made. i) At
-0.1 V after the equilibrium adsorption state of HSA/SnO_2
was attained. ii) At +0.5 V after potential scanning of
SnO_2 electrode with adsorbed protein from -0.1 to +0.5 V

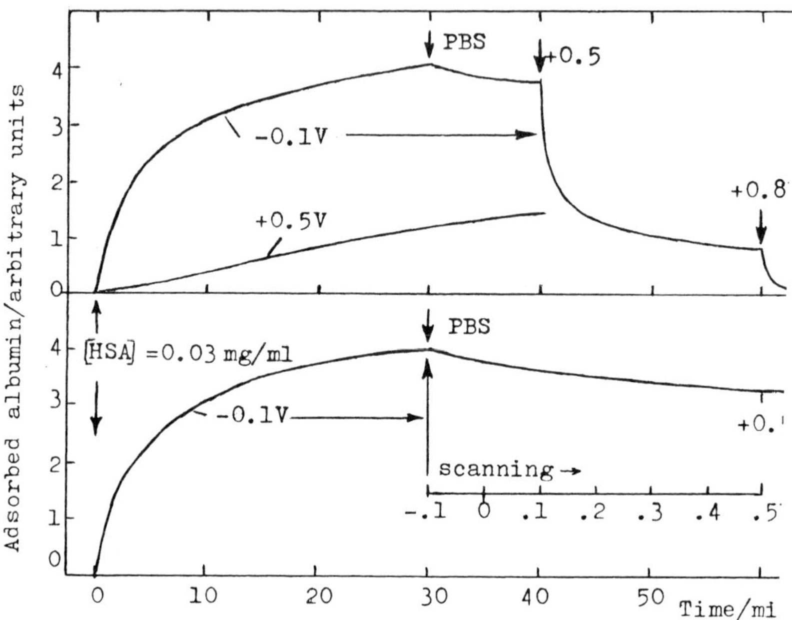

FIGURE 4.   Kinetics of HSA adsorption from 0.03 mg/ml
HSA solution flow (10 ml/min) and kinetics of desorptic
of HSA irreversibly adsorbed at −0.1 V after switching
the potential from −0.1 to +0.5 and from +0.5 to +0.8 ·
(upper curve). The effect of slow scanning (20 mv/min)
in the voltage range from −0.1 to +0.5 Volt on the protein desorption (lower curve). Pure PBS flow 10 ml/min.

with scanning rate 20 mV/min. The decay of fluorescence
anisotropy was found to have different time characteristics in these two cases. Interpretation of these data is,
however, ambigous.

If the conjugate is a solid spherical protein molecule with rigidly attached fluorescent label, and if the
pool of adsorbed molecules is homogeneous, the anisotropy
decay A(t) is described by equation (Dobretsov, 1989):

$$A(t) = (I_\| - I_\perp)/(I_\| + 2I_\perp) = A_0 exp(-t/\tau) \qquad (7)$$

Here $I_\|$ and $I_\perp$ are the fluorescence intensity components
parallel and perpendicular to the plane of excitation
light polarization; A_0 - initial anisotropy; τ - rotational correlation time. In this simple case the fluorescence anisotropy decay is independed of fluorescence lifetime. The rotational correlation time has the following
simple meaning:

$$\tau = Q\eta/kT \qquad (8)$$

where Q is the rotation sphere, η - the viscosity of the
medium, k - Boltzman's factor, and T - temperature.

In our case of the energetically heterogeneous system
the interpretation of the anisotropy decay is more complicated. The fluorescence intensities as well as A(t) show
decay kinetics that could not be approximated by one or
two exponents. The best physical value, that can be used to
describe the decay process of this type, is half-lifetime,
the time in which the signal decays from the initial level
to a one half of that. The value of half-lifetime of A(t)
represents a weighted mean correlation times for the rotations of HSA-PyS molecules as the whole, internal rotations of the fluorophore, domain motions in the adsorbed
molecules, and so on.

The half-lifetime value of fluorescence anisotropy
decay measured at the SnO_2 potential -0.1 V is approximately two times greater than that at +0.5 Volt:

$$\tau_{\frac{1}{2}}(-0.1\ V) = 74\pm8\ nsec; \quad \tau_{\frac{1}{2}}(+0.5\ V) = 39\pm5\ nsec$$

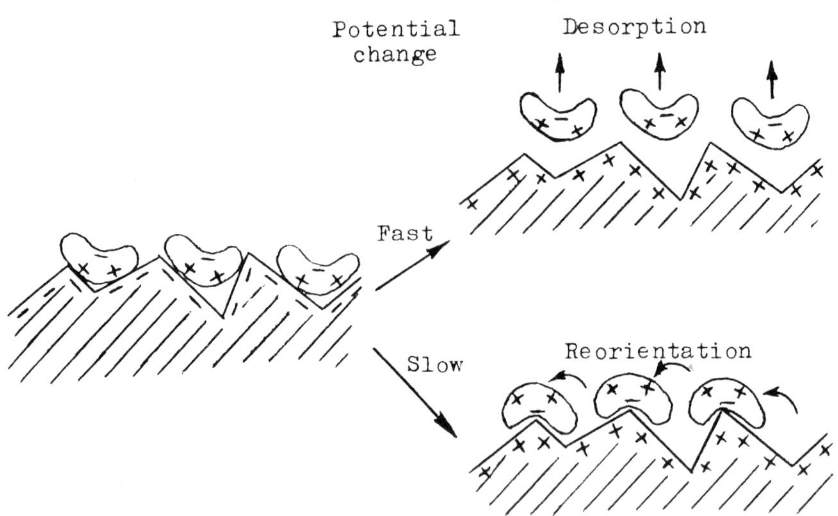

FIGURE 5.   Schematic illustration of the protein interactions with energetically heterogeneous surface at different surface charge. Desorption under the fast switching of the potential, and reorientation of adsorbed molecules under the slow potential scanning.

Many changes in the microenvironment of the fluores-
cent label may be responsible for the changes in $\tau_{\frac{1}{2}}$ values. The most probable changes during slow potential
scanning seems to be the reorientation of adsorbed molecules, as shows the Figure 5. For the sake of illustration, energetical heterogeneity of the surface is shown
in the scheme by geometrical in-homogeneity. In fact,
most of real surfaces have mechanical inhomogenieties the
dimension of which is compared with the size of globular
proteins (~10 nm).

Molecules of globular proteins have on its "surfaces"
positively and negatively charged partches (Andrade,
1985). The proposed model assumes that the preferable
orientation of the protein molecule in the event of adsorption is that positively charged partches are faced to the
negatively charged surface, and vice versa. When one suddenly changes the electric charge of the surface, attractive electrostatic forces change for repulsive ones. Mo-

lecules quickly desorb from the surface. However, in the
case of slow changes of potential, protein molecules have
time to reorient into suitable position and remain in
adsorbed state.

It should be emphasized that the present considera-
ton of the protein interaction with solid surface is based on many simplifications. Considering the reorientation one have to take into account that protein molecules
in the dense and diffuse parts of the double electric
layer (DEL) are in high gradients of potential, pH value,
ionic strenth, ion concentrations, etc. These aspects
should be studied in future works.

CONCLUSIONS

A new spectroelectrochemical technique based on a
combination of TIRF spectroscopy and electrochemical system, allowed external control of the potential of transparent working electrode, was used to study the effect of
surface charge on protein behaviour at the interface.
Amount of adsorbed protein and the rates of adsorption
and desorption were found strongly affected by the electrochemical polarization of the surface.

Desorption of the protein irreversibly adsorbed at
negative potential drastically enhances under fast changing of the potential from negative to positive. Slow
changes in the same potential range results in only very
limited desorption. The proposed model is based on assumption that under slow potential changes protein molecules
have time to reorient and remain in adsorbed state, while
under fast potential change molecules have no time for
the transition into a new adsorbed state.

The effect of electrochemical control could give new
insight into mechanisms of protein interactions with solid
surfaces as well as with biological and artificial membranes. The spectroelectrochemical technique gives new instrument to study these phenomena. The effect of electrochemical control could be used for developing new biocompatible materials, reusable biosensors, immunoassays, bio-

mimetic and molecular electronic systems.

ACKNOWLEDGEMENTS

The authors thank Professor Rafail F. Khairutdinov
for stimulating discussions and revision of the manuscript.

REFERENCES

Andrade JD (1985): Principles of protein adsorption. In:
   Surface and interfacial aspects of biomedical polymers.
   Andrade JD ed. New York: Plenum Press. Vol.2, p.3.

Asanov AN, Kulik EA, and Sevastianov VI (1987): Energetical heterogeneity of solid surfaces in interactions
   with protein molecules. Doklady Akad.Nauk USSR; 296:
   735-740.

Beissinger RL, and Leonard EF (1980): Immunoglobulin sorp-
   tion and desorption rates on quartz: Evidence for multiple sorbed states: ASAIO Journal 3: 160-175

Bohnert JL, and Horbett TA (1986): Changes in adsorbed
   fibrinogen and albumin interactions with polymers
   indicated by decrease in detergent eluability. J.Col-
   loid Interface Sci.111: 363-377

Burghardt TP, and Axelrod D (1981): Total internal reflec-
   tion/fluorescence photobleaching recovery study of serum albumin adsorption dynamics.Biophys.J.:33:455-467.

Fukumura H, and Hayashi K (1990): Time-resolved fluorescence anisotropy of labeled plasma proteins adsorbed
   on polymer surfaces. J.Colloid Interface Sci.135:435-
   443.

Goldman M (1968): Fluorescente antibody techniques. New
   York: Academic press;

Dobretsov GE (1989): Fluorescent probes in studies of
   cells, membranes and lipoproteins. Moscow: Nauka.

Harrik NJ (1967): Internal reflection spectroscopy.
   New York: Interscience Publishers.

Hlady V, Van Wagenen RA, and Andrade JD (1985): Total
   internal reflection intrinsic fluorescence (TIRIF)

spectroscopy applied to protein adsorption. In: <u>Surface and interfacial aspects of biomedical polymers</u>. Andrade JD editor. New York: Plenum Press. Vol. 2, p. 81.

Sawyer PN, Srinivasan S, Chopra PS, Martin JG, Lucas CB, Borrowes T, and Sauvage L. Electrochemistry of thrombosis. An aid to selection of prosthetic materials. <u>J.Biomed.Mater.Res.</u>4: 43-55.

THE DIRECT ELECTRON TRANSFER REACTIONS OF CYTOCHROME OXIDASE IMMOBILIZED INTO A MEMBRANE MODIFIED ELECTRODE

John K. Cullison and Fred M. Hawkridge
Department of Chemistry
Virginia Commonwealth University
Richmond, Virginia 23284 USA

Naotoshi Nakashima
Department of Industrial Chemistry
Faculty of Engineering
Nagasaki University
Nagasaki 852, JAPAN

Charles R. Hartzell
Alfred I. duPont Institute
P. O. Box 269
Wilmington, Delaware 19899 USA

INTRODUCTION

There have recently been a large number of papers in the literature dedicated to examining monolayers and films assembled onto solid substrate surfaces (e.g., Swalen et al.,1987; Chidsey and Loiacono, 1990). Some of this research has been directed toward a biochemical approach (Hafeman et al., 1981; Ishiguro and Nakanishi, 1984; Brian and McConnell, 1984; Fabianowski et al., 1989; Prime and Whitesides, 1991; Tarlov and Bowden, 1991). In this paper we describe a procedure for preparing bilayers on electrode surfaces containing the enzyme cytochrome c oxidase. This procedure is based on previous reports for preparing vesicle resident enzyme preparations (Casey, 1984). Electrochemical studies on these modified electrode surfaces indicate that direct electron transfer between bilayer resident cytochrome c oxidase and the electrode is occurring. Experimental results are also shown that suggest that heterogeneous catalytic communication between cytochrome c oxidase on the electrode is occurring with its solution resident biological redox partner, reduced cytochrome c. The homogeneous electron transfer reaction between cytochrome c and cytochrome oxidase had been studied earlier (Long et al., 1988; Long et al., 1986). The purpose of the research described here is to model the heterogeneous electron transfer reaction that occurs between cytochrome c and cytochrome oxidase in vivo.

Cytochrome oxidase is the terminal enzyme of the oxidative phosphorylation chain. It accepts electrons from cytochrome c at the outer surface of the inner membrane of the mitochondria and catalyzes the four electron reduction of dioxygen to water on the inner surface (Palmer, 1987).

$$4 \text{ cyt.c}(2+) + 4H^{+} + O_2 \rightleftharpoons 4 \text{ cyt.c}(3+) + 2H_2O$$

There is disagreement in the literature concerning the molecular weight of cytochrome oxidase and the number of subunits it contains. This controversy is a result of the confusion of how the enzyme is defined. At issue is whether the enzyme should be defined as the minimum number of subunits that co-purify with the enzyme and still retain the ability to catalytically reduce dioxygen to water. Alternatively, should it be defined as the minimum number of subunits that still allow the enzyme to reduce oxygen and pump protons. Adding to this confusion is the fact that different purification procedures give forms of the enzyme that contain different numbers of subunits. Molecular weight estimates between 160,000 and 200,000 daltons have been cited (Capaldi et al., 1983; Azzi, 1980; Yoshikawa et al., 1988).

Cytochrome oxidase contains four redox active sites. Two of the redox sites, Cu_b and cytochrome a_3 form the binuclear center where dioxygen binds to the enzyme. The Cu_b and cytochrome a_3 sites are located on the matrix side of the inner membrane of the mitochondria. The other two sites, Cu_a and cytochrome a. are positioned toward the cytosolic side of the inner membrane. Cytochrome a is believed to be the initial acceptor of electrons from reduced cytochrome c (Palmer, 1987). The formal potentials of these four redox sites have been found to be cytochrome a = 215 mV, cytochrome a_3 = 350 mV (Schroedl and Hartzell, 1977), Cu_b = 350 mV, and Cu_a = 190 mV (Anderson et al., 1976). The bovine enzyme also contains one zinc and one magnesium atom per oxidase molecule (Einarsdottir and Caughey, 1985; Einarsdottir and Caughey, 1984) and a third copper atom that is not redox active (Steffens et al., 1987; Yewey and Caughey, 1988).

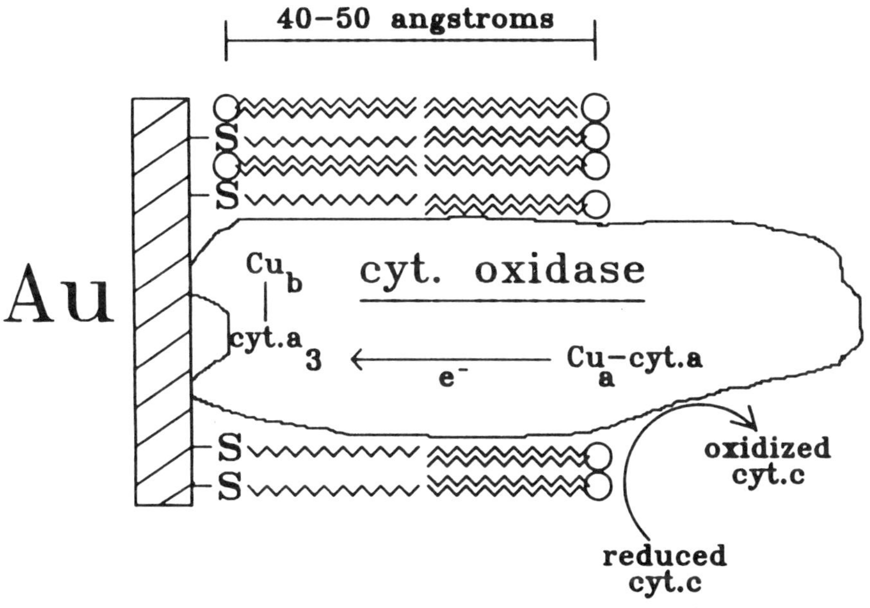

Fig.1 Model of Enzyme Modified Electrode

From electron microscopy studies

cytochrome oxidase has been shown to extend approximately 55 angstroms into the cytosolic side of the inner membrane of the mitochondria and no more than 20 angstroms into the matrix side of the membrane (Deatherage et al., 1982).

The initial step used in constructing these enzymes immobilized in bilayers on electrodes involves spontaneously adsorbing sub-monolayer levels of octadecyl mercaptan onto gold surfaces. This degree of surface coverage is intended to allow the oxidase to partition into the layer. The spontaneous assembly of thiols onto gold has received considerable attention (Swalen et al., 1987; Nuzzo and Allara, 1983; Porter et al., 1987). Octadecyl mercaptan forms very stable monolayers on gold and it is now believed that such alkane thiols are chemisorbed to the surface and that the chemisorbed species is a thiolate (Nuzzo et al., 1987; Bain et al., 1989).

Following the octadecyl mercaptan modification a cholate dialysis is performed. This procedure is similar to cholate dialysis procedures that have been used to incorporate cytochrome oxidase into vesicles in a unidirectional orientation (Casey, 1984; Hinkle et al., 1972; Zhang, 1984; Zhang, 1985). Briefly, cholate prevents amphiphiles from forming vesicles. As the cholate is removed from such solutions via dialysis, vesicles form and the enzyme incorporates unidirectionally into the vesicle bilayers (Casey, 1984). The aim here is to use these ideas to prepare bilayers on gold electrodes with cytochrome oxidase partitioned into these structures unidirectionally. Prior covalent modification of the gold electrode surface is intended to result in a more stable bilayer.

Figure 1 shows a cartoon model of the enzyme immobilized membrane that is described above. The surface of this electrode is more complex this cartoon. The surface of the electrode is not atomically smooth and the amphiphile orientation and packing density is not known in this system. However, this is a model that depicts the desired orientation of the cytochrome oxidase and the approximate dimensions expected for this modified electrode surface.

MATERIALS AND METHODS

In the initial experiments a gold foil electrode (Aldrich,99.99%) was used. The foil electrode was polished first with 1.0 micron (Buehler, Alpha Micropolish II deagglomerated Alumina) polish and then rinsed with water. The water used in all of the work reported here was purified with a Milli RO-4/Milli-Q system (Millipore Corp.) and it exhibited a resistivity of $18M\Omega cm^{-1}$ on delivery. The electrode was then polished with 0.3 micron Alumina polish, rinsed and finally polished with 0.05 micron alumina polish and rinsed. The electrode was then quickly placed under water to avoid contamination by organics from the atmosphere.

In more recent experiments vapor deposited gold electrodes (Evaporated Metal Films, Inc.) were used. These electrodes consisted of float glass covered with a thin film of CrO_2 with approximately 1000 angstroms of gold covering the CrO_2 layer. These electrodes were plasma cleaned (Harrick Plasma Cleaner) and were highly hydrophilic upon removal from the plasma cleaner. Clean gold surfaces are hydrophilic (Smith, 1980). The cleaned electrodes were very quickly removed from the plasma cleaner and placed under water to avoid contamination.

The vapor deposited gold electrodes were modified with 10^{-4} M octadecyl mercaptan (Aldrich,98%) in ethanol (USI Chemicals Co., 100%) for 1 minute and then immediately rinsed with ethanol and finally stored under water. The gold foil was modified in 10^{-4} M octadecyl mercaptan in ethanol for 5 minutes, rinsed with ethanol, and then with water, and finally stored under water.

Solutions containing 25 mg of L-α-phosphatidylethanolamine, dioleoyl (Sigma,99%) and 6 mg of L-α-phosphatidylcholine,dioleoyl (Sigma,99%) were placed on a roto-vap (Büchi, RE111 Rotovapor) and the storage solvent was removed by evaporation. The amphiphiles were washed twice with anhydrous ether (J.T.Baker, >98%) followed by evaporation. Then 50 mg of sodium deoxycholate (Sigma) were added along with 3 mL of 0.1 M phosphate buffer, pH = 7.4. The solution was then stirred gently in the dark at 4 celsius until the amphiphiles and cholate were dissolved. The resulting solution should be colorless.

Dialysis tubing (Spectrapor, MWCO: 3,500) was then soaked in water for a couple of hours, changing the water frequently, to remove glycerine and sulfides from the membrane. Then 2-5 mg of cytochrome oxidase were added to the buffer solution containing the amphiphile and deoxycholate. The octadecyl mercaptan modified gold electrodes were placed into the dialysis tubing, gold surface exposed, with the oxidase containing solution and dialyzed against 0.1 M phosphate buffer, pH = 7.4 for 4-6 days at 4 C. Control experiments were performed in an identical fashion except no cytochrome oxidase was added to the solution as described above.

The cytochrome oxidase was kindly given to our laboratory by Charles Hartzell and Professor Shinya Yoshikawa. The cytochrome oxidase provided to us by Charles Hartzell was isolated using a preparation developed by Hartzell and Beinert (Hartzell and Beinert, 1974). The cytochrome oxidase given to us by Professor Shinya Yoshikawa was isolated following an enzyme preparation developed by his laboratory (Yoshikawa et al., 1977). There was no difference between the two types of cytochrome oxidase in the experimental behavior observed .

Cytochrome c was reduced with sodium

dithionite and then passed through a desalting gel (Bio-Gel, P-6DG) to remove the unreacted dithionite and reaction products. Only the first half of the band was used to avoid any possible contamination.

The electrochemistry was performed using a standard three electrode cell configuration. The enzyme modified electrode was assembled into the cell under 0.1 M phosphate buffer, pH = 7.4 to eliminate any contact of the membrane with air. The reference electrode used was a 1 M KCl, Ag/AgCl electrode and all potentials are reported versus the Normal Hydrogen Electrode.

RESULTS AND DISCUSSION

Figure 2 shows data obtained from an initial experiment. In Figure 2A the solid line shows a cyclic voltammogram of an enzyme immobilized electrode alone using a gold foil electrode for the solid substrate. The cyclic voltammetric waves are bracketed around a potential of approximately 400 mV indicating that the cytochrome a_3 and/or the Cu_b site are communicating with the electrode. The position of these waves is also consistent with a unidirectional partitioning of the cytochrome oxidase into the membrane (i.e., the potential of the cytochrome a and/or the Cu_a sites are around 200 mV positive of these waves).

Using the model shown in Figure 1, it should be energetically more favorable for the oxidase to insert into the membrane with the cytochrome a_3 and Cu_b adjacent to the electrode based on free energy considerations. This is because cyochrome c oxidase has a considerable number of charged residues sticking out into solution in areas not normally contacting the membrane, which would not be stabilized in the hydrophobic interior of the membrane.

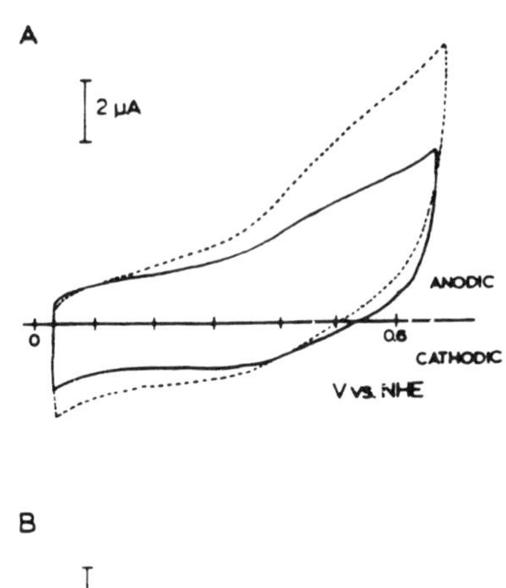

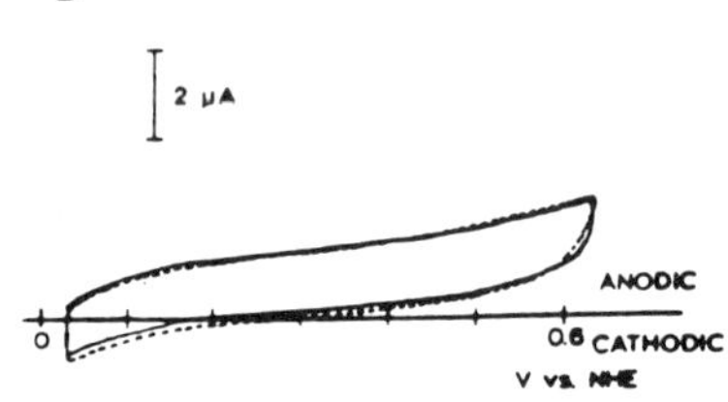

Figure 2. Cyclic Voltammetry of a Cytochrome c Oxidase Modified Bilayer Gold Electrode. The scan rate used is 50 mV/s and the buffer is 60 mM phosphate buffer, pH = 7.

A: <u>Solid line</u>: cyclic voltammogram of immobilized cytochrome oxidase in buffer alone.

<u>Dashed line</u>: cyclic voltammogram after the addition of a 50 μM solution of reduced cytochrome c.

B: <u>Solid line</u>: cyclic voltammogram in buffer alone following a control dialysis performed without adding cytochrome oxidase.

<u>Dashed line</u>: cyclic voltammogram following the addition of a 25 μM solution of reduced cytochrome c.

The dashed line in Figure 2A is the response observed following the addition of 50 μM solution of reduced cytochrome c. The anodic current increases at a potential well removed from cytochrome c's formal potential of 260 mV. This indicates a catalytic

oxidation of reduced cytochrome c, mimicking
the heterogeneous reaction between cytochrome
c and cytochrome c oxidase found in vivo.

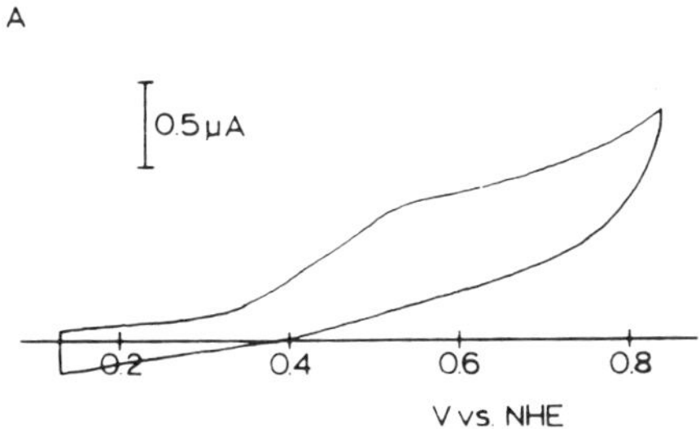

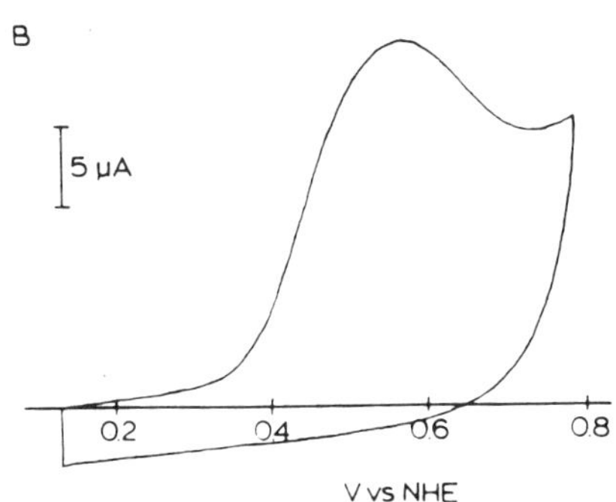

Figure 3. Cyclic Voltammetry of an Enzyme
Modified Vapor Deposited Gold Electrode.
The scan rate used was 50 mV/s and the buffer
was 0.1 M phosphate, pH = 7.4.
     A: Cyclic voltammogram of the enzyme
modified electrode in buffer alone.
     B: Cyclic Voltammogram after adding a
180 µM solution of reduced cytochrome c and
storing the cell at 4 celsius for four days.

     Figure 2B shows the results of the
control experiment performed without oxidase.
It is clear from this control experiment that
this membrane blocks the electrochemical
communication of the cytochrome c with the

electrode. These experiments show the best data obtained using the gold foil electrodes. These data represent the type of behavior observed in these experiments, however sometimes anomalous peaks are observed and we have attributed it to not removing all of the Alumina polish from the surface of the electrode.

Figure 3A shows the cyclic voltammetric response of an enzyme modified vapor deposited gold electrode in buffer alone. Again the formal potential is approximately 400 mV indicating that the cytochrome a_3 and Cu_b sites are oriented toward the electrode. Figure 3B shows the cyclic voltammetric response after the addition of a 180 μM solution of reduced cytochrome c and then storing the cell at 4 C for four days. There is over an order of magnitude increase in current after the addition of the reduced cytochrome c.

CONCLUSION

In conclusion this paper describes cyclic voltammetric results that are consistent with direct communication of immobilized cytochrome oxidase with a gold electrode. Catalytic currents following the addition of reduced cytochrome c to solution have also been observed. These voltammetric results are consistent with the redox properties and the electron transfer function of native cytochrome oxidase. Experiments using coupled optical absorption spectroscopy to further characterize these results are planned together with studies of effects arising from the addition of known inhibitors of cytochrome oxidase.

ACKNOWLEDGEMENTS

The authors gratefully acknowledge the financial support of this work by a grant from the National Science Foundation CHE 8520270. The kind gift of cytochrome oxidase from S.

Yoshikawa is also acknowledged.

REFERENCES

Anderson J.L., Kuwana T., and Hartzell C.R. (1976) <u>Biochem.</u> 15:3847.
Azzi A. (1980) <u>Biochim. Biophys. Acta</u> 594:231.
Bain C.D., Biebuyck, H.A., and Whitesides G.D. (1989) <u>Langmuir</u> 5:723.
Brian A.A., and McConnell (1984) <u>Proc. Natl. Acad. Sci. USA</u> 81:6159.
Capaldi R.A., Malatesta F., and Darley-Ushmar V.M. (1983) <u>Biochim. Biophys. Acta</u> 726:135.
Casey R.P. (1984) <u>Biochim. Biophys. Acta</u> 768:319.
Chidsey C.E.D., and Loiacono D.N. (1990) <u>Langmuir</u> 6:682.
Deartherage J.F., Henderson R., and Capaldi R.A. (1982) <u>J. Mol. Biol.</u> 158:487.
Einarsdottir O., and Caughey W.S. (1984) <u>Biochim. Biophys. Res. Commun.</u> 124:836.
Einarsdottir O., and Caughey W.S. (1985) <u>Biochim. Biophys. Res. Commun.</u> 129:840.
Fabianowski W., Coyle L.C., Weber B.A., Granata R.D., Castner D.G., Sadownik A., and Regen S.L. (1989) <u>Langmuir</u> 5:35.
Hafeman D.G., Tscharner V., and McConnell H.M. (1981) <u>Proc. Natl. Acad. Sci. USA</u> 78:4552.
Hartzell C.R. and Beinert H. (1974) <u>Biochim. Biophys. Acta</u> 368:318.
Hinkle P.C., Kim J.J., and Racker E. (1972) <u>J. Biol. Chem.</u> 247:1338.
Ishiguro T., and Nakanishi M. (1984) <u>J. Biochem.</u> 95:581.
Long R.C., Hawkridge F.M., and Hartzell C.R. (1986) <u>J. Electroanal. Chem.</u> 198:89.
Long R.C., Hawkridge F.M., Chlebowski J.F., and Hartzell C.R. (1988) <u>J. Electroanal. Chem.</u> 256:111.
Nuzzo R.G., and Allara D.L. (1983) <u>J. Am. Chem. Soc.</u> 105:4481.
Nuzzo R.G., Zegarski B.R., and Dubois L.H.

(1987) J. Am. Chem. Soc. 109:3559.

Palmer G. (1987) Pure and Appl. Chem. 59:749.

Porter M.D., Bright T.B., Allara D.L. and
  Chidsey C.E.D. (1987) J. Am. Chem. Soc.
  109:3559.

Prime K.L., and Whitesides G.M. (1991)
  Science 252:1164.

Schroedl N.A., and Hatzell C.R. (1977)
  Biochem. 16:1377.

Smith T. (1980) J. Colloid Interface Sci.
  75:51.

Steffens G.C.M., Biewald E., and Buse G.
  (1987) Eur. J. Biochem. 164:295.

Swalen J.D., Allara D.L., Andreade E.A.,
  Chandross E.A., Garoff S., Israelachvili
  J., McCarthy T.J., Murray R., Pease R.F.,
  Rabbolt J.F., Wynne K.J., and Yu H. (1987)
  Langmuir 3:932.

Tarlov M.J., and Bowden E.F. (1991) J. Am.
  Chem. Soc. 113:1847.

Yewey G.L. and Caughey W.S. (1988) Ann. N.Y.
  Acad. Sci. 550:22.

Yoshikawa S., Choc M.G., O'Toole M.C., and
  Caughey W.S. (1977) J. Biol. Chem.
  252:5498.

Yoshikawa S., Tera T., Takahashi Y.,
  Tsukihara T., and Caughey W.S. (1988)
  Proc. Natl. Acad. Sci. USA 85:1354.

Zhang Y., Georgevich G., and Capaldi R.A.
  (1984) Biochem. 23:5616.

Zhang Y., Capaldi R.A., Cullis P.R., and
  Madden T.D. (1985) Biochim. Biophys. Acta
  808:209.

THERMODYNAMIC AND ELECTROCHEMICAL STUDIES OF THE ELECTRON TRANSFER REACTIONS OF HEMOGLOBIN

Jennifer L. Detrich, Gabriel A. Erb,
David A. Beres and Lyman H. Rickard
Department of Chemistry
Millersville University
Millersville, PA 17520

INTRODUCTION

Heme containing proteins function as oxygen carriers, electron carriers or enzymes in biological systems. Of particular interest to this investigation is the ability of heme proteins to serve as electron carriers. The heme group, an iron atom bound to the four nitrogens of a porphyrin ring system, is the characteristic structural feature in common to all heme proteins. The iron atom serves as the site of oxidation or reduction during electron transfer. However, heme proteins exhibit significantly different thermodynamic and kinetic properties for electron transfer reactions. These differences between heme proteins must be due to the protein structure and the location of the heme group within the molecule. Although hemoglobin functions physiologically as an oxygen carrier rather than an electron carrier it is an ideal molecule to use as a model for the study of the electron transfer reactions of heme proteins. This is because it is readily available, has a moderate cost and has a known and documented structure.

The focus of this electrochemical investigation is to quantify and model the electron transfer reactions of hemoglobin at various electrode surfaces. Reported here are initial studies of the electrochemical determination of the relationship between pH and formal potential of hemoglobin and the kinetics of hemoglobin's heterogeneous electron transfer at an unmodified indium oxide electrode. Although hemoglobin has been extensively studied there have been relatively few electrochemical investigations. Early electrochemical studies of hemoglobin were done at mercury (Betso and Cover, 1972; Janchen et al, 1973; Kunetsov et al., 1977). Later electrochemical studies involved the use of mediators in homogeneous electron transfer (Kwee and Lund, 1974; Wheeler, 1985; Kwee, 1986; Durliat and Comtat, 1987; Durliat et al., 1988; Labrune et al. 1990; Song and Dong, 1988). The most recent electrochemical studies of hemoglobin have involved primarily heterogeneous electron transfer both at modified (Razumas et al., 1984; Wheeler, 1985; Song and Dong, 1988; Ye and Baldwin, 1988; Dong et al., 1989; Zhu and Dong, 1990a; Zhu and Dong, 1990b) and unmodified (Razumas et al., 1984; Wheeler, 1985; Durliat and Comtat, 1987; Song and Dong 1988, Durliat et al., 1988; Grubbs and Rickard 1989) electrode surfaces.

The structure of hemoglobin consists of four polypeptide chains with one heme group per chain located in crevices at the exterior of the molecule. Hemoglobin is an approximately spherical protein with dimensions of 64 x 55 x 50 angstroms. The iron atoms in the four heme groups form an irregular tetrahedron in which the iron atoms are separated by between 25 and 30 angstroms.

Initial electrochemical investigations of hemoglobin produced a faradaic response that was small compared to the background current. This resulted in data that was not satisfactory for the extraction of quantitative results. Therefore it was decided to use spectroelectrochemical techniques in which the system is probed electrochemically and monitored spectrophotometrically. Spectropotentiometry was used to determine the formal potential and n values. Methylene Blue was used as a mediator in these experiments in order to decrease the time required for each experiment. Single potential step chronoabsorptometry (with no mediator present), SPS/CA, was used for the heterogeneous electron transfer kinetics study.

METHODS AND MATERIALS

Horse hemoglobin from Sigma Chemical Company was purified by gel filtration on a column of Sephadex F-100-120 (Sigma Chemical Company). Tris(hydroxymethyl)aminomethane (99.8%)

and cacodylic acid (hydroxydimethylarsine oxide; 99%) were used as received from Fisher Scientific company. All solutions were prepared in distilled and deionized water exhibiting a resistivity of 8 Mohms/cm. All experiments were carried out in nonbinding tris/cacodylic acid buffer (Bowden et al., 1982a). Hemoglobin solutions were stirred in the cold under nitrogen (MG Industries prepurified grade, 99.998%) for 30 minutes before use. All other chemicals used in this work were reagent grade.

Electrochemical cells were of a conventional three electrode design. The optically transparent thin layer electrochemical (OTTLE) cell used for spectropotentiometric measurements has been described by Bowden, Cohen and Hawkridge (1982b) and the cell used for SPS/CA measurements has been described by King and Hawkridge (1987). Working electrodes were tin doped indium oxide deposited on glass from Donnelly Corporation, Holland, MI. Auxiliary electrodes were platinum wire. The reference electrodes used for all measurements were Ag/AgCl (1.0 M KCl). Reference electrodes were calibrated against quinhydrone. All reported potentials have been corrected to the normal hydrogen electrode. All measurements were taken at 30 $\pm$ 2°C.

Spectra were obtained using a Perkin-Elmer Lambda 7 Spectrophotometer interfaced to a Perkin-Elmer 3700 Data Station. Electrochemical experiments were carried out using an IBM EC/225 Voltametric Analyzer and an in-house constructed potentiostat.

Prior to each experiment the OTTLE was stored under vacuum for at least one hour to remove oxygen from the lucite cell body. The working electrode was pretreated by published procedures (Bowden et al., 1984) before each experiment. The working electrode was then preconditioned for 20 minutes in electrolyte buffer by voltammetric cycling over the potential range to be employed in the experiment.

Hemoglobin concentrations were determined by the reduced - minus - oxidized difference molar absorptivity, 9.7 x 10^4 $M^{-1}cm^{-1}$ (Wheeler, 1985). Deoxyhemoglobin (reduced form) was obtained by reduction with dithionite.

Data acquisition and data treatment procedures for spectropotentiometric methods have been described by Heineman et al. (1975). Procedures for potential step chrono-absorptometry have been described by Albertson et al. (1979) and Bancroft et al. (1981).

RESULTS AND DISCUSSION

Spectropotentiometric Measurements

Spectra obtained for a hemoglobin and methylene blue solution poised at various potentials are shown in Figure 1. Methemoglobin (oxidized form) and deoxyhemoglobin show absorbance maxima at 405 and 429 nm respectively. Spectra which are intermediate between the fully oxidized and fully reduced forms correspond to the intermediate potentials that give rise to a mixture of oxidized and reduced hemoglobin in the solution. Potentials were applied until the absorbance monitored at 405 or 429 was constant (change of less than 0.001 AU/min). Typically five to twenty minutes was required before steady-state equilibrium was established with oxidation requiring a longer time period than reduction. A potential sufficiently negative to cause complete reduction of the hemoglobin and any trace amount of oxygen was applied first. All other potentials were applied in a random order. Although hemoglobin could be completely electrochemically reduced in every experiment, total oxidation was achieved only intermittently. It was therefore necessary to calculate the absorbance of the totally oxidized hemoglobin from spectra taken to determine the concentration for each trial.

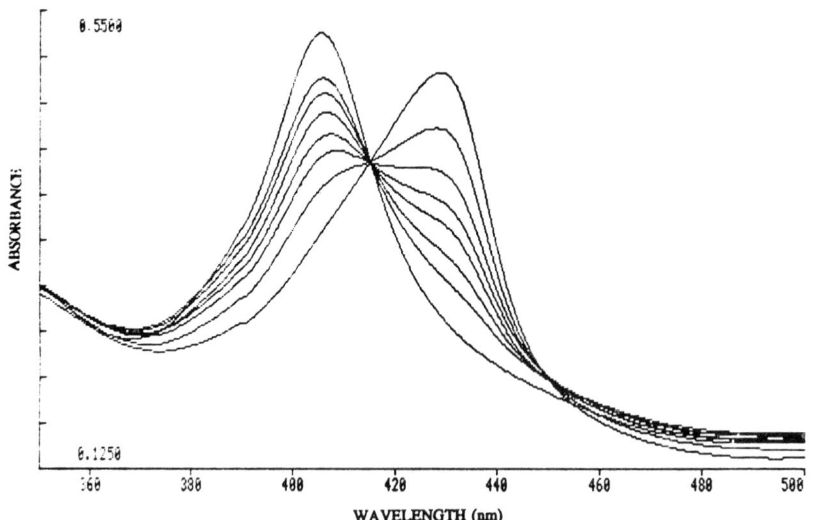

FIGURE 1. Spectra of 1.1 x 10^{-4}M hemoglobin and 2.0 x 10^{-4}M methylene blue mediator in Tris/Cacodylic Acid Buffer at pH 7.73. Applied potentials of: 7, 32, 47, 62, 77, and 92 mV vs NHE.

The isobestic point observed at 416 nm is evidence that only the methemoglobin and deoxyhemoglobin are present and that there is no oxyhemoglobin (oxygen bound form). The absorbance of methylene blue is insignificant in the spectral range and concentrations being used for the hemoglobin measurements. Absorbance changes due to methylene blue are observed in the 550 to 770 nm region.

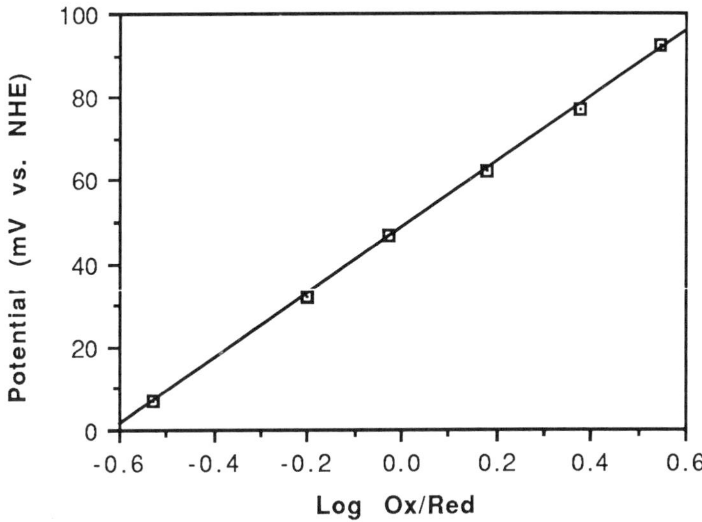

FIGURE 2. Nernst Plot of $E_{applied}$ vs. Log ([Ox]/[Red]) for data from Figure 1.

Figure 2 shows a plot of potential vs the log (Hb_{Ox}/Hb_{Red}) in which the ratio of the concentration of the oxidized to reduced hemoglobin is calculated from the absorbance data in Figure 1. The y-intercept of the line in Figure 2 represents the formal potential, $E^{0'}$ of hemoglobin and the slope can be used to calculate n, the electron stoichiometry. These measurements were repeated at various pH values to yield Figure 3 which shows the relationship between pH and formal potential for hemoglobin. At acidic pH values below 6.8 hemoglobin's formal potential is independent of pH. Above pH values of 6.8 the formal potential decreases with a slope of - 64 mV/pH. Over the pH range 5.96 to 8.10 hemoglobin's n value is observed to increase from 0.59 to 0.83.

The decrease in hemoglobin's formal potential and the increase in n value in increasingly alkaline media has been described in the literature (Taylor and Hastings, 1939, Antonini et al., 1963,

Bull and Hoffman, 1975, Santucci et al., 1986). These studies used chemical titration and the method of mixtures. A dependence of formal potential on pH has also been observed in other heme proteins such as myoglobin and cytochrome c (Antonini and Brunori, 1971; Faladijan et al., 1982; Koller and Hawdridge, 1988).

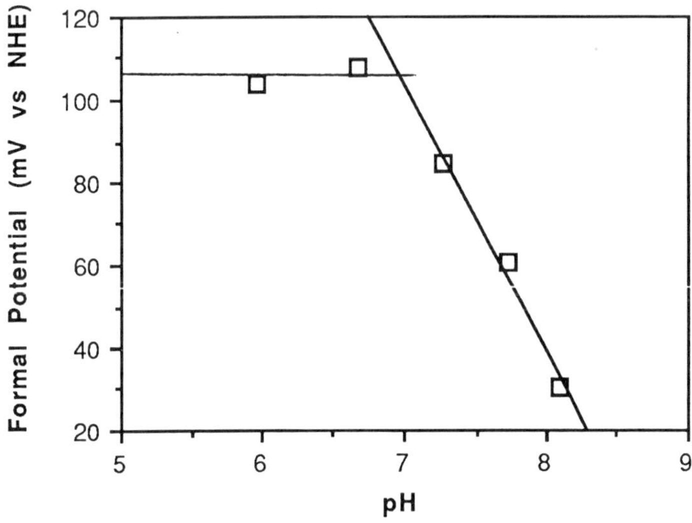

FIGURE 3. Dependence of $E^{0'}$ on pH of horse hemoglobin.

The relationship between pH and formal potential found in this study agrees quite well with the literature cited above. However, the individual values for formal potential and for the electron stoichiometry are not in agreement with the literature. The formal potential obtained for hemoglobin at pH 7.0, 104 mV, is more negative than the literature values for horse hemoglobin of 132, 139, 144, and 150 mV (Taylor, 1981 and references). Other electrochemical investigations of hemoglobins have also found formal potentials more negative than the accepted value. Betso (1972), and Scheller (1974, 1975) found polarographic half wave potentials for human hemoglobin in the range of -.36 to -.51 V (accepted formal potential = .150 V). Wheeler (1985) used spectropotentiometry at carbon fibers to determine values ranging between 126 and 147 mV. Formal potentials for bovine hemoglobin determined using spectropotentiometry at modified platinum electrodes have ranged from 32 to 133 mV (Song and Dong, 1988; Dong et al., 1989; Zhu and Dong, 1990). Literature values for bovine hemoglobin are 95 150 and 170 mV (Taylor, 1981 and references).

Electron stoichiometric values obtained in this investigation are more tenuous than the formal potentials. Values of n are a measure of the interaction between redox sites (heme groups) in the hemoglobin molecule. The values obtained from this study are less than one. An n value of less than one implies an anticooperative interaction between heme groups. Accepted values for n are pH dependent and range from 1.2 to 2.5 for hemoglobin (Antonini, 1964; Antonini and Brunori, 1971; Bull and Hoffman, 1975). These values mirror the cooperative effect between heme groups observed during the binding of dioxygen to hemoglobin. The n values for this investigation were reproducibly obtained from spectra like that shown in Figure 1 and with Nernst Plots having high correlation coefficients. An explanation for this discrepancy has not yet been found.

Several explanations have been proposed for the relationship between hemoglobin's formal potential and pH in alkaline solution. Since this relationship mirrors the oxygenation Bohr Effect which is observed when dioxygen binds to hemoglobin, it has been termed the oxidation alkaline Bohr Effect. The correlation between these two effects implies that the mechanism controlling them is the same (Antonini and Brunori 1971). The accepted literature value of - 60mV/pH for the effect suggests a change of one proton per heme group. Dissociation of the water molecule coordinated to the heme iron in the sixth position of aquo-methemoglobin to yield hydroxy-methemoglobin is thought to be responsible for the major portion of the relationship between formal potential and pH (Antonini and Brunori 1971). However, the dissociation of water does not account for the entire dependency observed and therefore it is expected that other acid linked groups must exist.

Several explanations have been offered for the change in cooperativity of hemoglobin with increasing pH. Antonini and Brunori (1971) state that the relationship is due to the differences associated with the redox potentials of the alpha and beta chains of hemoglobin. Perutz claims that there is a shift in the conformational equilibrium between the T (tense, unliganded) and the R (relaxed, liganded) forms of hemoglobin which causes the change in cooperativity (Santucci et al., 1986 and references). Santucci et al. (1986) reported data that indicates the existence of intermediate conformers between the R and T form in methemoglobin.

Potential Step Chronoabsorptometry

Initial measurements of the heterogeneous electron transfer rate parameters for the reduction of hemoglobin at an unmodified indium oxide electrode were performed using single step chronoabsorptometry. Figure 4 shows the results for a series of 60 second potential steps at five different overpotentials. The theoretical diffusion controlled absorbance at a given time can be calculated from (Albertson et al., 1979)

$$\Delta A = 2/\pi^{1/2} \; \Delta\varepsilon D^{1/2} C^o t^{1/2}$$

The data in Figure 1 are then normalized against these values of the diffusion controlled absorbance. The resulting normalized absorbance data are then fit to working curves from the literature (Albertson et al., 1979) to obtain the heterogeneous rate constant for each overpotential employed in the experiment. The relationship between the experimentally measured rate constants and overpotential is shown in Figure 5. The formal heterogeneous electron transfer rate constant, $k^{o'}$, is calculated from the y-intercept of this plot to yield an estimated value of 1×10^{-6}cm/sec. The slope of this plot is used to determine the electrochemical transfer coefficient, alpha, which is found to be 0.06.

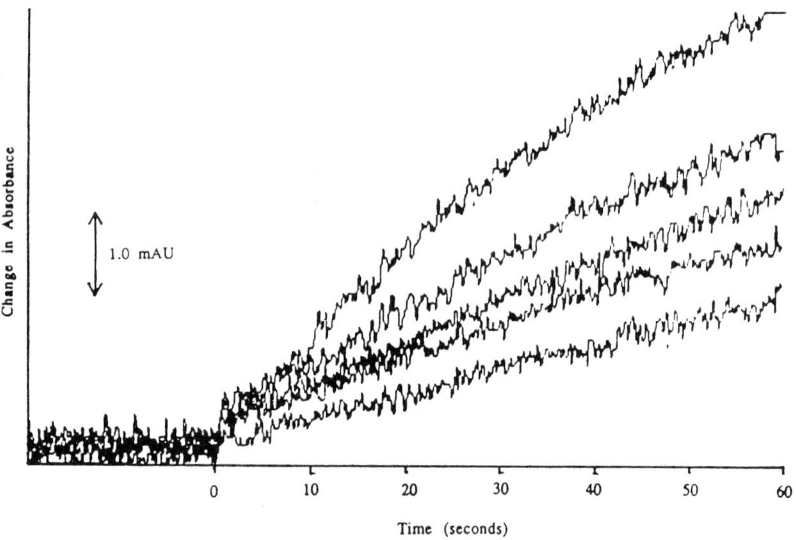

FIGURE 4. Potential step chronoabsorptometry of hemoglobin indium oxide. Solution: 1.0×10^{-4}M hemoglobin in tris/cacodylic acid buffer, pH 7.0. Overpotentials: 400, 500, 600, 700, 800 mV vs NHE.

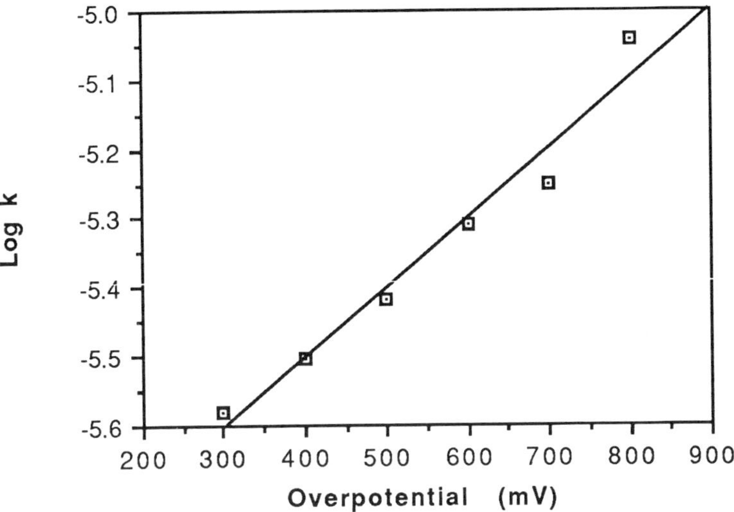

FIGURE 5. Dependence of rate constant on overpotential.

The first report of a formal heterogeneous electron transfer rate constant for hemoglobin showed irreversible kinetics (Razuma et al., 1984). The rate constant determined was 5.2×10^{-8} cm/sec and the transfer coefficient was 0.29 for electron transfer at a tin oxide electrode surface modified with tetraethylammonium chloride. Wheeler (1985) determined a formal rate constant of 5.1×10^{-5} cm/sec and transfer coefficient of 0.14 for electron transfer at an unmodified carbon fiber electrode. A platinum electrode modified with Brilliant Cresyl Blue gave a formal rate constant of 3.7×10^{-7} cm/sec (Dong et al, 1989) while one modified with Methylene Green exhibited much faster kinetics with a formal rate constant of 1.97×10^{-4} cm/sec (Zhu and Dong 1990). As can be seen from these results, hemoglobin's formal rate constant can vary over a wide range dependent on the type of electrode, modification of the electrode surface and other experimental conditions.

The application of the spectroelectrochemical techniques of spectropotentiometry and potential step chronoabsorptometry for the investigation of the thermodynamic and kinetic parameters of the electron transfer reactions of hemoglobin has been demonstrated. Future work will be directed toward optimizing solution conditions and electrode material for hemoglobin's electron transfer. Work will continue to resolve problems described in this investigation. Extension of the present work is expected to include the separation of hemoglobin into its alpha and beta chains and the study of their electron transfer reactions.

ACKNOWLEDGEMENTS

The authors would like to thank Dr. Edmond Bowden of North Carolina State University for the OTTLE cell and Dr. Fred Hawkridge of Virginia Commonwealth University for the tin-doped indium oxide material and many helpful discussions. The authors gratefully acknowledge the support of this work by Research Corporation, C-2667.

REFERENCES

Albertson DE, Blount HN and Hawkridge FM (1979): Spectroelectrochemical determination of heterogeneous electron transfer rate constants. *Anal Chem* 51: 556-560.

Antonini E and Brunori M (1971): *Hemoglobin and Myoglobin in Their Reactions with Ligands*, London: North-Holland Publishing.

Antonini E, Wyman J, Brunori M, Taylor JF, Rossi-Fanelli A and Caputo A (1964): Studies on the oxidation-reduction potentials of heme proteins. *J Biol Chem* 239: 907-912.

Bancroft EE, Blount HN and Hawkridge FM (1981): Single potential step chronoabsorptometric determination of heterogeneous electron transfer kinetic parameters of quasi-reversible processes. *Anal Chem* 53: 1862-1866.

Betso SR and Cover RE (1972): Electrochemical reduction of human methaemoglobin. *JCS Chem Comm*: 621.

Bowden EF, Hawkridge FM and Blount HN (1982a): The heterogeneous electron transfer properties of cytochrome c. In: *Electrochemical and Spectrochemical Studies of Biological Redox Components*, Kadish KM, ed. Washington, D.C.: American Chemical Society.

Bowden EF, Cohen DJ and Hawkridge FM (1982): Anaerobic thin-layer electrochemical cell for planar optically transparent electrodes. *Anal Chem* 54: 1005-1008.

Bowden EF, Hawkridge FM and Blount HN (1984): Interfacial electrochemistry of cytochrome c at tin oxide, indium oxide, gold and platinum electrodes. *J Electroanal Chem* 161: 355-376.

Bull C and Hoffman BM (1975): Redox equilibria of liganded form of methemoglobin. *Proc Nat Acad Sci* 72: 3382-3386.

Dong S, Zhu Y and Song S (1989): Electrode processes of hemoglobin at a platinum electrode covered by brilliant cresyl blue. *Bioelectrochem Bioenerg* 21: 233-243.

Durliat H and Comtat M (1987): Electrochemical reduction of methemoglobin either directly or with flavin mononucleotide as a mediator. *J Bio Chem* 262: 11497-11500.

Durliat H, Barrau MB and Comtat M (1988): FAD used as a mediator in the electron transfer between platinum and several biomolecules. *Bioelectrochem Bioenerg* 19: 413-423.

Grubbs WT and Rickard LH (1989): Hemoglobin electron transfer reactions. In: *Charge and Field Effects in Biosystems-2* Allen MJ, Cleary SF and Hawkridge FM, eds. New York: Plenum Press.

Haladjian J, Pilard R, Bianco P and Serre P (1982): Effect of pH on the electroactivity of horse heart cytochrome *c*. *Bioelectrochem Bioenerg* 9: 91-101.

Heineman WR, Norris BJ and Goelz JF (1975): Measurement of Enzyme $E^{\circ'}$ Values by Optically transparent thin layer electrochemical cells. *Anal Chem* 47: 79-84.

Janchen M, Scheller F, Prumke HJ, Mohr P, and Etzold G (1973): Das wechselstrom-polarographische verhalten von humanen methamoglobin und metmyoglobin in wassriger losung. *Studia Biophsica* 39: 1-8.

King BC and Hawkridge FM (1987): A study of the electron transfer and oxygen binding reaction of myoglobin. *J Electroanal Chem* 237: 81-92.

Kuznetsov BA, Shumakovich GP and Mestechkina NM (1977): The reduction mechanism of cytochrome c and methemoglobin on the mercury electrode. *Bioelectrochem Bioenerg* 4: 512-521.

Koller KB and Hawkridge FM (1988): The effects of temperature and electrolyte at acidic and alkaline pH on the electron transfer reactions of cytochrome *c* at In_2O_3 electrodes. *J Electroanal Chem* 239: 291-306.

Kwee S and Lund H (1974): Indirect electrolysis of macromolecules by means of pteridone mediators. *Bioelectrochem Bioenerg* 1: 87-95.

Kwee S (1986): A novel mediator for the investigation of the electrochemistry of metalloproteins. *Bioelectrochem Bioenerg* 16: 99-109.

Labrune P, Bergel A and Comtat M (1990): Indirect electrochemical reduction of methemoglobin: design of the process. *Biotechnol Bioeng* 36: 323-329.

Razumas VJ, Zapalskyte AA, and Kulys JJ (1984): Redox conversion of methemoglobin on SnO_2 electrodes. *studia biophysica* 103: 57-62.

Santucci T, Amiconi F, Alberto B, Brunori M and Ascoli F (1986): A potentiometric study on the redox properties of hemoglobin from camelus dromedarius. *Bioelectrochem Bioenerg* 15: 521-526.

Song S and Dong S (1988): Spectroelectrochemistry of the quasi-reversible reduction and oxidation of hemoglobin at a methylene blue adsorbed modified electrode. *Bioelectrochem Bioenerg* 19: 337-346.

Taylor JF and Hastings AB (1939): Oxidation-reduction potentials of the methemoglobin-hemoglobin system. *J Biol Chem* 131: 649-662.

Taylor JF (1981): Measurement of the oxidation-reduction equilibria of hemoglobin and myoglobin. In: *Hemoglobins*, Antonini E, Rossi-Bernardi L, Chiancone E, eds. New York: Academic Press.

Wheeler, JR (1985): *Spectroelectrochemical Investigations of Biological Redox Molecules*, dissertation Duke University, Durham, NC.

Ye J and Baldwin RP (1988): Catalytic reduction of myoglobin and hemoglobin at chemically modified electrodes containing methylene blue. *Anal Chem* 60: 2262-2268.

Zhu Y and Dong S (1990a): Rapid redox reaction of hemoglobin at methylene green modified platinum electrode. *Electrochim Acta* 35: 1139-1143.

Zhu Y and Dong S (1990b): Rapid oxidation and reduction of hemoglobin at a bifunctional dye of Janus Green modified electrode. *Bioelectrochem Bioenerg* 24: 23-31.

THE EFFECT OF ADSORBED IODINE ON THE ELECTRICAL CONDUCTIVITY OF PHOSPHOLIPID FILMS

Gordon L. Jendrasiak
Department of Radiation Oncology
East Carolina University School of Medicine

Thomas J. McIntosh
Department of Cell Biology
Duke University Medical Center

Gregory E. Madison
Department of Radiology
Andrews Air Force Base

Ralph Smith
Department of Radiation Oncology
East Carolina University School of Medicine

INTRODUCTION

The interaction of iodine with phospholipids has been studied by optical spectroscopic techniques by Bhowmik (1986) and Chatterjee (1988). Iodine's effect in increasing the electrical conductivity of lipids in solid state films has also been observed (Rosenberg, 1968). The addition of elemental iodine to the solution surrounding black lipid membranes (BLM) results in a large increase in the electrical conductivity of the membranes (Finkelstein, 1968, Liberman, 1968, Jendrasiak, 1969). The iodine induced electrical conduction is believed to occur perpendicular to the phospholipid bilayer for BLM, whereas in the solid state film work reported in this paper, we believe it occurs along the phospholipid head-group layers. Upon iodine adsorption by phospholipid films, optical absorption bands at approximately 294 nm and 365 nm have been observed (Bhowmik et al., 1967).

In this paper, we have measured the electrical conduction of dry films formed from EPC, DPPC, and certain other phospholipids as a function of the iodine adsorbed. This electrical conductivity is related to the magnitude of the 294 nm and 365 nm optical absorption bands developed upon iodine adsorption by EPC and by DPPC films. We tentatively ascribe the 294 nm and 365 nm absorption bands to the presence of I_3^- (Ross and Baldwin, 1966). Since the hydrocarbon chains in EPC have considerable unsaturation, whereas those in DPPC are completely saturated, the effect of unsaturation on the iodine induced electrical conductivity has thus been measured. We have also used x-ray diffraction techniques to study the structural effect iodine has on films formed from both these phosphatidylcholines as well as $DC_{8,9}PC$ and thus have correlated the structure and electrical behavior of the films.

METHODS AND MATERIALS

EPC (>99% pure), DPPC, egg phosphatidylethanolamine (EPE) and dipalmitoylphosphatidylethanolamine (DPPE) were obtained from Avanti Polar Lipids, Inc., Birmingham, Alabama. The purity of the lipids was checked by chromatography. Also, $DC_{8,9}PC$ was obtained from the United States Naval Research Laboratory. All other solvents and iodine were of the highest purities available.

Platinum electrodes of dimensions 0.5 cm by 0.5 cm, separated by 0.5 cm, were vacuum deposited in the center of a quartz plate (Suprasil #1, Engelhard Industries, Inc.) prior to lipid film deposition. The electrodes were formed by vacuum depositing titanium on the quartz and then the platinum. The lipids were deposited as films, from chloroform solution, on the quartz plates and the chloroform removed by vacuum. Complete drying of the lipid samples took several days and was determined by the absence of an electrical current, i.e. $<10^{-16}a$ in our case, which was the detection limit of our electrometer. The mass of lipid deposited on the quartz plate was typically 2 - 2.5 mg. The film mass and area along with lipid density were used to calculate an average film thickness, typically, 25-45 μ. The electrical potential was applied across the films by the vacuum-deposited electrodes. A quartz plate with a four-point probe arrangement was used to study the electrical contact resistance of the films (Streetman, 1980).

A Keithley electrometer, Model 617, was used to detect the current through the lipid films. A controlled D.C. voltage, usually 1v, was applied from the electrometer itself. An Omega 670 thermometer with Type T thermocouple was used to monitor the sample temperature. For electrical activation energy studies, the temperature of the lipid film was lowered by immersing a copper bar, in contact with the quartz plate, in liquid nitrogen at a controlled rate. The electric activation energy, E_A, was obtained by plotting the conductivity data using the equation $I=I_o \exp(-E_A/2kT)$. Here I is the electrical current at absolute temperature T, k is Boltzmann's constant and I_o is the current at room temperature, i.e. the starting temperature for the activation energy measurements.

The iodine released into the chamber containing the lipid samples was obtained by diluting a saturated solution of iodine in diethylene glycol. This solvent has a very low vapor pressure and in the absence of iodine does not affect the electrical properties of the lipid films. Dilutions of the saturated solution were used to obtain various "iodine vapor pressures" (Weast, 1983) in the chamber.

Numerous control measurements were made to assure that the electrical current observed was due to current through the lipid film with its adsorbed iodine rather than through other components of the experimental system.

Optical measurements of the lipid films were obtained with a Shimadzu, UV-160, recording spectrophotometer using the appropriate lipid film, not exposed to iodine, as a reference. The characteristic iodine peak at 512 nm as well as peaks at 294 and 365 nm were observed for the phosphatidylcholine films when exposed to iodine vapor. For actual quantification of the iodine adsorbed by the film, a phospholipid film, similar in weight to that used in the electrical studies was placed on a teflon substrate, dried and exposed to known amounts of iodine. The sample and substrate were then placed in chloroform and this solution measured for total iodine content. The peaks at 294 and 365 nm were measured; the 294 nm peak was chosen for the calibration. In this way a calibration curve of O.D. at 294 nm vs. iodine adsorbed was constructed. The iodine itself, in its interaction with the dry lipid film, was, to a great extent, converted to the species absorbing at 294 nm (and 365 nm, as well) which we tentatively ascribe to I_3^-. The appropriate lipid film used for the electrical conductivity measurements was then also dissolved in chloroform and its O.D. at 294 nm measured. By comparison to the calibration curve, the amount of iodine adsorbed to this lipid film at various relative iodine vapor pressures (RIVP), was thus determined. Adequate control measurements were run to assure that this measurement was an accurate measure of the iodine adsorbed during the electrical conductivity studies. Corrections were made for the small amounts of iodine adsorbed by system components other than the lipid film.

For x-ray diffraction analysis, oriented phospholipid multilayers were prepared by placing a small drop of lipid/chloroform solution on a flat piece of aluminum foil and evaporating the chloroform under a stream of nitrogen. The aluminum foil substrate was given a convex curvature by bending it around a Pasteur pipet as described previously (McIntosh et al., 1987). The specimen was then mounted in a controlled humidity chamber on a single-mirror (line-focussed) x-ray camera such that the x-ray beam was oriented at a grazing angle relative to the multilayers on the foil. The humidity chamber consisted of a copper canister with two mylar windows for passage of the x-ray beam. The relative humidity was controlled with a cup of saturated salt solution in the chamber. Diffraction patterns were obtained for a wide range of relative humidities. However, since the electrical conductivity measurements were performed at low water content, we concentrated on the low humidity region, either using "dri-rite" in the chamber for "dry" conditions, or else with a cup of saturated $CaCl_2$ solution, to maintain the relative humidity at 32%. For some experiments, the samples were exposed to 100% RIVP by the addition of iodine crystals to the chamber.

X-ray diffraction patterns were recorded on a stack of four sheets of Kodak DEF x-ray films. The films were processed by standard techniques and densitometered with a Joyce-Loebl microdensitometer as described previously (McIntosh and Holloway, 1987; McIntosh et al., 1989 a & b). After background subtraction, integrated intensities, I(h), were obtained for each order h by measuring the area under each diffraction peak. The intensities were corrected by a single factor of h due to the cylindrical curvature of the multilayers (Blaurock and Worthington, 1966, Herbette et al., 1977), so that $F(h) = \{hI(h)\}^{1/2}$.

Electron density profiles, $\rho(x)$, on a relative electron density scale were calculated from:
$$\rho(x) = (2/d) \sum \exp\{\phi(h)\} \cdot F(h)\cos(2\pi xh/d)$$
where x is the distance from the center of the bilayer, d is the lamellar repeat period, $\phi(h)$ is the phase angle for order h, and the sum is over h. Phase angles were determined by a sampling theorem analysis as described in detail previously (McIntosh and Holloway, 1987). All electron density profiles described in this paper are at a resolution of $d/2h_{max} \approx 6$ Å.

RESULTS

Fig. 1 shows the adsorption of I_2, for both EPC and DPPC films, as a function of RIVP from zero to 100%. The appropriate RIVP was obtained using the dilution technique described in the Methods and Materials section. The number of I_2 molecules adsorbed, per phospholipid molecule, was determined using the optical techniques described previously.

Fig. 2 displays the electrical conductivity of EPC and DPPC films, respectively, as a function of the number of I_2 molecules adsorbed by the films, per phospholipid molecule. The data points represent the average of measurements from a minimum of three different films in each case. The steady-state value of the electrical current at each adsorbed iodine value was attained within 48 hours. The current vs. voltage relationship for the films, exposed to iodine, was found to be linear from 1v to 100v, implying that the current obeys Ohm's Law. For $DC_{8,9}PC$ films, the electrical conductivity value was found to be about an order of magnitude less than for EPC, at maximum iodine adsorption for each film.

FIGURE 1. Adsorption isotherms for I_2 on EPC (●) and DPPC (■) films, respectively.

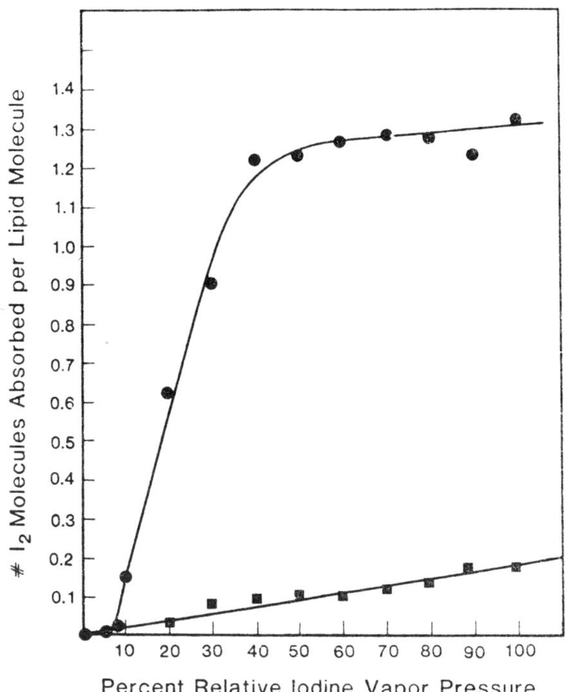

Electrical measurements of EPE and DPPE films were also made at maximum iodine adsorption. The electrical conductivity values for films of EPE containing unsaturated hydrocarbon chains were found to be some 3-4 orders of magnitude greater than were those for films containing saturated hydrocarbon chains (DPPE); this conductivity difference parallels that found for phosphatidylcholine films. The actual conductivity values are shown in Table 1.

TABLE 1. Electrical Properties of Lipids

Lipid	Average Electrical Conductivity at 100% RIVP	E_a (average)
EPC	$1.3 \times 10\text{-}3\,\Omega^{-1}cm^{-1}$	3.08 ev
DPPC	8.3×10^{-6}	4.12
$DC_{8,9}PC$	1.0×10^{-4}	4.54
EPE	1.1×10^{-7}	--
DPPE	2.2×10^{-11}	--

FIGURE 2. Electrical conductivity vs. number of I_2 molecules adsorbed on EPC (●) and DPPC (♦) films, respectively.

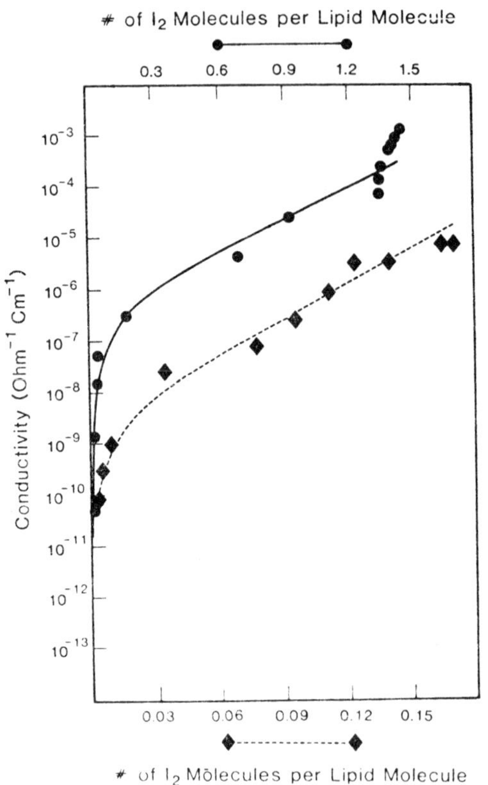

Fig. 3 shows typical optical absorption spectra for EPC and EPE (top) and DPPC and DPPE (bottom) films exposed to iodine vapor. Note the difference in O.D. scales for the saturated vis a vis unsaturated lipid. Each saturated lipid was exposed to the same iodine vapor pressure as its unsaturated counterpart. The conversion of I_2 to I_3^- is indicated by the appearance of absorption bands at 294 nm and 365 nm, upon iodine adsorption. Our optical measurements clearly indicate that films of phospholipids having unsaturated hydrocarbon chains adsorb much more iodine than do films of phospholipids having saturated chains.

Films of EPE show the formation of I_3^-, however the DPPE films, apparently because of their chalky-white appearance, scatter the light sufficiently so as to prevent observing any iodine bands. Other measurements do show the iodine adsorption by DPPE films to be quite small, relative to that for EPE films.

FIGURE 3. Optical absorption spectra for EPC and EPE (top, left and right, respectively) and DPPC and DPPE (bottom, left and right, respectively) films exposed to iodine vapor. Note O.D. scale difference for the saturated and unsaturated lipids.

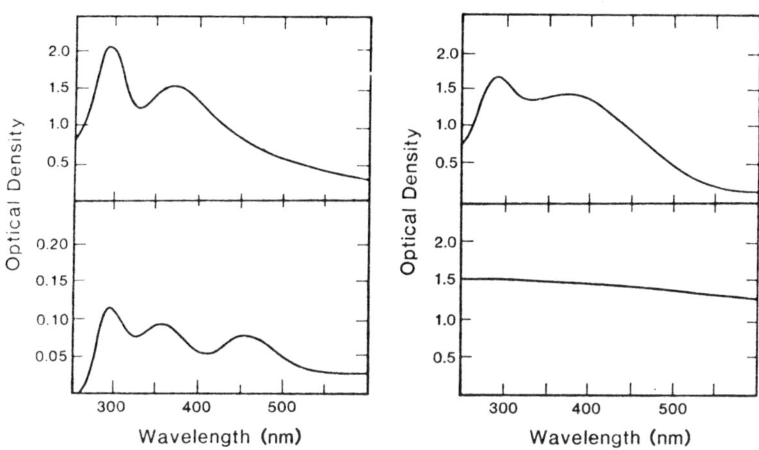

Fig. 4 shows representative electric activation energy plots for EPC, DPPC, and $DC_{8,9}PC$. Each film sample was cycled from room temperature to the lowest temperature at least three times. The activation energies were obtained from the slopes of the appropriate plots. The range of E_A obtained with at least three different films was 2.75 ev - 3.45 ev for EPC and 3.97 ev - 4.27 ev for DPPC. The average values of E_A are shown in Table I.

Four-point probe measurements indicated that the electrical contact resistances between the films and the electrodes were negligible in comparison to the lipid film resistance. Thus, the electrical measurements reported here represent the electrical properties of the bulk lipid films as the films adsorb iodine. This was further substantiated by tests using electrodes formed from other metals than those mentioned previously.

All x-ray diffraction patterns from EPC and DPPC, in the presence and absence of iodine, contained a series of low-angle reflections which indexed as orders of a single lamellar repeat period. This indicates that all samples contain multibilayers. The lamellar repeat periods and relative intensities of the diffraction orders were modified by exposure of the specimen to iodine. For samples in the

absence of iodine, the lamellar repeat periods were 56.3 Å for dry
DPPC and 50.5 Å for EPC at 32% relative humidity. Under these
humidity conditions, the presence of iodine crystals in the chamber
changed the lamellar repeat period to 65.1 Å for DPPC and 45.2 Å for
EPC. That is, iodine increased the repeat period for DPPC, but
decreased the repeat period for EPC bilayers. For $DC_{8,9}PC$, the
presence of iodine increases the repeat period from 56.7 Å to 63.1 Å.

FIGURE 4. Electric activation energy plots for EPC (●), DPPC (▲)
and $DC_{8,9}PC$ (♦) films.

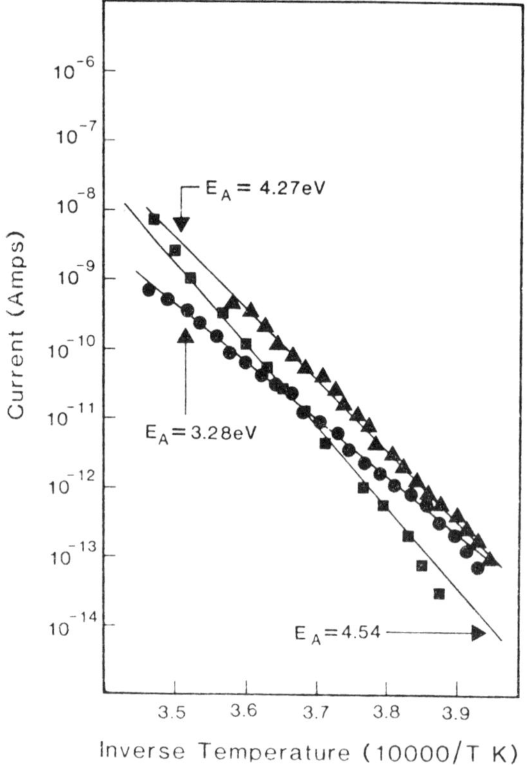

Electron density profiles for both dry DPPC and EPC at 32%
relative humidity in the presence and absence of iodine are shown in
Figs. 5 and 6, respectively. In the profile for dry DPPC, in the absence
of iodine, the high density peaks at ± 22 Å correspond to the lipid
head groups, the low density trough at 0 Å corresponds to the
localization of the terminal methyl groups in the bilayer center, the
medium density regions between the terminal trough and the head
group peaks correspond to the methylene chains, and the medium

density regions at the outer edges of the profiles correspond to the spaces between adjacent bilayers in the multilayer. The addition of iodine does not change the distance between head-group peaks across the bilayer, indicating that the thickness of the DPPC bilayer is not modified by the addition of iodine. However, the space between adjacent bilayers is increased by the incorporation of iodine.

FIGURE 5. Electron density profiles for dry DPPC bilayers in 0% RIVP (top) and 100% RIVP (bottom).

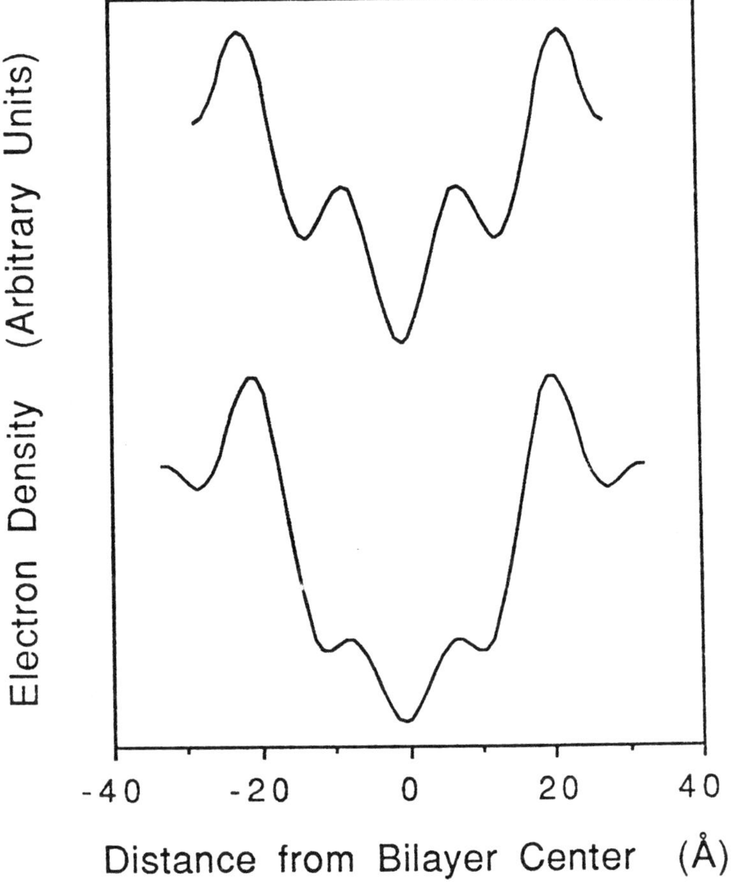

Moreover, compared to the control, DPPC profile, the presence of iodine increases the density in the head-group peaks relative to the depth of the terminal methyl trough. These profiles are both on arbitrary electron density scales, so that absolute electron densities cannot be obtained. However, the most likely explanation for the increase in width of the space between adjacent bilayers and the increased height of the head-group peak region compared to the terminal methyl trough is that iodine or I_3^- accumulates near the head-group of the bilayer. The presence of iodine or I_3^- in and around the head group would raise the electron density of the head group region relative to the hydrocarbon region of the bilayers, and could also sterically push apart adjacent bilayers. We emphasize that we consider this to be the most likely interpretation of these profiles, but that since the profiles are on arbitrary electron density scales, it is impossible to unequivocally localize the iodine.

FIGURE 6. Electron density profiles for EPC bilayers at 32% relative humidity in 0% RIVP (top) and 100% RIVP (bottom).

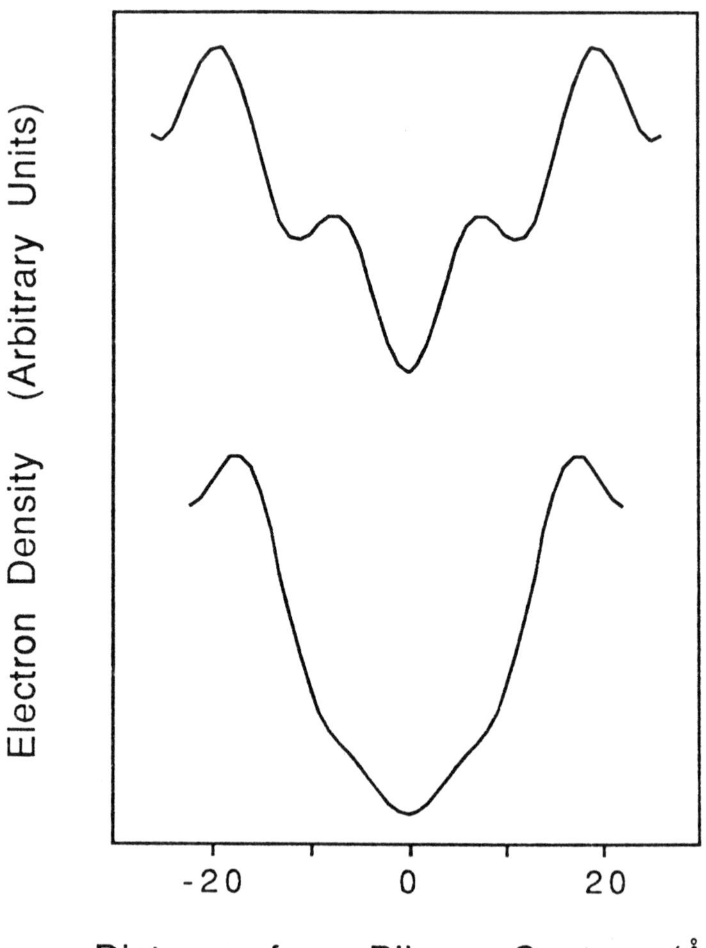

For the EPC profiles, in the absence of iodine, a terminal methyl trough is clearly present in the geometric center of the bilayer. In the presence of iodine, this trough is no longer visible and the distance between head group peaks has decreased by about 4 Å, meaning that the thickness of the hydrocarbon core of the bilayer has decreased due to the exposure to iodine. These observations imply that the structure of the hydrocarbon region of the EPC bilayer has been significantly modified by the iodine. It is impossible to localize iodine in the EPC profile since it is on an arbitrary electron density scale. The electron density profile for $DC_{8,9}PC$ is shown in Fig. 7. The appearance of a second, as yet unidentified lamellar phase upon   iodine exposure, however complicates the interpretation of this pattern.

FIGURE 7. Electron density profiles for dry $DC_{8,9}PC$ bilayers in 0% (top) and 100% RIVP (bottom).

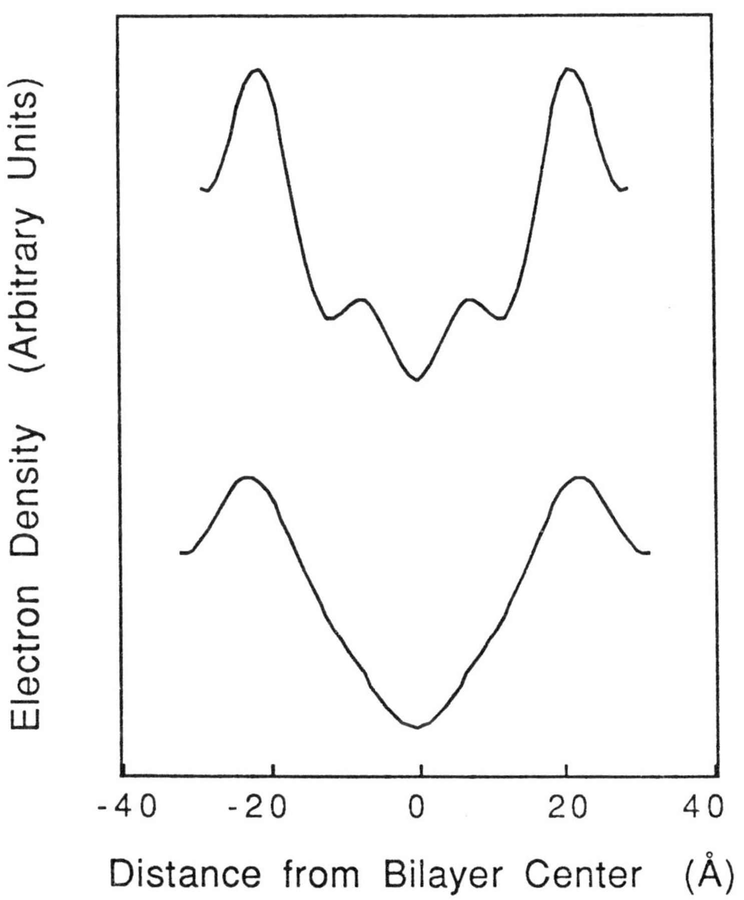

DISCUSSION

Our results show that the electrical conductivity of phospholipid films greatly increases upon exposure to iodine vapor. The most rapid increase occurs for the first few adsorbed I_2 molecules. The conductivity increase is a function of the amount of iodine adsorbed as evidenced by the increased formation of I_3^-.

The increase in the electrical conductivity is significantly greater, by some three (3) orders of magnitude, for EPC than for DPPC at the same RIVP. The conductivity of $DC_{8,9}PC$ is also orders of magnitude higher than for DPPC. Since we have found a similar conductivity difference for EPE as compared to DPPE, the presence of unsaturation in the phospholipid hydrocarbon chains results in a much higher electrical conductivity at the same RIVP than for phospholipids containing saturated chains. The electrical conductivities for the phosphatidylethanolamines of hydrocarbon chain composition similar to that for the phosphatidylcholines are however much less than for the phosphatidylcholines, and, therefore, both the head-group and the hydrocarbon chain composition play an important role in the electrical conductivity increase observed for phospholipids exposed to iodine vapor.

The x-ray diffraction results indicate that iodine reacts differently with DPPC than with EPC bilayers, i.e., iodine decreases the thickness of EPC bilayers, but does not decrease the thickness of DPPC bilayers. The most likely explanation for this result is that iodine interacts with the unsaturated hydrocarbon chains of EPC, modifying the structure of the hydrocarbon core of the bilayer. Iodine does not interact with the saturated chains of DPPC, although the x-ray patterns indicate that exposure to iodine does increase the density of the lipid head group region compared to the hydrocarbon chain region of the bilayer. The simplest explanation here is that iodine or I_3^- accumulates in the head group region of the bilayer. For $DC_{8,9}PC$ films, the situation is more complicated and further study is needed.

The iodine-exposed phospholipid films behave as electrical semiconductors in that the current follows the relationship $I = I_o \exp(-E_A/2kT)$, as discussed earlier. In comparing EPC and DPPC, the greater iodine adsorption corresponds to a higher electrical conductivity and to a lower E_A value. The E_A for $DC_{8,9}PC$, however, was found to be higher than that for DPPC. $DC_{8,9}PC$ has two triple bonds in each hydrocarbon chain and thus has a hydrocarbon chain unsaturation quite different from that of EPC. The electric activation energy of films formed from PC is thus a rather sensitive function of the nature of the hydrocarbon chain unsaturation.

Although our work does not delineate the conductivity mechanism for phospholipids exposed to iodine vapor, we speculate that upon adsorption of an I_2 molecule at the lipid polar head-group, I^+ and I^- entities are formed. The I^- would interact with other I_2 molecules adsorbed near the head-group to form I_3^-, energetically a favorable reaction (Popov, 1967). "Chains" of iodine molecules, adsorbed at the phospholipid polar head-groups, would then act as a pathway for an I^- ion to pass from I_2 to another nearby I_2, thus forming another I_3^- molecule. Such iodine chain formation in starch has been discussed (Thomas, 1959). Some workers (Berashn, 1961) have considered the iodine chain as a one-dimensional metal whereas others (Peticolas, 1963) have found results supporting an ionic conduction mechanism. The unsaturation of the hydrocarbon region of the lipid bilayers could play a role in total iodine adsorbed although in this scheme, it would have no direct action on the electrical conductivity. The validity and details of such a conductivity mechanism await further studies.

ACKNOWLEDGEMENTS: This research was funded in part by the North Carolina Biotechnology Center, grant number 87-6-00-212.

Special thanks go to Ms. Patricia Copenhaver for her excellent secretarial skills.

REFERENCES

Bersohn R and Isenberg I (1961): Metallic nature of the starch-iodine complex. *J Chem Phys* 35: 1640-3.

Bhowmik BB, Jendrasiak GL and Rosenberg B (1967): Charge transfer complexes of lipids with iodine. *Nature* 215: 842-3.

Bhowmik BB, Chatterjee I and Nandy P (1986): Liposome formation of egg lecithin and its interaction with iodine. *Chem Phys Lipids* 39: 271-7.

Blaurock AE and Worthington CR (1966): Treatment of low angle x-ray data from planar and concentric multilayered structures. *Biophysical J* 6: 305-312.

Chatterjee I, Nandy P and Bhowmik BB (1988): Nature of interaction of phospholipid liposome with iodine. *Chem Phys Lipids* 49: 57-63.

Finkelstein A and Case A (1968): Permeability and electrical properties of thin lipid membranes. *J Gen Physiol* 52: 145s-173s.

Herbette L, Marquardt J, Scarpa A and Blasie JK (1977): A direct analysis of a lamellar x-ray diffraction from hydrated oriented multilayers of fully functional sarcoplasmic reticulum. *Biophysical J* 20: 245-272.

Jendrasiak GL (1969): Effect of iodine on the electrical resistance of lipid bilayer membranes. *Chem Phys Lipids* 3: 98-101.

Liberman E and Topaly V (1968): Selective transport of ions through biomolecular phospholipid membranes. *Biochim Biophys Acta* 163: 125-136.

McIntosh TJ and Holloway PW (1987): Determination of the depth of bromine atoms in bilayers formed from bromolipid probes. *Biochemistry* 26: 1783-8.

McIntosh TJ, Magid AD and Simon SA (1987): Steric repulsion between phosphatidylcholine bilayers. *Biochemistry* 26: 7325-7332.

McIntosh TJ, Magid AD and Simon SA (1989a): Cholesterol modifies the short-range repulsive interactions between phosphatidylcholine membranes. *Biochemistry* 28: 17-25.

McIntosh TJ, Magid AD and Simon SA (1989b): Range of the solvation pressure between lipid membranes: Dependence on the packing density of solvent molecules. *Biochemistry* 28: 7904-7912.

Peticolas NL (1963): Helix coil transmission and electronic conductivity of amylose-iodine complex. *Nature* 197: 898-9.

Popov AI (1967): Polyhalogen complex ions, in Halogen Chemistry, Vol. 1, Ed. Viktor Gutman, Academic Press, London, New York, pgs. 225-264.

Rosenberg B and Jendrasiak GJ (1968): Semiconductive properties of lipids and their possible relationship to lipid bilayer conductivity. *Chem Phys Lipids* 2: 47-54.

Ross S and Baldwin VH Jr (1966): The interaction between iodine and micelles of amphipathic agents in aqueous and nonaqueous solutions. *J. Colloid and Interface Science* 21: 284-292.

Streetman BG: Solid State Electronic Devices, 2nd Edition, 1980, Prentice Hall, Englewood Cliffs, N.J., pg. 352.

Thomas JA and French D (1959): The starch-iodine-iodide interaction, Part I, Spectrophotometric Investigations. *J Amer Chem Soc* 82: 4144-7.

Weast R, Editor (1983-1984): CRC Handbook of Chemistry and Physics, D-202.

THE MOLECULAR ELECTROSTATICS OF GLYCOSPHINGOLIPIDS IN ORIENTED INTERFACES

Bruno Maggio
Department of Biochemistry & Molecular Biophysics, Medical College of Virginia, VCU Richmond, VA 23298-0614, U.S.A.

INTRODUCTION

Glycosphingolipids (GSLs) are abundant in nerve cell membranes. They are involved in cellular plasticity, recognition, receptor function and neurodegenerative processes (Ledeen et al., 1988). At the molecular level these lipids modify lipid-lipid and lipid-protein interactions. This has amplified consequences on the membrane structure, phase state and recognition. Figure 1 shows space-filling models of representative major GSLs. Different GSLs

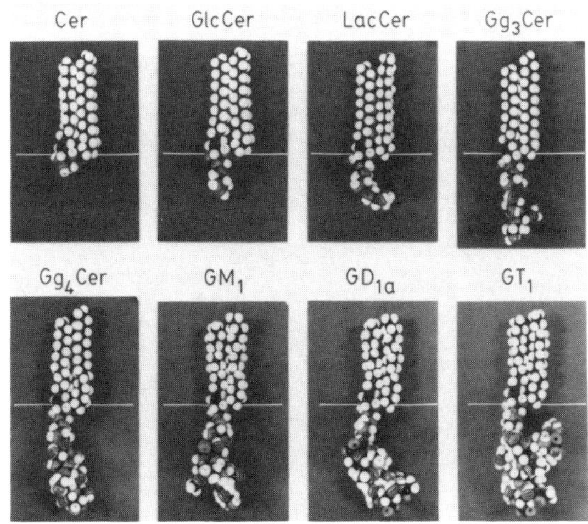

FIGURE 1. *Molecular models of glycosphingolipids*

have a polar head group that varies in the number and type of neutral carbohydrates and in the sialic acid content. The polar head group of GSLs has a dramatic effect on their membrane properties (Maggio et al., 1988).

The oligosaccharide chain of GSLs protrudes into the aqueous environment in a direction perpendicular, on average, to the membrane interface. The resultant molecular dipole determines a wide range of surface potential values for the different GSLs (Maggio et al., 1978a). This can affect the intermolecular packing (Maggio et al., 1986), the activity of phospholipases (Bianco et al., 1991) and intermembrane interactions (Maggio et al., 1989). In this paper we give details of the changes of surface electrostatics that are brought about by variations of the oligosaccharide chain of GSLs.

RESULTS AND DISCUSSION

Surface Electrostatic of GSLs on 145 mM NaCl

The hydrocarbon portion is relatively similar in most of the natural compounds. Although this portion has an effect on the monolayer and bilayer phase behavior, its influence is comparatively less than that induced by changes in the polar head group (Maggio et al., 1985a, 1985b, 1986; Curatolo, 1987). It has been shown that variations of the type, conformation and charge of the oligosaccharide chain modify the molecular packing, membrane topology, surface potential and micropolarity (Maggio et al., 1986; 1988). Fig. 2 shows the changes of the molecular area (upper part) and surface potential (lower part) with the surface pressure for representative GSLs with an increasing number of carbohydrate residues in the oligosaccharide chain, over unbuffered 145 mM NaCl, pH 5.6. Only minor variations occur over the pH range 4 to 12 but marked changes are induced below pH 2 due to protonation of sialosyl residues (Maggio et al., 1981).

The surface potential varies greatly with the lateral pressure (Fig. 2). In lipid monolayers, bilayers and natural membranes the "average" value of the lateral surface pressure is in the range of 30-35 mN/m (Cevc & Marsh, 1987). Due to the thermal energy, the latter can fluctuate by more than 10 mN/m. As a consequence, the electrostatic field gradient perpendicular to the membrane interface will also vary considerably depending on molecular packing.

The resultant molecular dipole moment is related to the measured surface potential by:

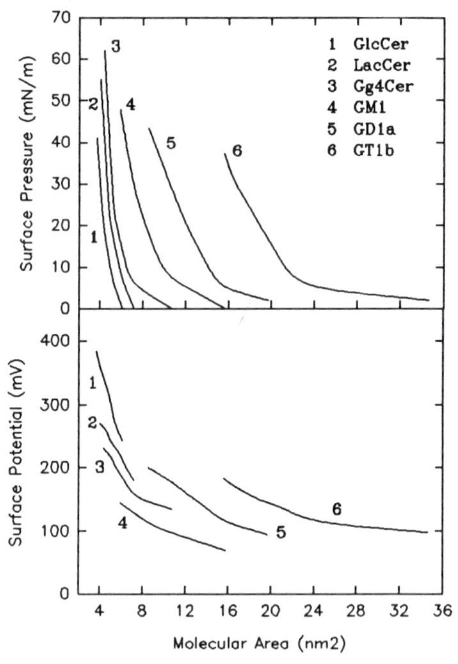

FIGURE 2. *Molecular area and surface potential of glycosphingolipids*

$$\Delta V = 4\,\pi\,n\,\mu \; + \; \phi_0 \qquad (1)$$

where n is the molecular surface density, μ the perpendicular resultant dipole moment of the molecule and ϕ_0 is the ionic double layer potential at the interface (Gaines, 1966). The latter is null for neutral GSLs but has a large negative magnitude for the acidic GSLs (Maggio et al., 1978a; 1981; McDaniel et al., 1986). ϕ_0 decays with distance from the surface in a manner that depends on the concentration and type of electrolyte in the aqueous medium (McLaughlin et al., 1989).

Recently, the electric dipole moments of several GSLs and phospholipids were reported, together with the estimated contribution of the polar head group to the local surface potential (Beitinger et al., 1989). The qualitative conclusions reached are in keeping with several of our earlier studies (Maggio et al., 1978a; 1981). However, the absolute values found for the molecular area and surface potential of gangliosides show considerable variations. This is due to differences in the ganglioside source and the method used for their purification (Fidelio et al., 1991). The molecular dipole moments reported recently (Beitinger et al., 1989) were obtained without substracting the ionic double layer ϕ_0 potential from the measured monolayer surface potential. Not surprisingly, the values reported vary greatly depending on which subphase was used (Beitinger et al., 1989). This is because ϕ_0 is a consequence not only of the molecular charge and interfacial molecular density but also of the ionic conditions in the aqueous medium.

Due to the orientation of the oligosaccharide chain (Maggio et al., 1978a, 1980), the negative charges in gangliosides are not located at the interface but are displaced at least 1 nm away into the water phase. Due to this, the definition of ϕ_0 at the hydrocarbon-water interface is not as straightforward as for acidic lipids with short polar head groups. When the plane of charges is 1 nm away into the water phase this potential is not as negative as for charges located at the interface (McLaughlin et al., 1989). Therefore, the simple Gouy-Chapmann equation previously used to evaluate this potential (Maggio et al., 1978a) is inadequate.

Nevertheless, the Gouy-Chapmann theory can describe the dependence of the double layer potential with the surface charge density and ion concentration at distances $\geq$ 2nm from the surface (McLaughlin et al., 1989). In addition, it was demonstrated that the ionic double layer potential profile at distances $\leq$ 2nm from the surface can be calculated with the Poisson-Boltzmann equation and making all the inherent assumptions of the Gouy-Chapmann theory (McLaughlin et al., 1989). This approach has been successfully applied to describe the electrophoretic behavior of bilayer vesicles containing mono- di- or tri-sialogangliosides (McDaniel et al., 1986). This also means that the effect of negatively charged sialic acid residues located slightly beyond 1 nm from the interface is essentially averaged at that distance. In this work, this approach was used to obtain ϕ_0 at the hydrocarbon-water interface of ganglioside monolayers and to evaluate the molecular dipole moments.

Once ϕ_0 is substracted, the resultant molecular dipole moment still contains several contributions. These belong to the fundamental dipoles in the polar head group, the water dipoles in the hydration shell and those of the hydrocarbon chains. The first two are not easily separated but the latter can be taken into account. The dipole moment contribution of closely packed fatty acyl chains at the air-water interface was determined to be + 0.35 D, positive toward the methyl end of the chain (Vogel & Mobius, 1988). Since the moments of two parallel closely packed chains are additive, the resultant hydrocarbon dipole (above a lateral surface pressure of 30 mN/m) should be + 0.70 D. This value has been used to obtain the polar head group dipole moments of GSLs (Beitinger et al., 1989). However, the hydrocarbon portion of GSLs does not contain two fatty acyl residues but ceramide which consists of a (C18 or C20) sphingosine base and an amide-linked fatty acyl chain (usually C18:0). We have previously determined the dipole moment contribution of the ceramide portion (Maggio et al., 1978a). Although similar to that of two fatty acyl chains, the dipole moment of closely packed ceramide is lower by about 0.1 D. Therefore, we used this value to obtain the dipolar contribution of the oligosaccharide chain (plus the hydration shell) to the overall molecular dipole of GSLs.

Fig. 3 (upper part) shows the variation of the molecular and polar head group dipole moments with the molecular area for different GSLs. However, the surface electrostatic potential rather than the dipole moment, is the magnitude more commonly used and the one that is expressed at the supramolecular level. Once the molecular and polar head group dipole moments have been obtained as described above, the local surface potential (without the ϕ_0 contribution) of the whole molecule and of the oligosaccharide portion can be recalculated (Beitinger et al., 1989). This is shown in Fig. 3 (lower part) as a function of the molecular areas of GSLs. It can be seen that the molecule surface potential

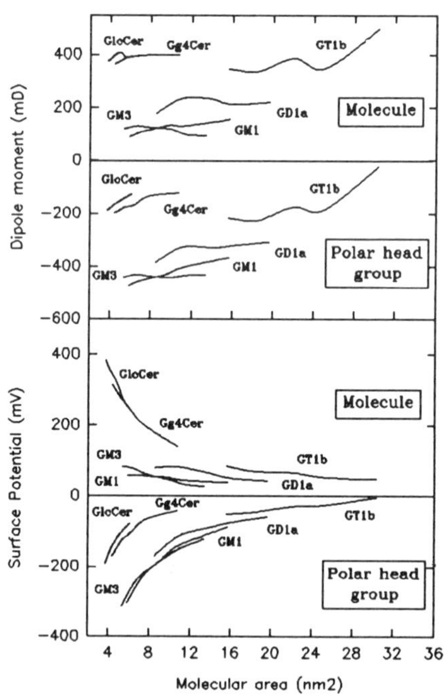

FIGURE 3. *Dipole moments (upper) and surface potential (lower) of glycosphingolipids*

(ceramide, plus oligosaccharide, plus hydration shell) of gangliosides is much lower than that of the neutral GSLs with a comparable number of carbohydrates. The polar head group surface potential is always negative for all the GSLs. This is in agreement with our previous approximate values obtained with the simple Gouy-Chapmann equation for ϕ_0 (Maggio et al., 1978a). Also, in qualitative agreement with our previous results, the polar head group potential is lower for monosialogangliosides but it increases (becomes less negative) for polysialogangliosidcs. The difference between mono- and polysialo-gangliosides can be accounted for by a different spatial orientation, with respect to the rest of the molecule, of the negative charge in the second and third sialosyl residue (Maggio et al., 1978a, 1980).

The values span a wide range of molecular areas and surface potential. The negative surface potential contributed by the oligosaccharide chain represents a depolarization of the interface at the local molecular level that depends on the molecular packing. In addition, the latter varies with the lateral surface pressure (itself a fluctuating parameter according to the, intermolecular interactions and the membrane phase state). The changes of the surface potential also depend on hydration-dehydration effects at the interface and on the presence of solutes that can alter the water structure in bulk and around the polar head group of GSLs (Fidelio et al., 1986; Bianco et al., 1988). Therefore, the surface electrostatics can change dynamically as a response to interactions occurring in the membrane plane and in the aqueous environment.

Changes of the oligosaccharide chain conformation, due to changes in the glycosidic linkages and carbohydrate sequence cause considerable variations of the surface electrostatics (Maggio et al., 1985a). A rigid L-bend (Koerner et al., 1984) is present in the oligosaccharide chain of globoside due to an α glycosidic linkage between the second and third carbohydrate residues. This induces larger intermolecular spacings and decreased surface potential compared to asialo-GM1 (all ß linkages) at comparable lateral surface pressures (Maggio et al., 1985b). On the other hand, the formation of internal carbohydrate esters (ganglioside lactones) between the sialosyl carboxylate and neighboring hydroxyl groups in gangliosides (Ando et al., 1989) brings about dramatic modifications of the electrostatic field at the interface. This has been recently

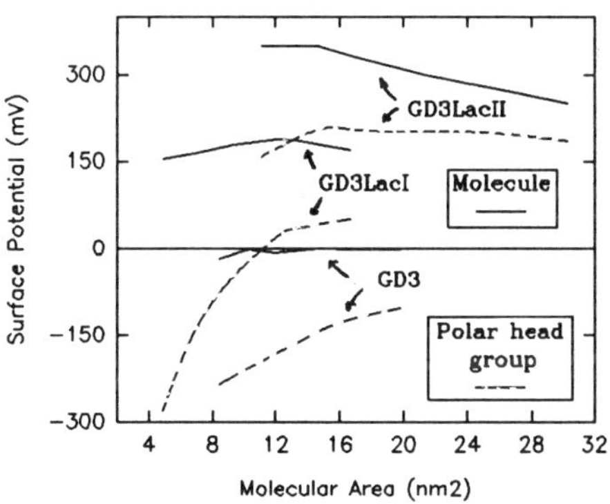

FIGURE 4. *Surface potential of ganglioside GD3 and GD3 lactones*

studied for ganglioside GD3 (Maggio et al., 1990). Fig 4 shows the molecule and polar head group surface potential of two GD3 lactones compared to the parent ganglioside. The large and negative local potential of the oligosaccharide chain of GD3 is decreased, eliminated, or even converted to positive values by the formation of a first (GD3LacI) or second (GD3LacII) lactone ring, depending on the molecular packing areas. This means that the molecular domain can be depolarized or hyperpolarized by interconversion among the ganglioside and its lactone forms. Ganglioside lactones occur naturally in biological membranes (Riboni et al., 1986) and are formed non-enzymatically from the parent ganglioside in acidic media (Ando et al., 1989). The occurrence of a lower surface pH compared to the bulk pH for negatively charged interfaces, due to ϕ_0, has been a long known fact (Gaines, 1966). The values of ϕ_0 obtained for gangliosides are in the range of -20 to -100 mV, depending on the molecular packing areas. As a consequence, the surface pH of these interfaces (for a bulk pH of 7.00) can reach values between 3-4; this is well within the range required for the ganglioside-lactone conversion (Ando et al., 1989). Also, due to ϕ_0, the pK_a of the acidic groups at the surface is increased by 2-4 units. Since this depends on the molecular surface density, different degrees of charge dissociation occur according to the molecular packing. These processes may constitute a rapid manner to vary the surface organization, electrostatics and recognition of ligands (Yu et al., 1985; Ando et al., 1989; Maggio and Yu, 1989). This would not require energetically more costly (and slower) changes of complex metabolic pathways to regulate surface events.

Effect of Ca^{++} on the Surface Electrostatics of GSLs

The surface pressure- and surface potential-area isotherms of representative GSLs in the presence of Ca^{++} at different concentration in the subphase (in addition to 145 mM NaCl at pH 5.6) are shown in Fig. 5. In the concentration range employed Ca^{++} induces only minor variations of the intermolecular packing. Our previous (Maggio et al., 1980; 1987) and present results are in agreement with the reported lack of high affinity or specific binding of Ca^{++} to gangliosides (McDaniel & McLaughlin, 1985; Langner et al., 1988).

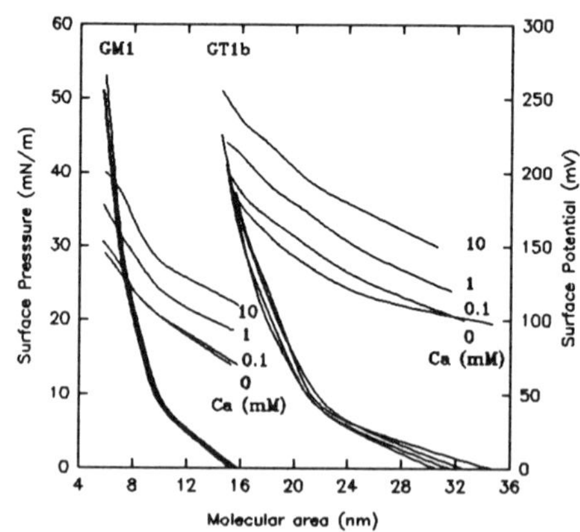

FIGURE 5. *Molecular area and surface potential of GM1 and GT1b in the presence of calcium ions*

The surface potential increases progressively in the presence of increasing concentrations of Ca^{++}. No such increase is observed for the same change of NaCl concentrations. The increase is due to the known preferential accumulation of the divalent cation in the proximity of any negatively charged surface due to the ionic double layer potential ϕ_0 (see eq. 1). According to the values for ϕ_0, the concentration of ions near the interface of closely packed acidic GSLs (calculated with the Grahame equation, see Israelachvili, 1985) can be of the order of 1-10 M for a bulk concentration of 145 mM NaCl and 1 mM $CaCl_2$. On relative terms the divalent cation is preferentially accumulated (Israelachvili, 1985).

Fig. 6 (upper part) shows the molecule and polar head group surface potential in the presence of 1 mM $CaCl_2$ as a function of the molecular area. These values were obtained after substracting the ionic double layer potential (see above). In this case, a ϕ_0' was calculated with the Grahame equation for monovalent and divalent ions (Israelachvili, 1985). The ϕ_0' was ascribed to a plane 1 nm away from the hydrocarbon-water surface. The value for ϕ_0 at the hydrocarbon-water interface was then obtained from ϕ_0' with the Poisson-Boltzmann eq. (Langner et al., 1989). Fig. 6 (upper part) should be compared with Fig. 3 (lower part). A direct comparison of the effect of 1 mM and 10 mM Ca^{++} on the molecule and polar head group surface potential is shown in Fig. 6 (lower part) for different GSLs at a lateral surface pressure of 30 mN/m; negligible changes are observed below 500 uM Ca^{++}. In the mM range, the presence of Ca^{++} induces a marked increase of the molecule and polar head group potentials for all gangliosides. For gangliosides GD1a and GT1b at close molecular packing, the negative contribution of their polar head group to the surface potential is much diminished or almost eliminated, depending on the Ca^{++} concentration. As a consequence, the local

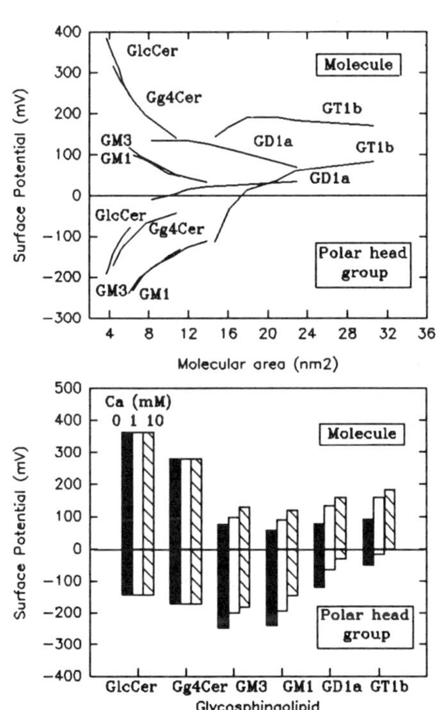

FIGURE 6. *Surface potential of glycosphingolipids in the presence of calcium ions*

depolarizing effect of the oligosaccharide chain is greatly decreased. The bulk Ca^{++} concentration at which these effects are observed is in the range of the intrinsic association constant for gangliosides and acidic phospholipids (McDaniel & McLaughlin, 1985). In the mM range Ca^{++} also modifies the micropolarity of interfaces with acidic GSLs (Montich et al., 1985). Therefore, besides molecular packing and polar head group conformation, the surface can be hyperpolarized or depolarized at the local molecular level by changes in concentration of ion second messengers such as Ca^{++}.

Variations in the magnitude of ϕ_0 depending on packing and on the surface charge density transmit a tangential stress and exert additional pressure on the molecules along the surface. The tangential pressure can vary by more than 10 mN/m, depending on the GSLs and molecular packing. Lateral pressure waves in this range can induce electrostatically driven phase transitions and membrane curvature effects (Cevc & Marsh, 1987; Maggio et al., 1988). These factors can also affect the repulsive pressure acting between approaching surfaces; this extends far into the aqueous medium (Israelachvili, 1985) and can mediate membrane-membrane interactions (Maggio & Yu, 1989).

Effect of GSLs-phospholipid interactions

GSLs exhibit different interactions with phospholipids depending on the polar head group of both lipids. This modifies the mean area and surface potential per molecule (Maggio et al., 1978b). Fig. 7 (upper part) shows the magnitude of the interactions at a surface pressure of 30 mN/m (as % deviations from the ideal behavior with no interactions) for mixtures of some GSLs (molar fraction 0.25) with dipalmitoyl phosphatidylcholine (dpPC).

The mean area and surface potential per molecule are increased in mixed monolayers with neutral GSLs or monosialogangliosides but are decreased with polysialogangliosides. This is a consequence of thermo dynamically favorable or

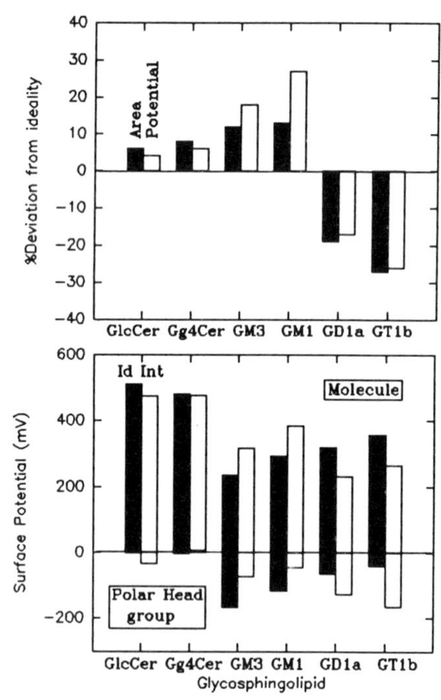

FIGURE 7. *Effect of interactions of glycosphingolipids with phosphatidylcholine on the surface potential*

unfavorable, respectively, dipolar interactions which in turn affect the molecular packing areas (Maggio et al., 1978b, 1980).

Fig. 7 (lower part) shows that the polar head group surface potential in monolayers with dpPC and monosialogangliosides is less negative than for the ideal mixture; this leads to an increase of the molecule surface potential. Conversely, with polysialo- gangliosides the polar head group potential is more negative than that for non-interacting molecules and this results in a decreased surface potential of the molecule. Thus, the local surface electrostatics in mixed lipid interfaces can be markedly modified by GSLs. Local hyperpolarization or depolarization can occur depending on the mutual interactions. This can have amplified consequences on the membrane topology, permeability and phospholipid degradation by enzymes that generate lipid second messengers (Maggio et al., 1988; Bianco et al., 1991)

CONCLUSIONS

The surface electrostatics of GSLs is markedly influenced by the properties of the oligosaccharide chain. The latter contributes a large negative electrostatic potential perpendicular to the surface. This can locally depolarize de interface in a manner that depends on, and in turn affects, the intermolecular packing. The conformation, the type of glycosidic linkages, the state of ionization and the presence of internal lactone rings affect dramatically the surface potential and the intermolecular organization.

The presence of Ca^{++} in the mM range induces variations of the molecule and polar head group surface potential due to its preferential accumulation in the ionic double layer. As a consequence, the depolarizing effect of the oligosaccharide chain is decreased and the overall molecule surface potential is reverted to more positive values. The presence of the ionic double layer potential causes that the surface pH of acidic GSLs is 2-4 units less than the bulk pH and the pK_a of ionizable groups is increased by similar values. This depends on the molecular packing and varies with the accumulation of electrolytes in the ionic double layer. The lateral surface pressure is modified by these electrostatic effects; the fluctuating tangential pressure thus generated along the surface is of the order of 2-10 mN/m, depending on the molecular packing and ϕ_o. These values are in the range of those required to induce electrostatically driven phase transition processes in the membrane. In addition, the large negative electrostatic field of the oligosaccharide chain should enhance the repulsive pressure between approaching surfaces in an aqueous medium. These effects can occur within physiological conditions of pH and electrolyte concentration and may represent rapid mechanisms for the transduction or amplification of surface events.

ACKNOWLEDGEMENTS

The research described was supported in part by grants from the National Multiple Sclerosis Society (FG-644A-1 and RG-2170-A-2) in the USA and by CONICOR and CONICET (Argentina). The continuous stimulus and support of Dr. R.K. Yu is gratefully acknowledged.

REFERENCES

Ando S, Yu RK, Scarsdale JN, Kusunoki S and Prestegard JH (1989): High resolution proton NMR studies of gangliosides. Structure of two types of GD3 lactones and their reactivity with monoclonal antibody. *J Biol Chem.* 264:3478-3483.

Beitinger H, Vogel V, Möbius D and Rahmann H (1989): Surface potential and electric dipole moments of ganglioside and phospholipid monolayers: contribution of the polar head group at the water/lipid interface. *Biochim Biophys Acta* 984:293-300.

Bianco ID, Fidelio GD and Maggio B (1988): Effect of glycerol on the molecular properties of cerebrosides, sulphatides and gangliosides in monolayers. *Biochem J* 251:613-616.

Bianco ID, Fidelio GD, Yu RK and Maggio B (1991): Degradation of dilauroylphosphatidylcholine by phospholipase A2 in monolayers containing glycosphingolipids. *Biochemistry* 30:1709-1714.

Cevc G and Marsh D (1987): *Phospholipid Bilayers. Cell Biology: A series of Monographs* (Bittar EE, ed.) vol. 5. Wiley-Interscience. New York.

Curatolo W. (1987): The physical properties of glycolipids. *Biochim Biophys Acta* 906:111-136.

Fidelio GD, Maggio B and Cumar FA (1986): Molecular parameters and physical state of neutral glycosphingolipids and gangliosides in monolayers at different temperatures. *Biochim Biophys Acta* 854:231-239.

Fidelio GD, Ariga T and Maggio B (1991): Molecular parameters of gangliosides in monolayers: comparative evaluation of suitable purification procedures. *J Biochem.* 101:111-116.

Gaines GL Jr (1966): Insoluble monolayers at liquid-gas interfaces. *Interscience monographs on physical chemistry* (Prigogine I, ed.) Interscience. New York.

Israelachvili J (1985): *Intermolecular and surface forces.* Academic Press. New York.

Ledeen RW, Hogan EL, Tettamanti G, Yates AJ and Yu RK (1988): *New trends in ganglioside research: neurochemical and neuroregenerative aspects.* FIDIA Research Series, vol. 14. Liviana Press. Padova. Italy.

Langner M, Winiski A., Eisemberg M, McLaughlin A and McLaughlin S (1988): The electrostatic potential adjacent to bilayer membranes containing either charged phospholipids or gangliosides. In: *New trends in ganglioside research: neurochemical and neuroregenerative aspects* (Ledeen RW, Hogan EL, Tettamanti G, Yates AJ and Yu RK, eds.) FIDIA Research Series, vol. 14, pp 121-131. Liviana Press. Padova. Italy.

Maggio B and Yu RK (1989): Interaction and fusion of unilamellar vesicles containing cerebrosides and sulfatides induced by myelin basic protein. *Chem Phys Lipids* 51:127-136.

Maggio B, Cumar FA and Caputto R (1978a): Surface behavior of gangliosides and related glycosphingolipids. *Biochem J* 171:559-565.

Maggio B, Cumar FA and Caputto R (1978b): Interaction of gangliosides with phospholipids and glycosphingolipids in mixed monolayers. *Biochem J* 175:1113-1118.

Maggio B, Cumar FA and Caputto R (1980): Configuration and interactions of the polar head group in gangliosides. *Biochem J* 189:435-440.

Maggio B, Cumar FA and Caputto R (1981): Molecular behavior of glycosphingolipids in interfaces. Possible participation in some properties of nerve membranes. *Biochim Biophys Acta* 650:69-87.

Maggio B, Ariga T, Sturtevant JM and Yu RK (1985a): Thermotropic behavior of glycosphingolipids in aqueous dispersions. *Biochemistry* 24:1084-1092.

Maggio B, Ariga T and Yu RK (1985b) Molecular parameters and conformation of globoside and asialo-GM1. *Arch Biochem Biophys.* 241:14-21.

Maggio B, Fidelio GD, Cumar FA and Yu RK (1986): Molecular interactions and thermotropic behavior of glycosphingolipids in model membrane systems. *Chem Phys Lipids* 42:49-63.

Maggio B, Sturtevant JM and Yu RK (1987): Effect of calcium ions on the thermotropic behavior of neutral and anionic glycosphingolipids. *Biochim Biophys Acta* 901:173-182.

Maggio B, Monferran CG, Montich GG and Bianco ID (1988) Effect of gangliosides and related glycosphingolipids on the molecular organization and physical properties of lipid-protein systems. In: *New trends in ganglioside research: neurochemical and neuroregenerative aspects* (Ledeen RW, Hogan EL, Tettamanti G, Yates AJ and Yu RK, eds.) FIDIA Research Series, vol. 14, pp 105-120. Liviana Press. Padova. Italy.

Maggio B, Ariga T and Yu RK (1990): Ganglioside GD3 lactones: polar head group mediated control of the intermolecular organization. *Biochemistry* 29:8729-8734.

McDaniel RV and McLaughlin S (1985) The interaction of calcium with gangliosides in bilayer membranes. *Biochim Biophys Acta* 819:153-160.

McDaniel RV, Sharp K, Brooks D, McLaughlin AC, Winiski AP, Cafiso D and McLaughlin S (1986): Electrokinetic and electrostatic properties of bilayers containing gangliosides GM1, GD1a or GT1. *Biophys J* 49:741-752.

McLaughlin S (1989): The electrostatic properties of membranes. *Ann Rev Biophys Biophys Chem* 18:113-136.

Montich GG, Bustos M, Maggio B and Cumar FA (1985): micropolarity of interfaces containing anionic and neutral glycosphingolipids as sensed by merocyanine 540. *Chem Phys Lipids* 38:319-326.

Vogel V and Möbius D (1987) Hydrated polar head groups in lipid monolayers: effective local dipole moments and dielectric properties. *Thin Solid Films* 159:73-81

DIRECT ENERGETIC INTERACTION OF ION TRANSPORT SYSTEMS IN BACTERIAL MEMBRANE

Armen A.Trchounian
Department of Biophysics, Biological Faculty of Yerevan State University,
375049 Yerevan, Armenia, USSR

INTRODUCTION

Bacteria accumulate considerable quantity of K^+ and create high distribution of K^+ between a cell and the medium through the genetically determined specialized membrane transport systems: the constitutive Trk in E.coli (Walderhaug et al.,1987) or Ktrl in S.faecalis (Harold and Kakinuma,1985) and the repressible. The K^+ uptake through the constitutive system is carried out together the secretion of H^+ via the proton pumps of membrane (Epstein and Schultz,1965; Durgaryan and Martirosov,1978; Martirosov and Trchounian,1981ab). How do these ion transport systems interact together?

Interactions between membrane transport systems acquire an important significance. According to Mitchell's theory (Mitchell,1966), primary transport systems - redox chain or the H^+-ATPase complex F_1F_0, which can extract energy from chemical compounds,- transfer H^+ through the membrane and generate the $\Delta\mu H$, then secondary systems use this energy for useful work in such processes as the ATP synthesis or ion transport. The thermodynamic value-$\Delta\mu H$ is the same for exclusive volume of a cell, and primary and secondary systems can be on distance in membrane and interact together indirectly via the $\Delta\mu H$. Transport systems in anaerobic bacteria which are able to transfer ions and other substances against transmembrane electrochemical gradients use energy from ATP under it's hydrolysis through the F_1F_0.

Many laboratories have collected data, which are not explained by Mitchell's theory. We can assume that primary and secondary transport systems interact together with formation of heterostructures within membrane and <u>direct transfer of energy</u> without the mediation of $\Delta\tilde{\mu}H$.

ION DISTRIBUTION BETWEEN A CELL AND THE MEDIUM AND $\Delta\tilde{\mu}H$ IN BACTERIA

Uneven ion distribution between a cell and the medium is high, but different for anaerobically and aerobically grown bacteria. Anaerobically grown <u>E.coli</u> (Martirosov and Trchounian,1986) and other bacteria (Trchounian et al.,1987ab) accumulate K^+ up to 0.8-1.2 M and create distribution of these cations between a cell and the medium higher than 10^3, at the same time such distributions in aerobically grown bacteria are lower (Table). Intensive accumulation of K^+ and high distribution of these cations between a cell and the medium are energy-dependent and need ATP or $\Delta\tilde{\mu}H$. Proton-motive force in anaerobically grown bacteria is equal to -100-160 mV and includes only the membrane potential ($\Delta\Psi$) under the <u>Δ</u>pH equal to zero, at the same time this force in aerobically grown bacteria is higher and equal to -160-200 mV (Martirosov et al.,1981; Martirosov and Trchounian,1986; Trchounian et al.,1987ab; Table).

TABLE. The values of K^+ distributions between a cell and the medium, K^+-equilibrium potentials and membrane potentials in anaerobically (A) and aerobically (B) grown bacteria.

Bacteria	Growth condi- tions	External K^+ activity for the mo- ment of max accumulati- on (mM)	Ratio of the internal to external K^+ activity for this moment	K^+-equilibrium potential (mV)	$\Delta\Psi$ (mV)
<u>E.coli</u>[1]	A	0.48	2530	207	142
<u>S.typhimurium</u>[2]	A	0.15	2460	207	110
<u>L.salivarius</u>[3]	A	0.23	1720	198	146
<u>E.coli</u>[1]	B	0.51	720	175	169
<u>E.coli</u>[4]	B	0.98	434	161	155
<u>S.typhimurium</u>[2]	B	0.19	585	169	166

1 (Martirosov and Trchounian,1986)
2 (Trchounian et al.,1987b)
3 (Trchounian et al.,1987a)
4 for cells treated with cyanide (Martirosov and Trchounian,1986)

These rezults permit us to conclude that the constitutive system for K^+ uptake, which creates high distribution of these cations between a cell and medium, operates as a pump with using of ATP energy in anaerobically grown and anaerobic bacteria and as an ionophore- uniport or symport with H^+ with using of the H in aerobically grown bacteria.

ION EXCHANGE IN BACTERIA AND ITS NATURE

Epstein and Schultz (1965) established that <u>E.coli</u> exchanges H^+ of a cell for K^+ of the medium and proposed an existence of a H^+-K^+-pump in bacteria. We have shown that anaerobically grown <u>E.coli</u> (Durgaryan and Martirosov,1978; Martirosov and Trchounian,1981,1982,1986) and other bacteria (Trchounian et al.,1987ab) carry out H^+-K^+-exchange. Such exchange has important pecularities: 1) it is inhibited by the N,N'-dicyclohexylcarbodiimide (DCCD); 2) it has two steps in gram-negative bacteria, the first of which is osmosensitive; 3) the DCCD-sensitive and osmodependent H^+-K^+-exchange has stable stoichiometry of cation fluxes, equal to 2. These data indicate that the $2H^+/K^+$-exchange in the first step is carried out through the same mechanism, functioning as a H^+K^+-pump. The second step of the DCCD-sensitive H^+-K^+-exchange in gram-negative bacteria has no stable stoichiometry and is carried out most probably as a H^+-K^+-antiport.

Such exchange in aerobically grown bacteria has no stable stoichiometry (Martirosov and Trchounian,1986; Trchounian et al.,1987b). Many other distingtions in the H^+-K^+-exchange between anaerobically and aerobically grown bacteria can be observed.

All the above results indicate that the character of ion exchange in bacteria depends on growth conditions and metabolism.

Interesting fact about the DCCD-sensitive $2H^+/K^+$-exchange in anaerobically grown and anaerobic bacteria put forward two questions: 1) why does K^+ uptake take place together with $2H^+$ secretion and 2) why does this exchange manifest a sensitivity to the DCCD? It had been assumed that this exchange was carried out through the F_1F_0 (Martirosov,1979). Such assumption did not coincide with idea that the F_1F_0 was a pure proton pump. We propose that the F_1F_0 and constitutive system for K^+ uptake <u>Trk</u> or (<u>Trk</u>-like) are associated together with formation a supercomplex, functioning as a H^+-K^+-pump (Martirosov and Trchounian,1982,1983; Trchounian et al.,1987). And analysis of the character of ion exchange in mutants of <u>E.coli</u> with defect in different subunits of the F_1F_0 and in K^+ transport is of prime consequence.

Study of the character of H^+-K^+-exchange in mutants of <u>E.coli</u> with defects in K^+ transport (the mutants were the gift of Prof. W.Epstein from the University of Chicago, USA) has shown following: the DCCD-sensitive $2H^+/K^+$-exchange is carried out through the Trk and the DCCD-sensitive unstable H^+-K^+-exchange in the second step in anaerobically grown bacteria - via "defective Trk" (mutations in trkA and trkD genes) or "X" (Martirosov and Trchounian,1981a).

Study of the character of H^+-K^+-exchange in unc-mutants of E.coli with defects in different subunits of the F_1F_0 (the mutants were the gift of Dr.Barbara Bachman from Genetic Center in New Haven, USA) has shown following: 1) defects in α, β and γ subunits of F_1 responsible for ATPase activity of the F_1F_0 (Dunn,1978; Kanazawa et al.,1978) destroy the $2H^+/K^+$-exchange through the Trk without any effects on the DCCD-sensitive H^+-K^+exchange with unstable stoichiometry via "defective Trk" (Martirosov and Trchounian, 1981b,1983); 2) defect in ε subunit of F_1 breaks the regulation of $2H^+/K^+$-exchange of the part of cell turgor without the influence on operation of mechanisms for ion exchange (Martirosov and Trchounian,1983); 3) under the defective a subunit of F_0 (m.m. 24 kDa) H^+-K^+-exchange loses the DCCD-sensitivity and the $2H^+/K^+$-exchange through the Trk in addition loses its osmosensitivity; under the defective b subunit of F_0 (m.m. 18 kDa) H^+-K^+-exchange loses the DCCD-sensitivity and the $2H^+/K^+$-exchange is blocked; the only mutation in the unc-operon producing the complete blocking of ion exchange is related to a defect in c subunit of F_0 (m.m. 8.4 kDa) which is the DCCD-sensitive and apparently the gate component of the F_1F_0 (Martirosov and Trchounian,1983). We want to note that the mutants with "defective Trk" as well as the mutants with defects in α subunit of F_0 utilize glucose, generate $\Delta\psi$ near -100-140 mV, which includes the DCCD-sensitive component (Martirosov et al.,1981), but carry out H^+-K^+-exchange with unstable stoichiometry for the DCCD-sensitive cation fluxes (Martirosov and Trchounian,1981ab).

Thus, gene-product of the unc-operon- the F_1F_0 in anaerobically grown bacteria has direct relationship to K^+ uptake, defects in F_1 have an effect on operation of the Trk and defects in F_0 affect functions of the Trk and "defective Trk" both. We are forced to confirm that the H^+-K^+-pump exchanged $2H^+/K^+$ consists from the F_1F_0 and Trk. In other words, these ion transport systems are associated into the united supercomplex functioning as the H^+-K^+-pump. Proposed by Epstein and Schultz (1965) H^+-K^+-pump acquires apparently molecular bases. We can assume also that the F_0 and "defective Trk" in anaerobically grown gram-negative bacteria carry out together unstable H^+-K^+-exchange.

THE H^+-K^+-PUMP IN BACTERIA: DIRECT ENERGETIC INTERACTION BETWEEN THE F_1F_0 AND K^+-TRANSPORT

Based on our data we propose that the H^+-K^+-pump exchanged $2H^+/K^+$ consists from independent ion-transport systems the F_1F_0 and _Trk_ (or _Trk_-like), which have possibility for intramembrane interactions and form a united supercomplex (Fig.). Unlike this model, these systems in aerobically grown bacteria operate separately and the _Trk_ functions as an ionophore - uniport or symport with H^+ using $\Delta\mu H$ (Fig.).

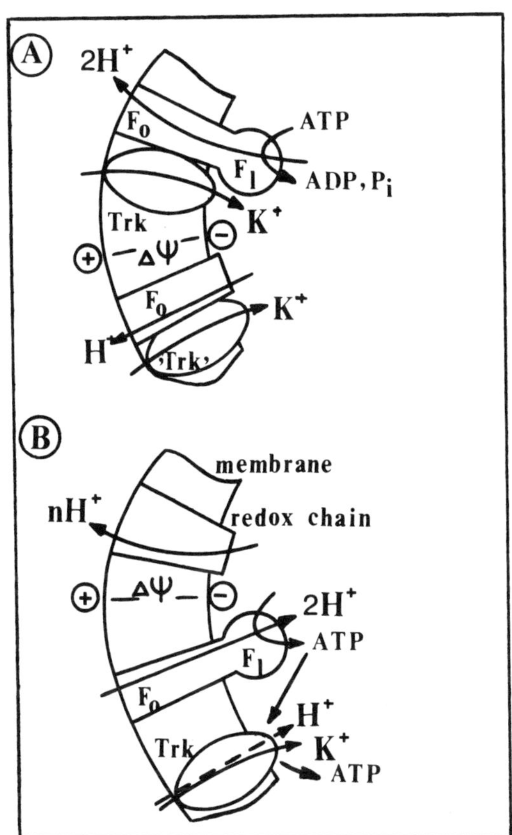

FIGURE. Proposed models for the H^+-K^+-exchange and interactions between the F_1F_0 and K^+-transport in anaerobically (A) and aerobically (B) grown bacteria.

Our assumption about direct interaction between the F_1F_0 and _Trk_ in anaerobically grown bacteria with formation of supercomplex, functioning as the H^+-K^+-pump, exchanged $2H^+/K^+$, has consequences, which can

be experimentally proved. 1) the H^+-K^+-pump must be reversal and in definite cinditions exchange $2H^+$ of the medium for K^+ of a cell with coupled synthesis of ATP. 2) H^+-K^+-pump must show K^+-dependent ATPase activity. 3) H^+-K^+ pump must display 8 typical subunits for the F_1F_0 (Foster
and Fillingame,1979; Leingruber et al.,1981) and additional proteins,
corresponding to Trk.

Taken into consideration that reverse of the H^+-K^+-pump must be observed under the values of $\Delta\mu H/F$ and $\Delta\mu K/F$ near to cell phosphate potential (52 kJ) and low turgor pressure, we have illustrated reversal of
the mechanism exchanging $2H^+/K^+$ with the coupled synthesis of ATP,
which is inhibited with the DCCD; the stoichiometry of entire pump cycle is $2H^+$-K^+-ATP (Martirosov and Trchounian,1982).

We have determined the DCCD-sensitive ATPase activity in protoplasts
(Martirosov et al.,1988) and isolated membranes (Trchounian and Ogandjanian,1989) of anaerobically grown bacteria dependent on K^+. Using mutants of E.coli with defects in different subunits of the F_1F_0 and K^+-
transport, we have shown also that this ATPase activity is observed only under structural integrity of the F_1F_0 and Trk.

Finally, we have shown that the F_1F_0 isolated from anaerobically
grown bacteria with sodium deoxycholate and ammonium sulphate (Leingruber et al.,1981) includes 8 typical and two additional subunits (Trchounian et al.,1991). These two additional proteins with m.m. 50-53 and 12-
15 kDa correspond as we assume to the Trk. Such preparations demonstrate
the DCCD-sensitive ATPase activity, increased strongly in 3 times with
changing of K^+ activity in the medium from zero to 100 mM.

Thus, we have all bases to confirm that the F_1F_0 and Trk system can
interact together within bacterial membrane and form a supercomplex,
functioning as the H^+-K^+-pump. Within such supercomplex direct transfer
of energy without $\quad$ H takes place.

Interaction of the F_1F_0 with Trk within membrane seems to clear, but
direct transfer of energy between these proteins doesn't remain. Transfer of energy is coupled usually with transport of reducing equivalents
($H^+ + e^-$) but the F_1F_0 and Trk have no redox locuses. Therefore, interesting question about third component in interaction between these proteins, provided reducing equivalents, came up. Bagramyan and Martirosov
(1989) have shown that E.coli, performed heteroenzymatic mixed fermentation, gives off H_2 with formate-hydrogen lyase under the $2H^+/K^+$-exchange
through the F_1F_0 and Trk, and proposed experimental model of direct

transduction of energy within such supercomplex via a dithiol-disulphide interchange with reaction of 2SH-- S-S + H_2. According to this model, S-S-link between proteins is a mediator and S-H-link - accumulator of energy. One H^+ from $3H^+$ transfered through the F_1F_0 is used for reducing of SH-group, electron for this is taken away from formate with formate-hydrogen lyase. Another SH-group is reducing by H which is taken away also from formate. Proposed model is possible, but doesn't explain ATP-ase activity in anaerobically grown <u>E.coli</u> (Martirosov et al.,1988; Trchounian and Ogandjanian,1989; Trchounian et al.,1991) and the $2H^+/K^+$-exchange with the F_1F_0 and <u>Trk</u>-like system in other bacteria performed homoenzymatic lactoacid fermentation (Trchounian et al.,1987a).

CONCLUSION

We have resulted arguments for the structural association of the F_1F_0 and K^+-transport into the united supercomplex with direct transfer of energy. We suggest that interaction between these proteins within membrane of anaerobically grown and anaerobic bacteria is performed directly without the mediation of $\Delta\mu H$.

ACKNOWLEDGEMENTS

I wish to thank Profs. S.M.Martirosov, Galina D.Mironova, D.N.Ostrovsky and L.S.Yaguzhinsky for active discussion and valuable advices and Drs. Elena S.Ogandjanian, Victoria A.Ter-Nikogossian, A.G.Vardanian and Karine A.Bagramyan for collaboration.

REFERENCES

Bagramyan KA and Martirosov SM (1989): Formation of an ion transport supercomplex in <u>Escherichia coli</u>. An experimental model of direct transduction of energy. <u>FEBS Lett.</u> 246: 149-152.

Dunn S (1978): Identification of the altered subunit in the inactive F_1-ATPase of an <u>Escherichia coli</u> <u>uncA</u> mutants. <u>Biochem.Biophys.Res.Commun.</u> 82: 596-601.

Durgaryan SS and Martirosov SM (1978a): An electrochemical study of energy-dependent potassium accumulation in <u>E.coli</u>. 1. Two steps in potassium accumulation and osmosensitivity. <u>Bioelectrochem.Bioenerg.</u> 5: 554-560.

Durgaryan SS and Martirosov SM (1978b): ... 3. Stoichiometry of H^+-K^+-exchange sensitive to N,N'-dicyclohexylcarbodiimide. <u>Bioelectrochem.Bi-</u>

oenerg. 5: 567-573.

Epstein W and Schultz SG (1965): Cation transport in Escherichia coli. 5. Regulation of cation content. J.Gen.Physiol. 49: 221-234.

Foster DL and Fillingame RH (1979): Energy transducing H^+-ATPase of Escherichia coli. Purification, reconstitution and subunit composition. J.Biol.Chem. 254: 8230-8236.

Harold FM and Kakinuma Y (1985): Primary and secondary transport of cations in bacteria. Annu.N.Y.Acad.Sci. 456: 375-383.

Leingruber RM, Jensen C and Abrams A (1981): Purification and characterization of the membrane adenosine triphosphatase complex from wild type and N,N'-dicyclohexylcarbodiimide-resistant strains of S.faecalis. J.Bacteriol. 148: 363-372.

Martirosov SM (1979): An electrochemical study of energy-dependent potassium accumulation in E.coli. 4. Regulation of the H^+-K^+-pump operation by a periplasmic protein. Bioelectrochem.Bioenerg. 6: 315-321.

Martirosov SM, Ogandjanian ES and Trchounian AA (1988): ... 12. K^+-dependent ATPase activity (Arguments for the structural association of H^+-ATPase with K^+-ionophore). Bioelectrochem.Bioenerg. 19: 353-357.

Martirosov SM, Petrosian LS, Trchounian AA et al. (1981): ... 8. Membrane potential (with comparison of S.faecalis). Bioelectrochem.Bioenerg. 8: 613-620.

Martirosov SM and Trchounian AA (1981a): ... 6. Identification of H^+-K^+-exchanging systems. Bioelectrochem.Bioenerg. 8: 597-603.

Martirosov SM and Trchounian AA (1981b): ... 7. On the structure of H^+-K^+-exchanging systems. Bioelectrochem.Bioenerg. 8: 605-611.

Martirosov SM and Trchounian AA (1982): ... 9. Reversal of the mechanism exchanging $2H^+$ for K^+ with the coupled synthesis of ATP. Bioelectrochem.Bioenerg. 9: 459-467.

Martirosov SM and Trchounian AA (1983): ... 10. Operation of transport systems exchanging H^+ for K^+ in unc-mutants. Bioelectrochem.Bioenerg. 11: 29-36.

Martirosov SM and Trchounian AA (1986): ... The Trk system in anaerobically and aerobically grown cells. Bioelectrochem.Bioenerg. 15: 417-426.

Mitchell P (1966): Chemiosmotic coupling in oxidative and photosynthetic phosphorylation. Biol.Rev. 11: 445-502.

Trchounian AA and Ogandjanian ES (1989): On the reason of the principle
of intramembranal interaction of transport systems in bacteria.
Stud.Biophys. 132: 231-234.

Trchounian AA, Ogandjanian ES and Martirosov SM (1987a): H^+-K^+-exchange
and $\Delta\mu H$ in Lactobacillus salivarius. Bioelectrochem.Bioenerg.
17: 503-508.

Trchounian AA, Ogandjanian ES and Mironova GD (1991): An electrochemical study of energy-deoendent potassium accumulation in E.coli.
13. On the interaction of the F_1F_0 with Trk. Bioelectrochem.Bio-
energ. 25.

Trchounian AA, Ter-Nikogossian VA and Martirosov SM (1987b): Energy-dependent K^+ uptake and H^+-K^+-exchange in Salmonella typhimurium.
Bioelectrochem.Bioenerg. 17: 183-192.

Walderhaug MO, Dosch DC and Epstein W (1987): Potassium transport in bacteria. In: Ion transport in prokaryotes. N.Y.: Acad.Press.

Ion and Electron Transport Properties of
Biological and Artifical Membranes

ELECTRONIC BEHAVIOR DIFFERENCES IN MUSCLE MEMBRANES
Milton J. Allen
Biophysical Laboratory, Department of Chemistry, Virginia
Commonwealth University, Richmond, Virginia 23284-2006

INTRODUCTION

The development of techniques for the study of charge transfer
processes in 'sheets' of cells from metabolically viable systems has
led to many interesting areas of investigation (Allen, 1989), (Allen
& Geffert, 1989a).

Previous studies on the electronic behavior of male CD-1 Swiss
Albino mice (Charles River) demonstrated properties which differed
somewhat from the characteristics of the recently investigated
female of this particular strain. Whether these sexual dissimilari-
ties exist in other strains of mice remains to be seen. However, be
as it may, the implications of this divergence in behavior at a
molecular and submolecular level is of great interest.

Some years ago it was found that the electrochemical parameters
of a disease process, namely muscular dystrophy, was reflected in
the metabolic viability of the erythrocytes of the afflicted individuals (Allen, 1971). Thus it was suggested that perhaps such a
phenomenon might occur in muscle tissue from an animal afflicted
with a neurological disorder. In this instance the shiverer
(shi/shi) strain of mouse was chosen for the studies. The shi/shi
is mutation of the myelin basic protein structural gene (MBP) which
results in the absence of myelin in the central nervous system. As
a result, the animal demonstrates a shivering gait which can be
observed within a few days of birth and persists until death at 50
to 100 days. Introduction of a wild type MBP gene into the genome

of a shi/shi results in a shiverer mouse homozygous for the MBP transgene. This animal is phenotypically and behaviorally normal ('cured'). Examination of the shi/shi diaphraghm muscle demonstrated significant differences in electronic behavior when compared to similiar muscle membranes obtained from cured mice.

For some time it has been known that insulin will bind to muscle tissue (Stadie, et al, 1950) and more recently, the extent of this binding, detectable from the electronic behavior of this tissue (Allen, 1985). Therefore in both of the above mentioned studies it was felt it would be of interest to examine the effects of insulin pretreatment on the muscle tissues used.

EXPERIMENTAL

Animals

The CD-1 strain mice consisted of both male and female contained in separate cages maintained under identical conditions. These animals were 40-55 days of age when used. The shiverer (shi/shi) and the cured mice investigated were restricted to the male of the strain and utilized between the ages of from 34-46 days.

A general procedure consisted of extirpating the diaphragm from the rib-cage and placing it in a petri dish containing 15ml Hanks basic salt solution (HBSS). The tissue was maintained in a viable state for up to 24 hr by refrigeration at 4°C. In instances where the diaphragm tissue was to be pretreated with insulin (IPT), a section of the excised tissue was stored overnight in 15ml HBSS containing 3 Units of Procine sodium insulin/ml. Prior to use, a sector of the membrane was removed and rinsed three times in successive containers of HBSS.

Electrochemical Cell

The cell used in these studies is described in the following figure.

FIGURE 1

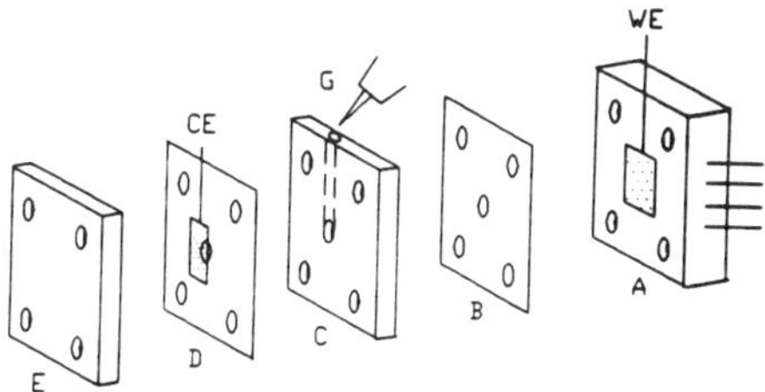

(A) thermoelectric temperature block 5X7.5X0.6cm, (B,D) 0.1cm thick silicone rubber gaskets with centrally located 0.3cm ID windows, (C) perspex compartment 5X7.5X0.4cm with 0.3cm window and .25cm ID channel for introduction of media and (G) S.C.E. reference electrode (RE) and salt bridge containing HBSS, (E) perspex cover plate 5X7.5X 0.6cm, (WE) membrane plated Pt electrode, (CE) Pt counter electrode. Reprinted with permission of Plenum Publishers from Charge and Field Effects in Bio-systems-2 pg. 116 (1989)

Instrumentation

The instrumentation utilized is depicted in Figure 2.

FIGURE 2

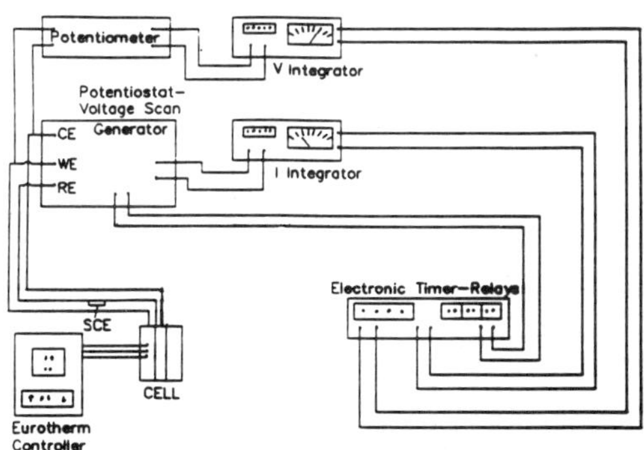

Schematic of Instrumentation

As shown in Figure 2 the voltage-scan generator combination consists of a high sensitivity Wenking LB75L potentiostat and a Wenking VSG72 voltage scan generator. A Kiethly Electrometer is used to obtain the voltage between the working (WE) and the counter (CE) electrodes. High sensitivity current and voltage integrators serve to produce the necessary information from which the charge and voltage was calculated. A programmable Eurotherm temperature controller served to maintain the desired temperature of 20° and 37±.01°C. To obtain Arrhenius plots, the controller was programmed to operate over a range of 20 to 44°C with the appropriate temperature increments and dwell periods.

Procedures

It was the objective in these studies to obtain the relative 'charge' produced as a result of imposed electrical stimulation, as well as the electron:hole (e:h) ratio and the resulting conductance of the various membrane systems under investigation. This was achieved by application of the techniques used earlier in studies of the Electrokinetic Behavior (EKB) of plant and mammalian membranes (Allen, 1984, 1985).

A small section of the excised diaphragm was removed and placed on a glass slide. A similar slide was placed over the section and gently compressed so as to restore the dimensions to that found in vivo. This expanded section was then 'plated' on the Pt WE and positioned on the temperature controlled block (A in Fig. 1). Sections C,D,E were then added with the CE interjected between D and E as shown. After bolting the unit, 100 μl of HBBS was added through the channel in C and the reference electrode salt bridge G inserted. It should be noted that prior to use, the platinum electrodes are pretreated by successive 3 min exposures to 50% aqua regia and conc. nitric acid, both at 50°C.

The total system was equilibrated to the chosen temperature, 20° or 37°, for 10 min. These temperatures were selected, as it was reported in a review of earlier studies (Allen, 1989a) that at the lower temperature, a change in the lipidity within the membrane is noted and at the latter temperature the activation of the various enzymes occur. Subsequent to equilibration the potential of the WE vs RE was noted and the system's potential clamped with the potentiostat. This was followed by a 5 sec +300mV ramp pulse (60mV/sec) succeeded by a 5 min relaxation period. A total of six such pulse were averaged to yield the charge (μC) and conductance (μS).

A similar experiment was performed on a second section from the same diaphragm using 5 sec -300mV ramp pulses. It was now possible to obtain the e:h ratio by relating the charge obtained from this experiment with that produced by the +300mV pulsing procedure.

Arrhenius plots were acquired according to the protocol previously reported (Allen, 1989b). Briefly, using a Eurotherm temperature range of 20 to 44°C and a program of 4°C ramps/30 sec intercepted by 15 min dwell periods yielded well defined transitions. At the end of each dwell period, a 5 sec +300mV ramp pulse was again initiated. From the current/voltage obtained it was possible to calculate the conductance at each temperature. Two points were taken from the transitional slope and the corresponding charge values (μC) used to calculate the activation energy (Ea).

RESULTS AND DISCUSSION

Electrokinetic behavior (CD1 mouse strain)

A total of twelve male and a similar number of female CD1 mice were examined for their behavior at 20°C. An equivalent number of animals were used for the experiments performed at 37°C. The data described in Table 1 has been corrected for the contribution of the background electrolyte in the cell (C in Figure 1).

The results obtained indicated a reversal in the e:h ratio, which based on the number of animals examined, was significant. This is especially noticeable at 37°C where enzyme activity is enhanced. Of significance too is the fact that the differences are due mainly to the 'e' values between the male and female of this mouse strain. This suggests that the major alteration exists

TABLE 1. EKB behavior of male ($\hat{\male}$) and female ($\female$)
CD1 mouse diaphragm membranes

	+ Pulsed (e↑)*			- Pulsed (h = e↓)*	
	μC	μS		μC	μS
20°C ($\hat{\male}$)	10.92	7.02		8.46	4.86
SD	(1.67)	(0.78)	e:h = 1.29:1	(1.31)	(1.08)
($\female$)	7.41	5.26		9.91	5.75
SD	(1.18)	(0.88)	e:h = 1:1.34	(1.06)	(0.72)
37°C ($\hat{\male}$)	18.02	11.13		13.26	7.84
SD	(2.33)	(1.16)	e:h = 1.36:1	1.21	(2.09)
($\female$)	9.20	6.49		13.60	7.39
SD	(1.66)	(0.94)	e:h = 1:1.48	(2.07)	(1.33)

*
e↑ represents electron withdrawal and h = e↓ the electron
acceptor sites. SD indicates the standard deviation for
the mean values listed.

between the electron sites or negative charges existing at the
surface of and/or within the membrane.

Further verification for this diverse behavior was obtained from
the results of Arrhenius plots in which the diaphragm membranes from
male and female produced values of 21.9 KJ and 37.2 KJ respectively.

Over the years a large number of publications have appeared on
the biochemistry, physiology and clinical aspects of insulin binding
effects _in vivo_ and _in vitro._ However to date very few studies have
been made on the electrical phenomena associated with the effects
evinced as a result of insulin binding to appropriate bind sites on
the surface and to the internal structures of the membrane. It was
fully realized that with such a complex system only a phenomeno-
logical presentation would occur. This in due course will perhaps
lead to an explanation of the significance of the various electrical
data in biochemical an/or physiological terms. It was on these

conditions that this aspect of the study of the electrical behavior of the electrical characteristics of the male and female CD1 strain was approached. The information obtained from the different sexes is described in Table 2.

TABLE 2. EKB behavior of male and female CD1 after pretreatment of diaphragm with insulin

	+ Pulsed (e↑)			- Pulsed (h = e↓)	
	μC	μS		μC	μS
20°C (♂)	10.90	6.50		12.26	5.99
SD	1.33	1.20	e:h = 1:1.12	0.76	0.47
(♀)	10.91	7.28		9.98	5.62
SD	1.76	1.52	e:h = 1.09:1	0.60	0.27
37°C (♂)	12.44	8.41		11.11	6.23
SD	1.66	1.40	e:h = 1.12:1	1.94	0.66
(♀)	15.21	10.16		16.82	7.36
SD	1.03	1.11	e:h = 1:1.11	1.62	1.36

To summarize, it appears when comparing the information in Table 2 with that of Table 1, the results indicate that there is certainly an effect of insulin pretreatment on the male and female of this species. At 20°C the male appears to generate more 'holes' or electron acceptors due to insulin pretreatment, whereas the female more negative binding sites. At 37°C, undoubtedly because of induced enzyme activity at this temperature, the male presents a depressed 'e' and 'h' value as compared to the female which projects a picture of augmentation.

Additional evidence for the effects of insulin resulted from the Arrhenius plots, performed as described earlier. which gave values of 13.2KJ for the male and 23.9KJ for the female of the CD1 strain. These values were significantly lower than obtained with the untreated diaphragm membranes.

Electrokinetic behavior (shi/shi strain mice)

A total of six shi/shi and a similar number of 'cured' shi/shi were used in each of these studies. Due to the limited number of available mice all studies were restricted to experiments performed at 37°C. The results are described in Table 3 together with the data obtained as a result of pretreatment with insulin.

The observed differences in related charge (μC) values between the shi/shi and the 'cured' shi/shi clearly conveys the fact that the lack of the myelin sheath in the central nervous system of the shi/shi, which would allow the random transfer of charge, may explain the higher value for this particular subject. Conversely in the 'cured' sample, where there is a direct preset pathway for the transmission of charge, the μC value is lower. Of interest are the essentially nonexistent differences in electron acceptors, (h), between the two types of membranes examined.

TABLE 3. EKB behavior of shi/shi and "cured" shi/shi vs diaphragm sample pretreated with insulin(IPT)

	+ Pulsed (e↑)			− Pulsed (h = e↓)	
	μC	μS		μC	μS
shi/shi	18.71	10.65		14.18	6.98
SD	0.50	0.28	e:h = 1.32:1	0.72	0.61
'cured' shi/shi	13.76	8.78		13.54	6.50
SD	0.25	0.25	e:h = 1.02:1	1.09	0.64
IPT shi/shi	22.33	12.51		13.32	7.15
SD	0.78	1.02	e:h = 1.68:1	1.06	0.88
'cured' IPT shi/shi	12.18	8.24		17.41	8.89
SD	0.59	.66	e:h = 1:1.43	1.11	0.93

As a result of IPT there is an increase in the μC for the + pulsed shi/shi and a small decrease in μC for the 'cured' specimen. With the − pulsed systems no significant differences were noted in μC values except in the case of the IPT 'cured' shi/shi. However in examining the e:h ratios, noteworthy differences do exist.

The Ea values obtained from Arrhenius plots obtained for each of the systems described in Table 3 were 27.9KJ vs 35.7 KJ for the shi/shi vs IPT shi/shi and 39.8KJ vs 43.9KJ for the 'cured' shi/shi under similar conditions.

In continuing studies on the 'electronic behavior' of mammalian diaphragn muscle membranes, it has now been established that sexual differences in electrical charcteristics can be observed between males and females within the same strain of mice. Ensuing investigations on a myelin defficient strain of mice (shiverers) demonstrated that it was possible to distinguish, electrically, the shiverer from its genetically 'cured' counterpart. Observations in both experimental studies indicated divergencies between the, insulin pretreated and their respective controls. This suggested that insulin binding had occured as a result of exposure to this hormone.

ACKNOWLEDGEMENT

The shiverer mice and their 'cured' counterparts were kindly supplied by Dr. Carol Readhead, Department of Biology, California Institute of Technology. The insulin used in these studies was generously provided by the Lilly Research Laboratories.

REFERENCES

Allen MJ (1989): Bioelectrochemistry-Before and After. In: Electrochemistry Past and Present, Stock JT and Orna MV, eds. Washington, D.C.: ACS Symposium Series 390.

Allen MJ and Gefferet G (1989): Electronic properties of dystrophic muscle membrane systems. In: Charge and Field Effects in Biosystems-2, Allen MJ, Cleary SF and Hawkridge FM, eds. New York: Plenum Press.

Allen MJ (1985): The 'solid state' approach to the study of membrane behavior. In: Water and Ions in Biological Systems, Pullman A, Vasilescu V and Packer L, eds. Bucharest: Union of Societies for Medical Sciences.

Allen MJ (1984): The electro-genetic characteristics of plant membranes. In: <u>Charge and Field Effects in Biosystems-1</u>, Allen MJ and Usherwood PNR, eds. Tonbridge Wells, Kent: Abacus Press.

Allen MJ (1971): A voltammetric study of metabolizing erythrocytes. <u>Collection Czechoslov. Chem</u> <u>Commun</u> 36: 658-663.

Stadie WC, Haugaard N, Marsh JB, Hills GB (1950): The chemical combination of insulin with muscle diaphragm of normal rat. <u>Am J Med Sci</u> 218: 265-274.

THE HOMEOSTATIC EFFECT OF ELECTRICALLY NON-COMPENSATED HYDROXYL ($OH^{\ominus}$), (NEGATIVE HYDRO-AIRIONS) ON PHOSPHORYLATING RESPIRATION IN HIGHLY NATIVE MITOCHONDRIA.

Andrew Babsky
Department of Physiology
Lvov State University

Elena Grigorenko
Elena Okon
Marie Kondrashova
Laboratory of Energy Supply of Physiological
Functions Institute of Theoretical and
Experimental Biophysics Academy Sci.USSR.

INTRODUCTION

The light negative airions are known as very effective regulator of state of organism. The mechanism of their therapeutic effect is still not clear. The pioneers in this field proposed bioelectrical theory suggesting electrical interaction between airion charges and tissue components (Tchijevsky, 1941; Vasil'ev and Tchijevsky, 1933). More famons among recent views is the serotonin concept (Krueger, 1959). According to it all favourable effects of negative airions are attributed to serotonin elimination by stimuation of its oxidation. In our opinion the most experiments with hydroairions in isolated biological preparations do not correspond to the conditions of their action in organism. In in vitro experiments much greater doses than in vivo are necessary to observe any effect and this effect is inhibition. It is difficult to obtain in vitro most significant effect of negative airions: i.e. stimulation of functions. We think that the reason is that the tissue preparations of intact animals are used and that many preparations are far from the native state. In the meantime the effect of airions in organism is manifested in individuals with pathology and not in healthy ones. Some artifacts during isolation of preparations also disturb observation of physiological effects of airions.

In our investigation we used biological preparations obtained from non-intact animals and under conditions providing better conservation of the native state. In these cases we have observed airion effects corresponding to their therapeutic action in organism (Andreeva et al., 1989; Kondrashova and Grigorenko, 1985; Kondrashova et al., 1981; 1982a,b)

As negative airions we used hydroairions, i.e. hydroxyl separated from spraying water due to balloelectric effect. This is electrically non-compensated hydroxyl. We sign it $OH^{\ominus}$. It may be considered as a "flying cathode". The negative airions of this kind are rather natural, similar to those hydroairions which arise near running water. The negatively charged oxygen generated by electroeffluvial method and often used in experiments is more artificial as it may be contaminated by toxic oxidized products. These contaminations may also dusturb observation in vitro airion effects related to their action in vivo. Using hydroairions: $OH^{\ominus}$ is more appropriate for revealing stimulating effects of negative airions.

We found that native mitochondria (MCH) isolated in concentrated suspension without washing serve as a suitable preparation for observation of $OH^{\ominus}$ effect. These organelles keep a certain integrity and at the same time they are available for analysis of biochemical and biophysical mechanisms. Our method of MCH isolation provides conservation of the native state of MCH in living cell in the form of aggregations (Katz et al., 1985; Kondrashova and Grigorenko 1984, 1985; Kondrashova et al.1987). In such preparations the effect of $OH^{\ominus}$ is well observed both after $OH^{\ominus}$ influence on the animal and on experimental media before (or after) addition of MCH. In MCH we found pronounced normalization of phosphorylating respiration altered by stress of animal. The sign of effect depends upon the degree of stress-induced alteration of respiration. The initial phase is hyperactivation of succinate oxidation, the second phase is inhibition of succinate oxidation. Under 24 hour stress hyperactivation is observed in liver MCH whereas inhibition is observed in brain MCH. Correspond i ngly $OH^{\ominus}$ treatment of stressed animal results in a decrease of succinate supported respiration in liver MCH and an increase of the same in brain MCH. Any effect of $OH^{\ominus}$ on respiration in MCH of intact animals was not observed. As a more complex objects than MCH we used heart muscle preparations. Also two sign effect of $OH^{\ominus}$ was observed in these preparations as judged by their contractility and ion currents. In acute stress state immediately after isolation the decrease of hyperactivated parameters was observed, whereas their increase was induced in inhibited tired preparations. Our investigation in isolated biological preparations coincides well with numerous data in organism, showing that negative airions abolish pathological alterations without pronounced effect on normal individuals.

In this work we continue the study of $OH^{\ominus}$ effect in native MCH. In order to provide even better conservation their native state we used homogenates and short time periods after its preparation. As factor inducind alteration of intact state of animal we used administration of physiological dose of adrenaline. Oxidation of succinate was investigated as this most powerful process supports selectively intensive energy supply of active functional states (Kondrashova, 1989; 1991; Kondrashova et al., 1988).

MATERIAL AND METHODS.

Fed male non linear albino rats (200-220g) were used. Liver homogenate (HMG) was prepared with teflone nomogenizator in medium containing 300 mM sucrose, 1mM EDTA, 10 mM Tris-HCL, pH 7.4. 1ml medium per 1g tissue was taken. The incubation medium contained 150 mM sucrose, 50 mM KCL, 1 mM KH_2PO_4, 5 mM Tris-HCL, pH7.4; t^o 26 ^{o}C. Respiration was measured polarographically in 1ml cuvette containing 4.5-6.0 mg protein (Chance and Williams, 1956). 3 mM succinate was used as substrate and two additions of ADP per 200 μM for phosphorylating oxidation stimulation. The results of respiration measurments are given in diagrams. 10μg/100g of adrenaline was administrated intraperitoneally, 15 min before decapitation. An equal volume of 0.9% NaCl was injected into control animals.

Hydroairions were generated by spraying water using Mikulin's hydroaeroionizer. It produces 250.000 of light negative airions in 1 cm^3 at the point of their application with the ratio of negative ions to positive 10:1. The flow of ionized air was blown for 50-60 sec over the surface of 1ml cuvette for respiration measurment with stirring. After that cuvette was closed, HMG was added and respiration was measured. Measurments were began immediately after obtaining HMG and continued no longer than for 40-50 min of its storage.

RESULTS

The highly native MCH. According to conventional methods of isolation of MCH procedures between decapitation and obtaining MCH take 40-60 min and even more. Using HMG shorten considerably this period of time. This and possibly the presence of some natural components of MCH environment in HMG provided obtaining even more native state of MCH than those were mentioned above. It may be called the highly native state. It is obserwed only in freshly prepared HMG in the first 15-20 min and is characterized by lower than usual rate of phosporylating respiration (state 3) together with also low non-phosphorylating respiration (state 4). This low rate of state 3 respiration is not a result of damage and inhibition but on contrary it is due to a quite latent respiration activity. This latent activity is realized during storage and transforms into moderate increase of respiration.

Dual, homeostatic effect of OH^- in hyperactivated and latent MCH. In highly native MCH in HMG adrenaline stimulation of succinate oxidation is even greater than described earlier (Babsky et al., 1985; Kondrashova and Babsky, 1986; Shostakovskaja and Babsky, 1984). Such exsample is given in Fig.1.I.

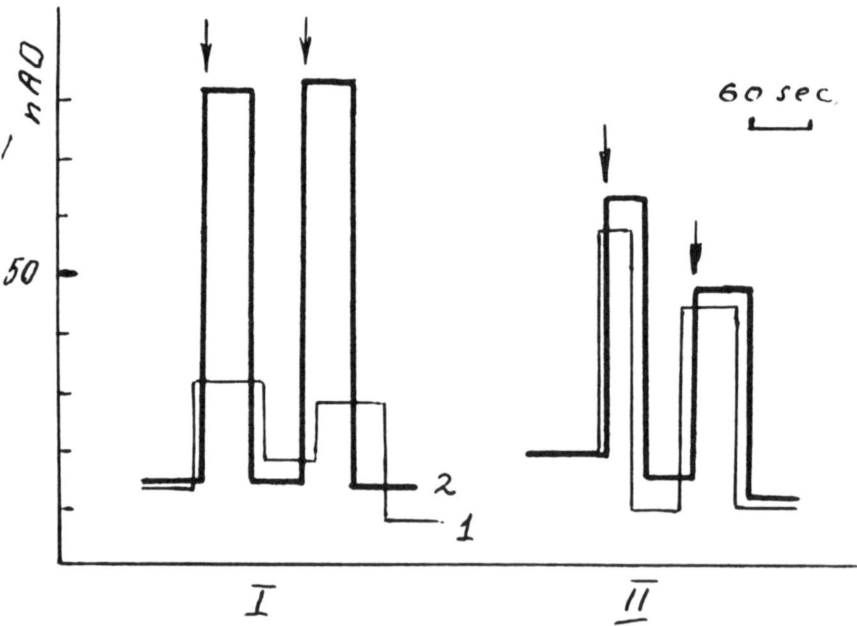

FIGURE 1. The effect of adrenaline and OH^- on the respiration of highly native mitochondria.
1 - Intact MCH, 2 - adrenaline MCH
I No OH^- II OH^- treatment. Arrows indicate ADP additions.
The polarographic registration of respiration is given in diagrams MCH response to two following ADP additions was registered. The

respiration of intact MCH is very low - 30 nAO. Under these conditions, providing the highly native state of MCH very high level of respiration with adrenaline was observed, more than 80. Thus adrenaline stimulation of state 3 respiration is more than 250%. The effect of $OH^{\ominus}$ is also well pronounced in highly native MCH (Fig 1.II). $OH^{\ominus}$ treatment decreases hyperactivated adrenaline stimulated respiration to the level around 55 nAO between responses to the first and second ADP addition. As it is shown in Fig 1, II the same level is reached under $OH^{\ominus}$ treatment by intact MCH. In this case $OH^{\ominus}$ increase very low rate of respiration per 100%. The attention should be attracted to that final levels of both adrenaline stimulated and intact MCH respiration a f ter OH^{-} treatment drow together very closely. It is level 60 nAO for the first ADP addition and level 45 nAO for the second ADP addition. In both cases respiration of intact MCH remains to be slightly lower than that of adrenaline stimulated. However their initial difference is practically completely abolished by $OH^{\ominus}$. OH^{-} stimulation of latent respiration is not identical to activation of inhibited respiration described earlier. As we mentioned low respiration of highly native MCH is not inhibited but latent. In the course of storage latent respiration is increased in contrast to inhibited respiration which is declined. This dual effect of $OH^{\ominus}$ consisting in both decrease of hyperactivated and increase of latent respiration was found only in the highly native MCH. We think that establishment of the same intermediate level of respiration under $OH^{\ominus}$ trea t ment indicates the special properties of corresponding state of MCH. Possiby this is more stable state than hyperactivated or deeply latent. Th e refore $OH^{\ominus}$ treatment corrects divergence providing maintenance of more stable, normal level. This effect of $OH^{\ominus}$ we called homeostatic.

<u>The loss of $OH^{\ominus}$ effect during storage of MCH</u>. With the loss of the latency of MCH both adrenaline and $OH^{\ominus}$ effects declined and disappeared. This is shown in Fig 2.

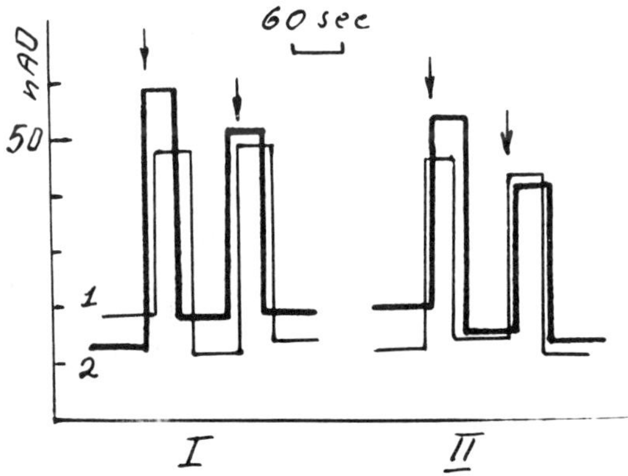

FIGURE 2. The loss of adrenaline and $OH^{\ominus}$ effect in mitochondria having lost their latency. Indications as in Fig 1.
Respiration of intact MCH presented in Fig 2,I is higher than those in Fig 1,I. This evidences loss of MCH latency in spite of they still posses good parameters of coupled respiration - high respiratory control and ADP:0 ratio. However together with the loss of latency of MCH effects of adrenaline (Fig 2,I) and of $OH^{\ominus}$ (Fig 2,II) also dissapear.
The restoration of the native state of MCH by $OH^{\ominus}$.

In Fig 3 the results are given of experiments with several preparations of different degrees of the native state.

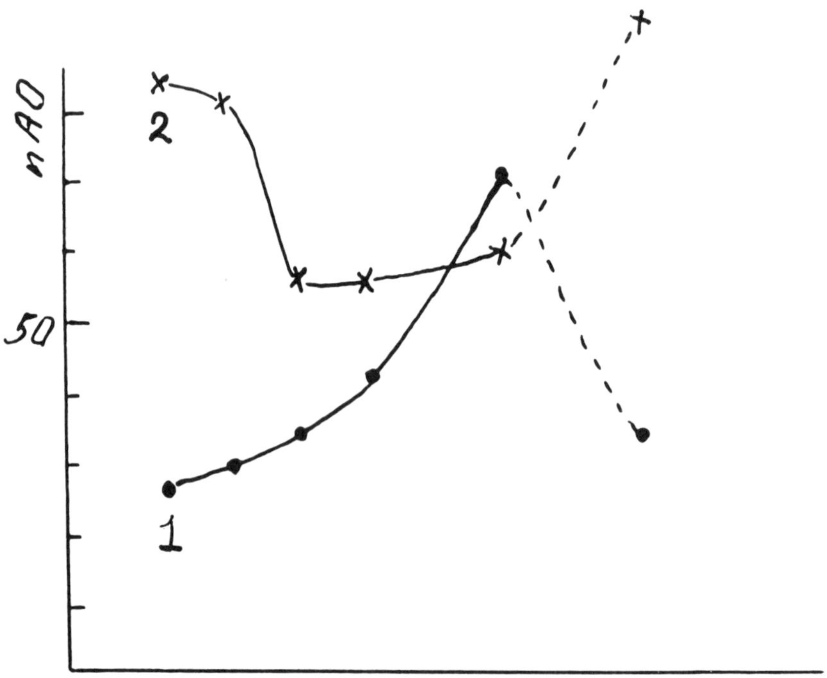

FIGURE 3. The dependence of adrenaline and $OH^{\ominus}$ effect on the latency of mitochondria.
1 - Intact MCH, 2 - adrenaline MCH.
Dot numbers from the left 1-6. 1-5 No $OH^{\ominus}$, 6 $OH^{\ominus}$ treatment.
The lower curve presents phosphorylating respiration of intact MCH, the upper - those after adrenaline administration. The observed stimulation of respiration by adrenaline strongly depends upon native state of MCH. It is well pronounced in highly native MCH with low level of respiration (pairs 1,2 of dots) decreasing with the loss of latency (pairs 3,4 of dots). More prolonged

storage of MCH results in more greater increase of intact MCH respiration reaching adrenaline stimulated level (pair 5 of dots). In this case respiration of adrenaline stimulated MCH instead of rise decreases. Noticeable is that in this high rate ra n ge $OH^{\ominus}$ diminishes hyperactive respiration MCH of intact animal as in the case of adrenaline stimulated respiration. Simultaneously stimulation of respiration by adrenaline is also appeared (pair 6).

Thus $OH^{\ominus}$ treatment provides restoration of native properties of MCH which have lost their latency during storage. These data well coincide with our previous observation showing the ability of $OH^{\ominus}$ to restore MCH aggregation dissipation during storage (Kondrashova et al., 1987).

DISCUSSION

The investigation of highly native MCH in HMG revealed new homeostatic effect of $O\,H^{\ominus}$ on phosphorylating oxidation of succinate: decrease of hyperactivated respiration and increase of latent respiration. Mo r eover $OH^{\ominus}$ treatment restored loss of latent respiration and reversed normal response of MCH to adrenaline (Fig 3) These effects are mediated through water as MCH were added in medium pretreated by $OH^{\ominus}$. We found earlier that pretreatment of medium prevents dissipation of aggregation of MCH and loss of their native respiratory responses. These effects were shown to be mediated through water and bound water rearrangements followed by structural interconversions of protein and MCH (Kondrashova et al., 1987; Tschegoleva and Oksuk, 1991). This mechanism supports bioelectric theory of airion action (Tchijevsky, 1941; Vasil'ev and Tchijevsky, 1933). According to Vasil'ev negative airion effect is like cathode (Vasil'ev, 1953). He considered very weak action of cathode which does not induce well known cathode-depression but on contrary quite opposite anode-like stimulation of inhibited functions (Vasil'ev, 1925,; 1933; Arschavsky, 1960). The separated hydroxyl ions, $OH^{\ominus}$ serve as the most appropriate candidate for very mild catode as it is flying and pulsating. The impulse influence is quite adequate for living tissue which is regulated by nerve impulses. The mild cathode can induce dual effect in vivo. On the one hand it can eliminate deep supression of cathode-depression nature. On the other hand it can initiate activity under hyperpolarization inhibition. Dual effect depends upon the state of tissue target.

In all cases the electrical interaction is the primary step of airion action whereas metabolic changes are the secondary step. Serotonin elimination should be considered as one of importance among metabolic changes (Krueger, 1959).

The described here and elsewhere metabolic regulation of phosphorylating succinate oxidation is also very important part of negative airions effect in vivo. The states of respiration of MCH observed in our experiments here and elsewhere may be presented in Fig.4.

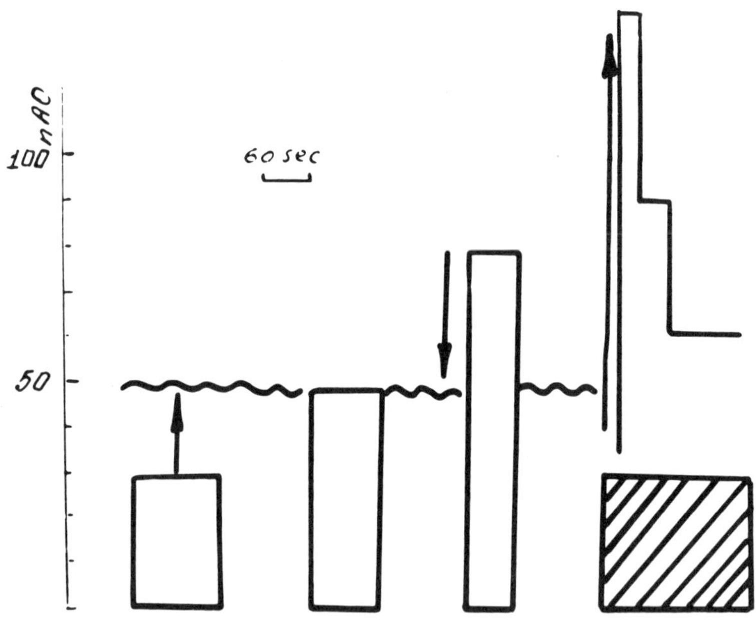

FIGURE 4. Different states of mitochondria and their response to $OH^{\ominus}$ (arrows).
From the left to right the following states are given:
L - Latent (highly native)
A - Common active
HA - Hyperactive
I - Inhibited
L and A states correspond to normal state in vivo,
HA - to excitation,
I - to stress inhibition.
Transitions between these states occur with increase of irritation in vivo or with damage of MCH in vitro. The corresponding effects of $OH^{\ominus}$ in these states are:
L - Mild activation,
A - No effect
HA - Restriction of hyperactivity

I - Sharp activation

Mild activation and restriction of hyperactivity cover the range of dual homeostatic effect described in this paper. As it was shown in Fig 1 $OH^{\ominus}$ treatment induces transition of L and HA respiration to general intermediate apparently stable normal level.

Sharp activation is out of homeostatic range of responses. Activation of stress-inhibited MCH may induce burst of respiration (Kondrashova et al., 1986). This acute response of respiration is related to some pathological events in vivo.

Obtaining native states of MCH permits observation ofseveral effects of low, physiological doses of $OH^{\ominus}$. This gives new approach to understanding biophysical and biochemical mechanism of biological action of airions.

SUMMARY

The effect was investigated of incubation medium treated by the flow of negative hydroairions (electrically non-compensated hydroxyl, $OH^{\ominus}$, obtained by spraying water, on respiration of mitochondria in rat liver homogenate. The strong effect of $OH^{\ominus}$ was shown on respiration coupled with ATP synthesis and not on non-phosphorylating respiration. It was found that the effect depends strongly upon the state of isolated MCH. $OH^{\ominus}$ effect is observed only in freshly prepared HMG and it dissappears in the course of their storage although they still keep good parameters of energy coupling. The found effect is dual. Hyperactive respiration of MCH obtained from animals after adrenaline administration is decreased by $OH^{\ominus}$. Low rate respiration of MCH of intact animals is increased by $OH^{\ominus}$. As a result respiration of MCH of active and intact animals drows together at the same middle level. This dual effect is called homeostatic. It shows that $OH^{\ominus}$ correct divergence providing maintenance more stable and normal level.

REFERENCES

Andreeva LA, Bakaneva VF, Grigorenko EV, Kondrashova MN (1989): The influence of negative hydroairions on electromechanical coupling in auricle of Rana Ridibunda. *Biofizika* 34:306-309.

Arschavsky IA (1960): Perielectroton as a base and mechanism of excitation and inhibition. *Biofizika* 5:143-151.

Babsky AM, Kondrashova MN, Shostakovskaja IV (1985): Action and afteraction of adrenaline on respiration in mitochondria. *Physiol. J.* (Kiev) 31:301-306.

Chawce B, Williams GR(1956): The respiratory chain and oxidative phosphorylation. *Adv Enzymol.* 17:4567-4576.

Katz J, Wals PA, Golden S, Raijman L(1985):Mitochondrial-reticular cytostructure in liver cells. *Biochem.J.* 214:795-813.

Kondrashova MN(1989): The structural-kinetic organization of tricarboxylic acid cycle under active function of mitochondria. *Biofizika* 34:450-458.

Kondrashova MN (1991): Interaction of transamination and oxidation of carboxylic acids in different functional states of tissues. *Biochimia* 56:388-405.

Kondrashova MN and Babsky AM(1986): The dose dependence of stimulation of respiration by adrenaline. *Ukrainian Bioch. J.* 58:49-54.

Kondrashova MN and Grigorenko EV(1984): Natural mitochondria well conserve their state in organism. *3 Europ. Bioenergetic Conf.* Hannover, Cogress Ed. 3B:711-712.

Kondrashova MN and Grigorenko EV (1985):Manifestation of stress at the level of mitochondria, their stimulation by hormones and control by hydroairions. *J.Gen.Biol.(Rus).* 46:516-526.

Kondrashova MN, Guzar IB, Grigorenko EV(1986):Burst of saccinate oxidation in stress state of mitochondria. *4 Europ. Bioenergetic Conf.*, Prague 4:402.

Kondrashova MN, Grigorenko EV, Guzar IB, Okon EB(1981): The elimination by negative airions stress induced respiration changes in mitochondria. *Biofizika* 26:687-691.

Kondrashova MN, Grigorenko EV, Guzar IB, Okon EB (1982a): Hyperactive metabolic state of mitochondria under stress. *2 Europ. Bioenergetic Conf.*, Lyon, L.B.T.M.- C.N.R.S., Villeurbanne 2:589-590.

Kondrashova MN, Grigorenko EV, Kosenko EA(1988) Rapid cycle of substrate oxidation under activation of energy metabolism. In:*V Europ.Bioenergetic Conf.*,Aberystwyth,p.297.

Kondrashova MN, Grigorenko EV, Temnov AV, Okon EB et al (1987): The effect of negative hydroairions on the structure and functional properties of mitochondria. *Biofizika* 32:313-322.

Kondrashova MN, Guzar IB, Grigorenko EV, Okon EB(1982b) The airion effect on the relationship between forward and reversed electron transter in mitochondria under stress. *Biofizika* 27:76-81

Krueger AP(1959):Gaseous ion effect on trachea. *J.Gen.Physiol.* 42:5-12

Shostakovskaja IV, Babsky AM(1984): The effect of adrenaline on Ca^{2+} transport and oxidative phosphorylation in mitochondria. *Ukrainian bioch.J.*56:57-62

Tchijevsky AL(1941):Airion biological action. Acta medica *Scandinavica* 61:129-137.

Tschegoleva TYu, Oksuk OD(1991):The effect of OH^- on dielectric permeability in biopolymere solutions. *Biofizika* in press.

Vasil'ev LL(1925): Dual theory of inhibition. In: *New in reflexology and physiology of nervous system.* Moscow, Gosizdat, 1:15-32.

Vasil'ev LL(1953): *Theory and practice of medical treatment by ionized air.*Leningrad 190p.

Vasil'ev LL, Tchijevsky AL(1933): Hypothesis of organic electroexchange. In: *Problems of ionification.* Voronez, I:219-232.

SELECTION RULES ON HELICITY DURING DISCRETE TRANSITIONS OF
THE GENOME CONFORMATIONAL STATE IN INTACT AND X-RAYED
CELLS OF E.COLI IN MILLIMETER RANGE OF ELECTROMAGNETIC
FIELD

I.Ya.Belyaev, V.S.Shcheglov, Ye.D.Alipov
Scientific Research Collective "Otklik"
Moscow Engineering Physics Institute

ABSTRACT

The method of viscosity anomalous time dependence
(VATD) was applied to study the influence of extremely high
frequency (EHF) electromagnetic radiation (EMR) on the
genome conformational state (GCS) of E.coli cells. The
cells were exposed to EMR at resonant frequencies of the
two earlier discovered frequency bands (41.25-41.50 GHz and
51.62-51.84 GHz) in which non-thermal microwaves suppressed
resonantly the reparation of X-ray induced changes in GCS.
The use of a unit with quarter-wave plates to obtain
circularly polarized EMR confirmed the existence of
selection rules on helicity during discrete transitions of
GCS in the millimeter range of the electromagnetic field.
At the resonant frequency of the second resonance (51.76
GHz) right-handed polarized microwaves effectively
influence GCS of X-rayed cells, while left-handed polarized
EMR is virtually ineffective and linear polarization's
effectiveness lies between the two circular polarizations.
Conversely, left-handed polarized EHF EMR is effective when
X-rayed cells are exposed to microwaves at the resonant
frequency of the first resonance (41.32 GHz). It was shown
that the difference in the effectiveness of the two
circularly polarized components of EMR increases when the

ellipticity coefficient goes down from 1.2 to 1.05.

It was established that the sign of effective circularly polarized EHF EMR does not depend on the sequence in which the cells are exposed to microwaves and X-rays. Yet when intact cells are exposed to microwaves at each of the two resonant frequencies, the effectiveness of the circularly polarized components is inverted. The registered effects are non-thermal and manifest themselves at the power density of 1 $\mu W/cm^2$.

The results confirm the role of the cell genome in the resonance response of cells to low-intensity millimeter waves and are described from the standpoint of the physical concept on the role of the electromagnetic field in the functioning of cells.

INTRODUCTION

Research in the biological effect of low-intensity millimeter waves occupies a special place among problems of electromagnetic biology because many questions remain unclear in despite of long and intensive studies (Motzkin et al., 1983; Postow and Swicord, 1986; Grundler et al., 1988) which show the ability of EHF EMR to generate biological effects resonantly, including gene expression (Lukashevsky and Belyaev, 1990). For instance, the reproducibility of observed effects presents a special problem (Sitko, 1989). There is no physical model of interaction between external EHF EMR and biological systems. On the other hand, a physical concept was suggested as far back as 1968 according to which in the millimeter range of the electromagnetic field a coherence of biological processes in multicellular organisms is ensured (Frolich, 1968). But resonant effects of EHF EMR can also be observed when it affects single cell microorganisms. It can be assumed therefore that the internal electromagnetic field of the millimeter range ensures coherence of main cellular functions — catalytic, membrane and DNA. Proceeding from the assumption of a possible role of chromosomal DNA in this biological coherence we had previously studied the influence of EHF EMR on the GCS of E.coli cells (Belyaev et al., 1990, 1991a). "Genome conformational state" is understood as spatial-topological organization of the entire chromosomal DNA which is ensured, among other things, by the supercoiling of DNA and DNA-protein bonds. In our studies we applied the VATD technique which demonstrates changes in the GCS. These changes are identified after lysis of exposured cells and partial proteolysis of proteins tied to DNA.We managed to reveal two frequency ranges in which non-thermal microwaves resonantly block reparation of induced by X-rays changes in the GCS. Manifestation of such resonant effects can be explained according to a physical concept of discrete states in living systems in the millimeter range of the electromagnetic field (Sitko at al., 1989; Keilmann, 1986). The concept assumes that transitions between discrete states obey the rules of selection, for instance, the rules of selection on helicity. If this is so, left- and right-handed polarized radiation of the same frequency must produce different effects. The assumption was confirmed experimentally (Belyaev et al.,1990, 1991b). It turned out that left-handed polarized EMR effectively influenced X-rayed cells at all the three studied frequencies of the first resonance (41.25-41.50 GHz), while right-handed polarized EMR had virtually no effect. A reverse effect was observed when cells were exposed to microwaves at frequencies of the second resonance (51.62-51.84 GHz) where right-handed polarized EHF EMR was effective . The insignificant effect of "ineffective" circular polarization can be attributed to the ellipticity of the applied EMR. The unit with spiral waveguides used for irradiation provided an ellipticity coefficient (K) of not less than 1.1.

The goal of this study was to determine the same effect of circularly polarized millimeter waves generated by a unit with a smaller ellipticity coefficient.

MATERIALS AND METHODS

A G4-141 generator was used to obtain linearly polarized millimeter waves. Three samples were exposed simultaneously to linearly left- and right-handed polarized radiation through three waveguides connected to the generator. Linearly polarized microwaves were converted to circularly polarized microwaves by means of mica quarter-wave plates inserted in the waveguides.The waveguides for circularly polarized EMR ended in cone-shaped horns with an area of 20 cm^2. The pyramid-shaped horn for linearly polarized EMR had the same area. The geometry and conditions of exposure to microwaves and X-rays were described in detail earlier (Belyaev et al., 1990, 1991 a-b). The K along the beam axis of circularly polarized EMR in the object's zone did not exceed 1.05±0.05.

Changes in the GCS of exposured cells were studied with the VATD technique in cell lysates.It is based on radial migration of high polymer DNA molecules in a high gradient hydrodynamic field of a rotational viscosimeter (Shafer et al., 1974). Radial migration of chromosomal DNA toward the revolving rotor causes anomalous changes of viscosity that can be registered by measuring the rotor rotation period (T). Molecules lighter than 10^6 daltons do not contribute to VATD and it is thus determined exclusively by the hydrodynamic behaviour of chromosomal DNA macromolecules. For their part, the hydrodynamic parameters of chromosomes depend on the nativity and supercoiling of chromosomal DNA, and on the degree of proteolysis of DNA-tied proteins. The VATD was studied as a dependence of the T , which is proportionate to viscosity, on the measurement time. The measurement time starts with the switching on of the viscosimeter's electromagnets that create a constant force moment revolving the rotor. Conditions of lysis were selected to minimize damage to DNA and ensure the proteolysis of DNA-tied proteins sufficient for registration of VATD curves. The VATD curves were measured three times in each variant of the experiment. Student's criterion was used to compare the experimental data.

RESULTS AND DISCUSSION

Figure 1 shows several VATD curves obtained by averaging out three single-type curves measured in one standard experiment. The T was measured 10 to 20 times, from the moment the electromagnetic field which rotates the rotor is switched on, to obtain a VATD curve. Viscosity and therefore T increases at the initial stage of measurement due to radial migration of DNA macromolecules to the rotor. Viscosity decreases again when chromosomal DNA is deposited on the rotor's surface. The maximum rotation period of the rotor (T_{max}) is the VATD parameter which is most sensitive to changes in the GCS.

Compared to the control level, T_{max} diminishes considerably upon exposure of E.coli cells to 20 Gy of X-rays that induce about one DNA single-strand break per genome. The decline in maximum viscosity may be due to changes in the conformational state caused by single-strand breaks of DNA or changes in the quantity and set of proteins that remain tied to DNA after lysis. The proteins that form the spatial-topological organization of chromosomal DNA in cells and lysates and are deposited on the rotor together with DNA during the VATD measurement will be described in a separate paper.

During incubation of cells in an M-9 buffer after exposure to X-rays, the maximum viscosity of the VATD curve again tends toward the control level. The process reflects reparation of radiation-induced changes in the GCS. Yet if cells are exposed to right-handed polarized EMR at 51.76 GHz during the post-irradiation incubation stage, the reparation of the GCS is inhibited. Given the same conditions of exposure, left-handed polarized microwaves are virtually ineffective. The effectiveness of linearly polarized EHF EMR was average between the two circular polarization.

When the PD was increased by an order (from 10 $\mu W/cm^2$ to 100 $\mu W/cm^2$), the effect of differently polarized microwaves on X-rayed cells did not change (Table 1). The dependence of the microwave effect on intact cells (not

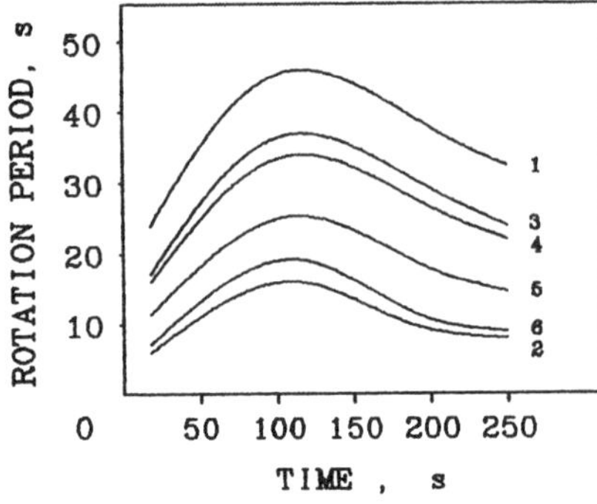

FIGURE 1. VATD of cell lysates after exposure of E.coli cells to X-rays (20 Gy) and EMR (51.76 GHz; 10 $\mu W/cm^2$): 1 - Control; 2 - X-rays; 3 - X-rays and incubation (90 min); 4,5,6 - left-handed, linearly and right-handed polarized EMR during incubation.

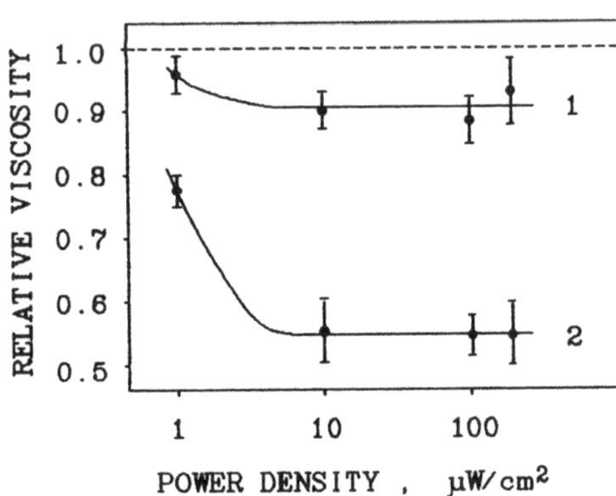

FIGURE 2. Relative viscosity dependence on PD of circularly polarized EMR:
1,2 - left- and right-handed polarized EMR.

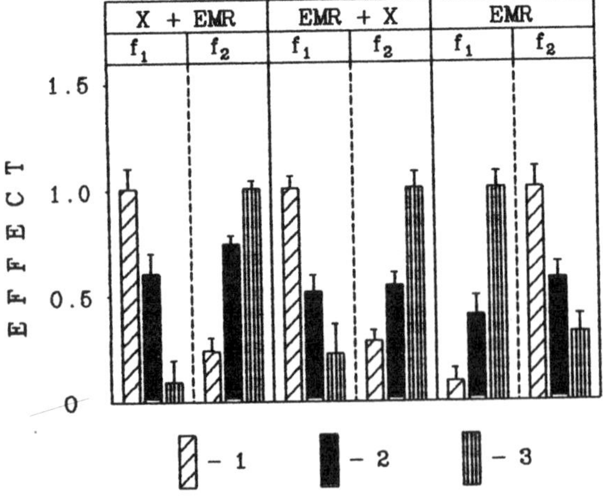

FIGURE 3. Effect (ε_1;ε_2) of E.coli cells exposure to EMR (f_1=41.32 GHz; f_2= 51.76 GHz; PD = 100 $\mu W/cm^2$) and X - rays (X) with different sequence of irradiation:
1,2,3- left-handed, linearly and right-handed polarized EMR.

exposed to X-rays) on polarization was totally different. The effect of right-handed polarized EMR did not differ from the control level, while left-handed polarized EMR proved effective.

Test number	Exposure of cells	EHF EMR polari- zation	$\overline{T}_{max} \pm SE,$ s	Significance level
1	Control	–	42.6±0.8	P(1,3)<0.001 P(1,7)>0.05
2	X-rays	–	14.1±1.0	P(2,4)<0.02
3	X-rays and incubation (90 min)	–	34.1±0.4	P(3,2)<0.00005 P(3,6)<0.03
4	X-rays and	right	18.3±0.2	P(4,5)<0.0005
5	incubation	linear	22.4±0.4	P(5,6)<0.002
6	with EMR	left	30.4±0.9	P(6,4)<0.0003
7	Only	right	39.7±0.9	P(7,8)<0.04
8	EMR	linear	36.2±0.6	P(8,9)<0.007
9	action	left	30.4±0.9	P(9,7)<0.003

TABLE 1. Examination of 10-minute effect of polarized EHF EMR (PD = 100 μW/cm^2; f = 51.76 GHz) on E.coli cells X-rayed in a dose of 20 Gy and intact cells. SE – standard error.

The exposure of E.coli cells to 20 Gy of X-rays inverted the sign of the effective circularly polarized EMR with a resonant frequency of 51.76 GHz. Inversion also took place when intact cells were exposed to microwaves at another resonant frequency, i.e. 41.32 GHz. In this case the GCS was strongly affected by right-handed polarized microwaves (Figure 2). With the PD of 1 to 200 μW/cm^2, the effect of left-handed polarized EMR was virtually no

122

different from the control level. The relative viscosity was determined as $\eta = \bar{T}_{max\ EMR} / \bar{T}_{max\ CONT}$, where $\bar{T}_{max\ EMR}$ is the average T_{max} in lysates of E.coli cells exposed to EHF EMR, and $\bar{T}_{max\ CONT}$ is the average T_{max} in lysates of intact cells. As can be seen from Figure 2, the power dependence of microwave effect has a plateau with the 10-200 $\mu W/cm^2$ range. At the minimal investigated PD of 1 $\mu W/cm^2$ the exposure of cells to microwaves resulted in a statistically significant decline in the maximum viscosity. The curve's threshold appears to be at less than 1 $\mu W/cm^2$. The threshold dependence of the EMR effectiveness on power density is characteristic of non-thermal resonant effect of millimeter waves on biological systems (Postow and Swicord, 1986; Grundler et al., 1988; Lukashevsky and Belyaev, 1990). We had earlier identified a similar dependence when the effect of linearly polarized EMR at a resonant frequency of 41.32 GHz was studied (Belyaev et al., 1990, 1991 a). The PD of 1 $\mu W/cm^2$ was effective, and its increase from 10 to 100 $\mu W/cm^2$ produced no significant changes in the GCS of exposed E.coli cells.

The presence of a highly quality resonant response ($\Delta f/f \sim 10^{-3}$) to the effect of millimeter waves of so low intensities is a strong argument in favour of a quantum character of interaction between EHF EMR and cells (Sitko et al., 1988). It seems to us that these repeated results should be taken into account when some norms of radiation hygiene are revised.

The effect of circularly polarized EMR at 41.32 GHz, as in the case of the second resonance, depended on whether E.coli cells were exposed to X-rays. Our research showed that when X-rayed cells were exposed to the EMR of 1-200 $\mu W/cm^2$ the reparation of the GCS was inhibited by left-handed polarized microwaves, while right-handed polarization was ineffective.

The histogram on Figure 3 present the results of 6 identical experiments that sum up the effect of polarized EMR with resonant frequencies of two different resonances. The effectiveness of EMR influence on reparation in X-rayed at 20 Gy cells was evaluated according to the formula:

$$\varepsilon_1 = \frac{\overline{T}_{max\ X+I} - \overline{T}_{max\ X+EMR}}{\overline{T}_{max\ X+I} - \overline{T}_{max\ X+EFF}} \quad ,$$

where $\overline{T}_{max\ X+EFF}$ is the average T_{max} in lysates of cells that were exposed to EHF EMR with effective circularly polarized component at a given frequency during or before post-irradiation incubation; $\overline{T}_{max\ X+I}$ is the average T_{max} in cells lysates after exposure to X-rays and subsequent incubation for 90 minutes; and $\overline{T}_{max\ X+EMR}$ is the average T_{max} in lysates of cells that were exposed to EHF EMR during or before post-irradiation incubation. The effect of microwaves on the GCS of intact cells was determined according to the formula below:

$$\varepsilon_2 = \frac{\overline{T}_{max\ CONT} - \overline{T}_{max\ EMR}}{\overline{T}_{max\ CONT} - \overline{T}_{max\ EFF}} \quad ,$$

where $\overline{T}_{max\ CONT}$ is the average T_{max} in lysates of intact E.coli cells; $\overline{T}_{max\ EMR}$ is the average T_{max} in lysates of cells exposed to EMR of a definite frequency and polarization; and $\overline{T}_{max\ EFF}$ is the average T_{max} in lysates of cells exposed to EMR of circular polarization which is effective at a given frequency.

It is clear from the histogram that the effect of EMR of a certain circular polarization depends both on the resonance frequency and on whether E.coli cells were exposed to X-rays. Incidentally, the sequence of cell exposure to microwaves and X-rays did not affect the sign of the effective circularly polarized component of EMR. In all the three types of exposure a linear polarization's effectiveness was average between the two circular polarizations. The effect of ineffective circularly polarized component was either insignificant or absent altogether.

The principal result of this part of our research was that it proved experimentally that differences in the effect of left- and right-handed circularly polarized EMR at the same frequency increased when its ellipticity coefficient declined. Thus the totality of previous experiments on the unit with K = 1.2±0.1 , the average effectiveness of the ineffective component amounted to ε_2=0.56±0.07 and ε_1=0.41±0.05 when EMR affected intact cells (11 experiments) and X-rayed cells (16 experiments) (Belyaev et al., 1990, 1991b). The present study was conducted on a unit with K = 1.05±0.05 and corresponding effects amounted to 0.24±0.03 (21 experiments in which intact cells were exposed to microwaves) and 0.19±0.03 (4 experiments with combined exposure). The increase in the relative effectiveness of circularly polarized components with decrease of the K coefficient were significant in both cases (p<0.003 and p<0.05 respectively). Our results suggest that a further decrease of K can render one of the circularly polarized microwaves totally ineffective, while ensuring high effectiveness of the one with an opposite sign.

Our assumption concerning the existence of selection rules on helicity during discrete transitions in living systems in the millimeter range of the electromagnetic field was once again confirmed experimentally. Obviously, the findings cannot be explained away by the concept of the "thermal" biological effect of millimeter waves that is behind hygienic standard setting (Elder, 1987). On the other hand, they fit into the assumption arising from the physical concept of a quantum mechanism of interaction between low-intensity EHF EMR and biological systems. Indeed, if a system has discrete energy levels, one can induce transitions between them if the frequencies of a transition and the external electromagnetic field coincide.

If these transitions are governed by the rules of selection on helicity we may expect that only one circularly polarized component of EMR will be effective for frequencies of different resonances (corresponding to different transitions). In this connection, the question of real targets (their sets) of the resonant EMR effect and sources of the internal electromagnetic field of the millimeter band appears very important. As is clear from the very essence of the VATD technique, an essential role in interaction between millimeter waves and cells is played by chromosomal DNA. The fact is confirmed by the identified inversion of the sign of effective polarization when cells are exposed to 20 Gy of X-rays. The dose is too small to cause noticeable damage in any other cell structures apart from DNA. Let us note here that the findings give rise to an assumption that there must be a dose in the 0-20 Gy interval which makes the effectiveness of the two components of EMR identical. This dose dependence of effect which demonstrates the genome's role in cell response to EHF EMR will be analyzed in a separate paper. We shall also describe a physical model of generation of the internal electromagnetic field of the millimeter band in elementary genetic processes such as transcription, replication, recombination and repair. According to the model, the presence of right- and left-handed conformational states of DNA (B- and Z-forms) in the cell genome makes a significant contribution to the interaction between cells and EHF EMR. The model suggests, inter alia, that the combined effect of circularly polarized EMR components of the same resonance frequency is not additive. The assumption is being tested experimentally.

ACKNOWLEDGMENTS

The authors are thankful to S.P.Sitko for discussing their finding.

REFERENCES

Belyaev IYa, Alipov YeD, Shcheglov VS and Lystsov VN (1990): Effect of millimetre waves on radiation-induced reparation of genome conformational state. In: <u>Workshop on DNA Repair on Mutagenesis Induced by Radiation. Proceedings</u>, Krasavin YeA, ed. Dubna: JINR Press (in Russian).

Belyaev IYa, Alipov YeD, Shcheglov VS and Lystsov VN (1991a): Resonance effect of microwaves on the genome conformational state of E.coli cells. _Z Naturforsch_, in press.

Belyaev IYa, Shcheglov VS and Alipov YeD (1991b): Existence of selection rules on helicity during discrete transitions of genome conformational state of E.coli cells exposed to low-level millimetre radiation. _Bioelectrochem Bioenerg_, in press.

Frolich H (1968): Long range coherence and energy storage in biological systems. _Int J Quantum Chem_ 2: 641-652.

Grundler W, Jentzsch U, Keilmann F and Putterlik V (1988): Resonant cellular effects of low intensity microwaves. In: _Biological Coherence and Response to External Stimuli_, Frolich H, ed. Berlin: Springer-Verlag.

Keilmann F (1986): Triplet-selective chemistry: a possible cause of biological microwave sensitivity. _Z Naturforsch_ 41c: 795-798.

Lukashevsky KV and Belyaev IYa (1990): Switching of prophage lambda genes in Escherichia coli by millimetre waves. _Med Sci Res_ 18: 955-957.

Motzkin SM, Benes L, Block N, Israel B, May N, Kuriyel J, Birenbaum L, Rosenthal S and Han Q (1983): Effects of low-level millimeter waves on cellular and subcellular systems. In: _Coherent Exitations in Biological Systems_, Frolich H and Kremer F, eds. Berlin: Springer-Verlag.

Postow E, Swicord ML (1986): "Window" effects in the millimeter-wave region. In: _CRC Handbook of Biological effects of electromagnetic fields_, Polk C and Postow E, eds. Boca Raton, CRC Press.

Shafer RH, Laiken N and Zimm BH (1974): Radial migration of DNA molecules in cylindrical flow. I. Theory of the free-draining model. _Biophys Chem_ 2: 180-188.

Sitko SP, Andreev EA and Dobronravova IS (1988): The whole as a result of self-organization. _J Biolog Phys_ 16: 71-74.

Sitko SP (1989): Why Devyatkov-Grundler resonances are not reproduced always? _Dokl Acad Nauk UkrSSR_ 4b: 74-77 (in Russian).

VISUALIZATION OF IONIC CHANNELS IN A LIPID MEMBRANE BY MEANS OF A SCANNING TUNNELLING MICROSCOPE AND FUTURE POSSIBILITIES FOR APPLICATION

Oleg V. Kolomytkin
Institute of Biophysics, Academy of Sciences of the USSR,
Pushchino, 142292, USSR

Alexander O. Golubok, Serge Y. Tipisev, Svetlana A. Vinogradova
Analytical Instrumentation Institute, Academy of Sciences
of the USSR, Leningrad, 198103, USSR

INTRODUCTION

Initially intented for the study of crystalline electroconductive surfaces, a scanning tunneling microscope (STM) has found an application in studies of the structure and electronic properties of organic molecules and biological objects (Binnig and Rohrer, 1986). Since 1985, when the work of Barto et al. on STM-visualization of viruses was published, there appeared a number of papers on STM studies of various organic and biological objects including simple organic molecules, lipid membranes, proteins, viruses, ets.

The design of a scanning tunneling microscope is the next. A metalic tip is placed above an electroconductive surface. The distance between the STM-tip and the surface is very low ($\sim$1 nm) in oder to record tunneling current between the tip and the surface. The surface is scanned by the tip. The scanning can be at constant tunneling current mode. In this case the values of Z-coordinate perpendicular to the surface is measured and imaged in a computer display. The scanning can be at constant Z-coordinate. In this case the tunneling current is measured and dispayed on osciloscope. The theory and design of the scanning tunneling microscopy is described in Refs. (Binnig et al., 1986, Rohler, 1990).

The advantages of STM are a high resolution ($\sim$0.1 nm), the ease of operation, the ability to operate in various media and a high locality of the method that enables one to work with monolayer films and smallest quantities of the substance (up to individual molecules). Besides, in studies of biological objects the investigators try to use them intact when their functional activity is not impaired. In this respect, STM studies in natural medium, i.e. aqueous or electrolyte solution, may appear essential.

Analysis of literature suggests that STM studies of biological objects are at the early stage. Most of the papers available demonstrate just the potentialities of STM. Unlike STM studies of solids where reliable experimental procedures and theoretical approaches to data interpretation are developed, in studies of molecular and biological objects many questions remain open. There is no unambiguous answer as to the mechanism of electron transfer across the organic molecules and their complexes in the STM experiment, no standard procedures for preparation and investigation of species, no criteria for reliable interpretation of experimental results.

At present the process of data accumulation is in progress. The results that are now avalable represent a collection of separate data on a range of objects obtained by different procedures on various substrates: conventional "decorating" with metals as it is done in electron microscopy (Amrein et al., 1988, Keller et al., 1990, Guckenberger et al., 1988), special chemical treatment (Hameroff et al., 1989, Cricenti et al., 1989), ets. The attempts are made to generalize and examine the problems of STM application in biology, model building and to discuss the mechanism of tunneling (Smith et al., 1987, Salmeron et al., 1990, Smith et al., 1990, Stemmer et al., 1988).

One of the most important trends in which certain progress has been made is the study of thin organic films and biological membranes. Thin films of organic substances on the surface of a solid substrate, in particular Langmuir-Blodgett films, represent an object that is of interest from the viewpoint of physics as well as biology. From the physical viewpoint such objects are thin-layer films with certain electrical, mechanical and optical properties, and as it was noted by some authors can be used for constructing electronic devices, biosensors and data storage systems at the molecular level (Al-Mohammad et al., 1990). From the biological viewpoint, the lipid Langmuir-Blodgett films are artificial analogs of natural biological membranes that form cellular walls and are very essential for vital activity of biological organisms. Artificial lipid membranes are used in biophysics for modeling the processes occurring in natural cellular membranes.

Smith et al. in 1987 obtained a molecular-level resolution on the lipid layer of cadmium arachidate. It was found that the lipid film packing is partially ordered and has a triclinic elementary cell. The parameters of the cell and a mean molecular density have been determined (a molecule per 1.94 nm^2). Using video recording the authors observed the dynamics of the bilayer. A lateral flow of the film was recorded to occur at a rate of 0.4 nm/s. In some sites of the film (mainly on the heterogeneities of graphite substrate) there was disordered molecule packing which shifted relative to stationary graphite lattice and could disintegrate into separate cadmium arachidate molecules. Images of Langmuir-Blodgett films were noisy, as compared to graphite images. Sharp drops in contrast and resolution occured. It could be because of molecular fluctuations inside the film as well as heterogenety of the film and possibly a weak linkage

with a substrate. Some time later similar results were
obtained for cadmium arachidate bilayer (Braun et al.,
198 , Horber et al., 1988, Lang et al., 1988).

Thin films of fatty acids on the substrates from Si,
Au, graphite and WSe_2 were studied (Braun et al., 198 ,
Horber et al., 1988, Fuchs et al., 1990). The molecular-
level resolution was obtained only for graphite and WSe_2.
These results indicate a strong effect of a substrate on
the STM visualization of the organic molecules.

Mono and multilayers of dipalmitoylphosphatidylcholine
on various substrates (Si, Au, Sn) were studied (Eng et al.,
1988). The authors observed different behaviour of films
on various substrates, however in none of the cases they
were able to obtain a molecular-level resolution.

In our work we undertook a further very important
step towards investigation of artificial membranes as
analogs of natural cellular walls. We showed the possibility
of using a scanning tunneling microscope to investigate the
structure of lipids and ionic channels at the molecular
level. To do this ionic channels (half-pores) formed by
antibiotic gramicidin A in lipid films were studied. Electrical properties of gramicidin A channels have been
discribed (Haydon et al., 1972). The structure of a gramicidin A channel is shown in Fig.1.

MATERIALS AND METHODS

Glycerinmonooleate was obtained from Fluka AG chemical
Company. Gramicidin A was from Sigma chemical Company.
Dipalmitoylphosphatidylcholine was synthesised and testified
biochemically by the method described in Ref. (Kates, 1972).

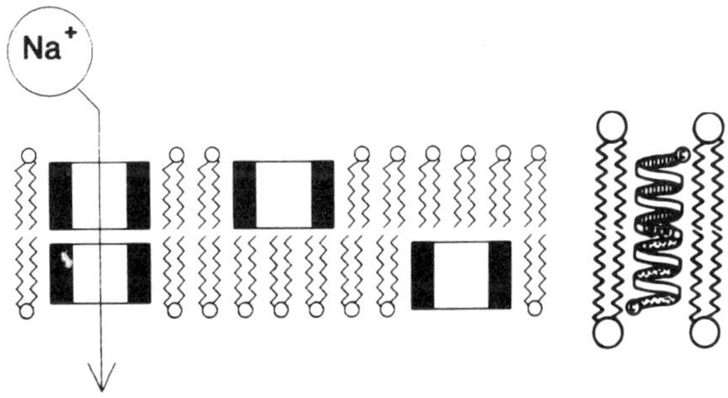

FIGURE 1. The structure of a gramicidin A channel. A dimer
of gramicidin A in the $\beta_{3,3}^{6}$ -helical conformation is an
ionic channel (Urry et al., 1971).

The surface of highly oriented pyrolitic graphite was made hydrophilic by means of electrochemical oxidation as described in Ref. (Lang et al., 1988).

The scanning tunneling microscope operated in air at ambient pressure was made at the Analytical Instrumentation Institute (USSR) (Golubok et al., 1989). Tungsten tips were formed by both electrochemical polishing and mechanical cutting methods. The tip was fixed and the investigated sample was placed at a scanner. The rough approach of a sample to the tip was made by means of an inertial piezo-walker. The piezoelectric tube scanner was 30 mm long and of 12 mm external and 10 mm internal diameter. Scanning tunneling microscope measurements were performed at constant tunneling current as well as at the constant tunneling distant mode. The maximum frame size was $5{\times}5\ \mu m^2$. The minimum time for measuring in one point at constant tunnelling current mode was 0.2 ms. The number of points in one frame was 150X150. Maximum frame frequency under constant average distance mode was 20 Hz. Operating, measuring and data analysis were made by means of the computer. At constant average distance mode the image was displayed on the oscilloscope screen. Images of a pyrolitic graphite surface with atomic resolution were used to calibrate the X and Y scales, Z scale calibration was made by optical interferometery (Golubok et al., 1989).

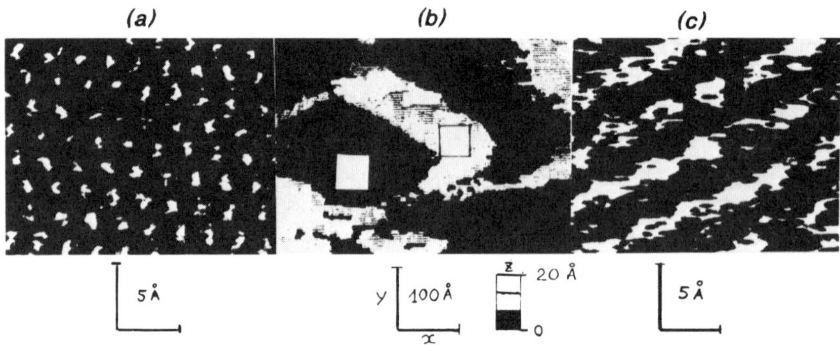

FIGURE 2. Scanning tunneling microscope image of glycerin-monooleate monolayer on a graphite substrate.
(b) islands of lipid monolayer on oxidized graphite surface are seen. Areas in two square frames in (b) were scanned with high resolution to obtain images of a pyrolitic graphite substrate (a) and glycerinmonooleate monolayer (c). Fast scanning in the constant tunneling distant mode was used. One image was obtained in 1 s. Tunneling current was registered under voltage 300 mV. White areas in pictures mean high tunneling current in comparison with black areas.

RESULTS AND DISCUSSION

The lipid monolayers containing gramicidin A were
formed at air-water interface and deposited on an oxidised
pyrolitic graphite substrate by the Langmuir-Blodgett
technique (Lang et al., 1988, Smith et al., 1987). At air-
water interface a monolayer of lipid and gramicidin A was
formed. Then the oxidised graphite substrate was slowly
removed from water to air. As a result a surface of the
oxidised graphite became covered by the monolayer of the
lipid and gramicidin A.

The lipid bilayers containing gramicidin A were formed
on a nonoxidised graphite by the similar technique. In
this case a nonoxidised graphite was immersed under water
and then removed from water to air. These samples were used
in experiments.

Pure lipid films on graphite were studied in the
control experiments as well. It is seen in Fig.2 that
islands of lipid monolayer on oxidized graphite surface
were present. Spatial modulation of conductivity was
observed when scanning was performed above lipid monolayers
(Fig.2c). The images presented in Fig.2 are well repro-
duceble. It is seen from Fig.2 that the spatial period
of the lipid monolayer structures was several times more
than the graphite period. The area per one white spot in
Fig.2c (0.175 nm^2) is very close to the area of one lipid

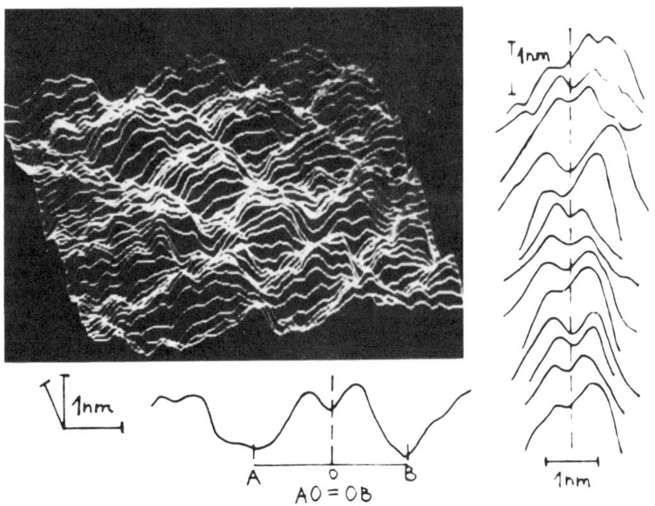

FIGURE 3. Side-view of the structure formed at relatively
high concentration of gramicidin A in the glycerinmonooleate monolayer (gramicidin A : glycerin -monooleate is 1:1
in weight). The scanned area is 7.5 10 nm . The constant
tunneling current mode was used. One image was obtained in
10 s. To the right cross sections of individual objects are
presented.

molecule at air-water interface known from surface pressure
measurements (Phillips et al., 1968). We conclude therefore
that the molecular structure of the lipid monolayer
appeared in the scanning tunneling microscope image.
Molecular resolution was obtained also when scanning was
performed above the dipalmitoylphosphatidylcholine mono-
and bilayer (Kolomytkin et al., 1991).

The main purpose of the present work was visualization
of ionic channels formed by gramicidin A in the lipid membranes. We have obtained scanning tunneling microscope images
of monolayers formed of lipid with gramicidin A.

Quasi-regular, densely packed structure was obtained
under high concentration of gramicidin A. Tipical image of
such region is given in Fig.3.

It is important to point out that we never found such
objects when we were studying pure lipid layers. This fact
and also the size of the objects (2$\pm$0.5 nm) lead to the
conclusion that these objects are STM images of gramicidin
A molecules build into lipid monolayer.

The cross-sections of individual molecules are given
in Fig.3, to the right. One can see that the characteristic
size of a gramicidin A molecule is 2$\pm$0.5 nm. Moreover, in
the central part of the STM image the cavity is present of
0.4$\pm$0.05 nm in halfwidth (diameter at 50% depth).

We also tried to obtained the images of individual
gramicidin A molecules in lipid monolayer with low concentration of antibiotic. In these experiments we saw as a
rule the structures of pure glycerinmonooleate monolayer.
Total area of 4,500 nm^2 was visualized. But six times we
met regions with significant local rise of tunneling
current. They looked like hills with an average diameter

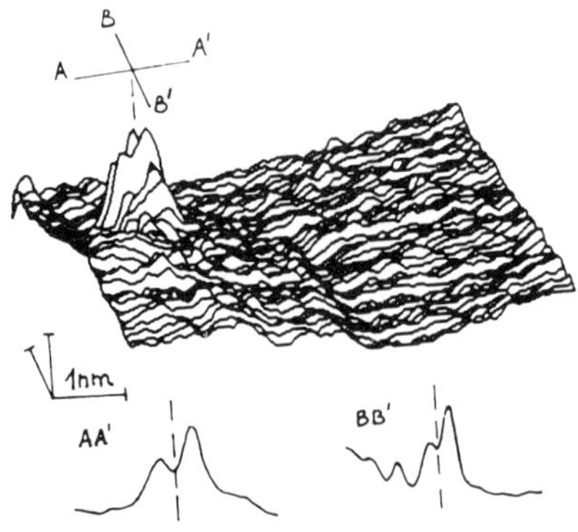

FIGURE 4. Side-view of the structure formed at low concentration of gramicidin A in the glycerin-monooleate monolayer
(0.1 nM gramicidin A was in solution when the monolayer was
formed on the water-air interface). The scanned area is
7.5$\pm$10 nm^2. Scan parameters are the same as the ones in
Fig.3. Cross sections of individual object are presented.

of 2±0.5 nm (Fig.4) on the images obtained at constant
tunneling current mode. The size and the shape of these
hills were similar to the same characteristic of the
objects obtained under high concentration of gramicidin A
(Fig.3). Therefore we are inclined to think that the
object shown in Fig.4 is an individual molecule of gramicidin A in the lipid monolayer.

It shoud be noted that parameters of the tunneling
current and voltage necessary for obtaining of good
quality images of gramicidin A and lipid molecules differ
of each other. Therefore it is difficult to obtain good
images of gramicidin A and lipid molecules at the same
picture. But the control experiment with different scaning
parameters showed that tightly packed lipid molecules were
present in space between gramicidin A molecules. Analysis
of the pictures showed that the total area of spots imaged
in Fig.3 was approximately 50% of the area of the monolayer.
The films were made of a mixture of gramicidin A and lipid
in weight ratio 1:1. Therefore the obtained result supported
the conclusion that gramicidin molecules were imaged in
Figs.3,4.

To confirm our conclusions the experiments with
another type of lipid layer, dipalmitoylphosphatidylcholine
bilayer, have been carried out.

Tipical result is given in Fig.5, where molecules
combined into clusters of chain type with average intermolecular distance of 3.0±0.1 nm are visualized. It could
be seen from cross-sections given to the right part of

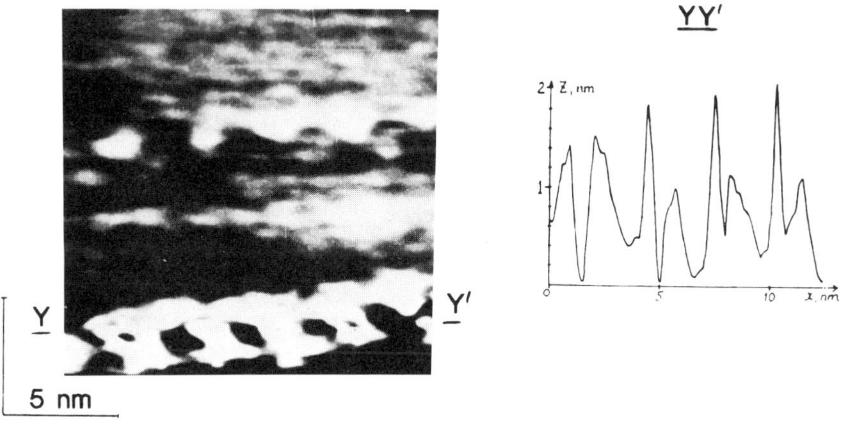

FIGURE 5. Scanning tunneling microscope image of "chain-
type" clusters of gramicidin A molecules in lipid bilayer
formed from dipalmitoylphosphatidylcholine on a graphite
substrate. Scans were performed with a frame size
15×15 nm^2, in 0.15 nm steps at a bias voltage of 30 mV.
Constant tunneling current of 0.3 nA was employed. One
image was obtained in 10 s. The ratio of lipid:gramicidin A
in the bilayer was 20:1 in weight. Top view is shown to the
left. Cross-section in line Y-Y' is shown to the right.

134

Fig.5 that the cavity exists in the center of the STM image
of gramicidin A molecules.

The images obtained in repeated scanning in the
smaller frame which corresponds to the left edge molecule
from this chain cluster is given in Fig.6.

Let us compare the obtained images with a molecular
model of gramicidin A channel. Urry proposed two
$\beta_{3,3}^{6}$ -helixes dimerized by head-to-head hydrogen bonding as
the dominant channel structure in a bimolecular lipid
membrane (Urry, 1972). Each $\beta_{3,3}^{6}$ -helix is located in one
lipid monolayer. In Fig.6, below the molecular model of a
gramicidin A channel is shown. It is seen that hydrophobic
butt-end of the gramicidin "half-pore" in a channel has
triangular form because of tryptophan residues. A pore
0.4 nm in diameter is located in the center of $\beta_{3,3}^{6}$ -helix.
We can see that the scanning tunneling microscope image of
gramicidin molecule shown in Fig.6 has similar form and
size as hydrophobic butt-end of the gramicidin $\beta_{3,3}^{6}$ -helix.

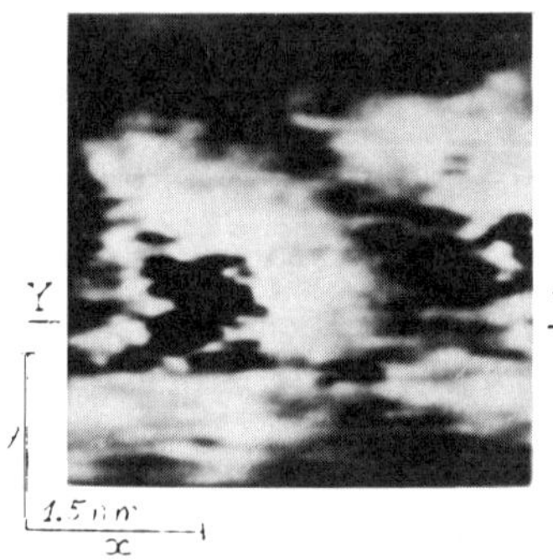

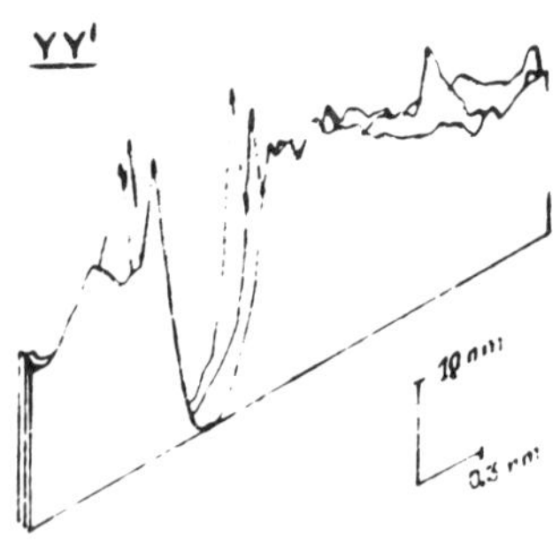

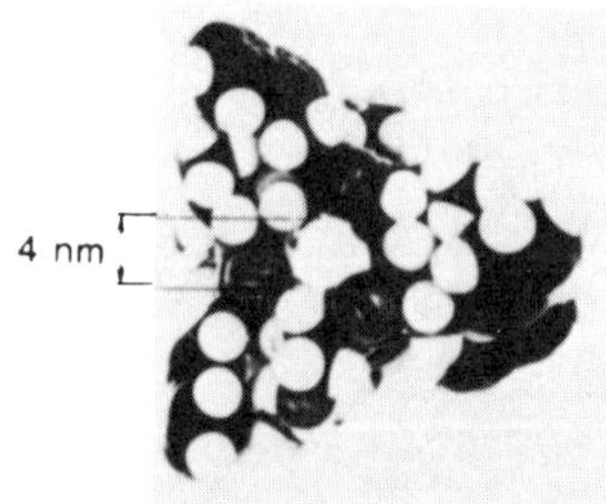

**Molecular model
of gramicidin A in
the $\beta_{3,3}^{6}$-helical
conformation
(Urry D.W., 1972).**

**Constant tunneling current
of 0.3 nA was employed.
Voltage 30 mV. Time 10 s.
Lipid/gramicidin A =
20:1 in weight.**

FIGURE 6. Scanning tunneling microscope image and molecular
model of a single gramicidin A molecule. Scanning tunneling
microscope top-view image of gramicidin A molecule in the
dipalmitoylphosphatidylcholine bilayer is shown to the left.
The scanned area is 4×4 nm^2. One cross section in YY' plane
is shown to the right. Two other cross sections are the
result of neighbouring line scans. Below is shown channel
view (helix axis perspective) of Corey-Pauling-Koltun
molecular model of gramicidin A in the $\beta_{3,3}^{6}$ -helical
conformation taken from Ref. (Urry, 1972).

Location and dimension of black area in Fig.6,a and the channel pore in Fig.6,c are quite similar to each other. Therefore it is possible to conclude that the image of the gramicidin $\beta^{\delta}_{3,3}$-helix was obtained.

It is interesting that gramicidin A molecules can be combined into clusters (Fig.3,5). It is possible that existence of such clusters should be taken into account when time parameters of channel gate functioning in a lipid bilayer is estimated (Stark et al., 1986).

It is known that STM is able to operate in water (Lindsay et al., 1988). Therefore one could expect that the direct experiments of both ionic channel visualization and conductivity measurements may be carried out by means of STM.

Let us consider a number of problems that are specific in STM studies of biological objects.

The first is the problem of electroconductivity and electron transfer. Numerous investigations with the use of other physical methods have shown that most of biological materials and many organic films are dielectrics. There is no answer to the question now why tunnelig current can flow across such thick area as lipid layers with incorporated peptides. Solution of this problem requires further theoretical and experimental study of the nature of contrast of the STM images of proteolipid films and other biological objects.

The second group of problems is associated with the interaction of biological objects with the tip and substrate as well as with their mobility and instability determined by their surface activity and conformational fluctuations.

Besides general requirements for substrates (smoothness, stability and electroconductivity) additional requirements are imposed that are, generaly speaking, mutually exclusive. On the one hand the substrate should be sufficiently inert. On the other hand the substrate should sufficienly strongly bind the molecules investigated. A strong interaction of the tip with the object (both electrical and mechanical, in particular, due to a close approach of the tip to the sample up to a contact, owing to a low conductivity of objects) results in that adsorbates weakly associated with the surface begin to be deformed or move along the substrate under the action of the tip (Coombs et al., 1988, Fuch et al., 1990, Smith et al., 1987, Salmeron et al., 1990). Therefore the object should be tightly bound to the substrate. But a strong surface attraction may result in the conformational changing of the molecule (Stemmer et al., 1988).

Besides, the effects related with the STM tip may manifest themselves as "astigmatism", distortion and multiple images of a single object. Control of STM tip quality is quite important (Stemmer et al., 1988). The tip should be tested not only above the smooth surfaces but also above the surfaces with characteristic relief with nonuniformities close to expected dimensions of the objects studied.

The third group of problems concerns interpretation of visualized structures. These problems are closely related to the mechanism of electron transfer.

The problems mentioned are to be solved in further studies.

In spite of the the currently existing problems we can compare the peculiarities and characteristic sizes of molecular models with STM images. The obtained results demonstrate that the scanning tunneling microscope can be used for high spatial resolution study of the structure of the molecules forming ionic channels in a lipid membrane.

It is known that same important membrane proteins can be isolated and incorporated into a lipid membrane. For example earlier we were able to reconstruct the postsynaptic gamma-aminobutiric acid receptor isolated from rat brain in lipid membrane (Kolomytkin et al., 1986, Kolomytkin et al., 1989). Therefore the presented paper opens a new way for structural investigation of these membrane proteins by means of the scanning tunneling microscope.

REFERENCES

Al-Mohammad A, Smith C, Al-Saffar IS and Slifkin MA (1990): Thin organic films for electronics applications. Thin Solid Films 189(1): 175-181.

Amrein M, Stasiak A, Gross H, Stoll E and Travaglini G (1988): Scunning tunneling microscopy images of metal-coated bacteriophages and uncoated double-stranded DNA. Science 240: 514-516.

Bando H, Kashiwaya S, Tokumoto H, Anzai N, Kinoshita N and Kajimura K (1990): Tunneling spectroscopy on an organic superconductors (BEDT-TTF)2Cu(NCS)2. J.Vac.Sci. Technol. A8(1): 479-483.

Baro AM, Miranda R, Alaman J, Garsia N, Binnig G, Rohrer H, Gerber Ch and Carrascosa JL (1985): Determination of surfase topography of biological specimens at high resolution by scanning tunneling microscopy. Nature 315: 253-254.

Binnig G, and Rohrer H (1986): Scanning tunneling micros-copy. IBM J. Res. Develop. 30: 355-369.

Braun HG, Fuchs H and Schrepp W (198): Surface structure investigation of Langmuir-Blodgett films. Thin Solid Films 159: 301-314.

Coombs JH, Pethica JB and Welland ME (1988): Scanning tunneling microscopy of thin organic films. Thin Solid Films 159:293-299.

Cricenti A, Selci S, Felici A, Generosi R, Gori E, Djaczenko W and Chiarotti G (1989): Molecular structure of DNA by scanning tunneling microscopy. Science 245: 1226-1227.

Eng L, Hidber HR, Rosenthaler L, Staufer U, Wiesendanger R, Gunterodt H and Tamm L (1988): Summary Abstract: Dipalmitoilphosphatidylcholine-Langmuir-Blodgett films on various substrates: Si(111), Au, Sn. J. Vac. Sci. Technol. A6(2): 358-362.

Fuchs H, Akari S and Dransfeld K (1990): Molecular reso-lution of Langmuir-Blodgett monolayers on tungsten diselenide by scanning tunneling microscopy. Z. Phys. B.-Cond. Matter, 80(3): 389-392.

Golubok AO, Davydov DN, Masalov SA, Nakhabtsev DV and

Timofeev VA (1989): Scanning tunneling microscope observation of graphite surface at ambient pressure. Surface, physics, chemistry, mechanics 3: 146-149.

Guckenberger R, Kosslinger C, Gatz R, Breu H, Levai N and Baummeister W (1988): A scanning tunneling microscope (STM) for biological applications: design and performance. Ultramicroscopy 25: 111-112.

Hameroff SR, Simic-Krstic J, Kelley MF, Volker MA, He JD, Dereniak EL, McCuskey RS and Schneiker CW (1989): Conditions for microtubule stabilization. J. Vac. Sci. Technol. A7(4): 2890-2894.

Haydon DA and Hladky SB (1972): Ion transport across thin lipid membranes: a critical discussion of mechanism in selected systems. Quarterly reviews of Biophysics 5: 187-283.

Horber JKH, Lang CA, Hansch TW, Heckl WM and Mohwald H (1988): Scanning tunneling microscopy of lipid films and embedded biomolecules. Chem. Phys. Lett. 145(2): 151-158.

Kates M (1972): Techniques of lipidology. North-Holland Publishing Company, Amsterdam, London/American Elsevier Publishing Co., Inc., New York.

Keller RW, Dunlap D, Bustamante C, Keller DJ, Garcia RG, Gray C and Maestre MF (1990): Scanning tunneling microscopy images of metal-coated bacteriophages and uncoated double-stranded DNA. J. Vac. Sci. Technol. A8(1): 706-712.

Kolomytkin OV, Golubok AO, Davydov DN, Timofeev VA, Vinogradova SA and Tipisev SYa (1991): Ionic channels in Langmuir-Blodgett films imaged by a scanning tunneling microscope. Biophys. J. 59: 889-893.

Kolomytkin OV and Kuznetsov VI (1986): Incorporation of GABA receptor into planar bilayer. Studia biophysica 115: 157-164.

Kolomytkin OV, Kuznetsov VI and Akoev IG (1989): Microwaves affect the function of reconstituted and native receptor membranes of the brain. In: Charge and Field Effects in biosystems-2. Allen MJ, Cleary SF and Hawkridge FM, eds. New York and London: Plenum Press.

Lang CA, Horber JKH, Hansh TW, Hechl W and Mohwald H. (1988): Scanning tunneling microscopy of Langmuir-Blodgett films on graphite. J. Vac. Sci. Technol. A6(2): 368-370.

Lindsay SM and Barris B (1988): Imaging deoxyribose nucleic acid molecules on a metal surface under water by scanning tunneling microscopy. J. Vac. Sci. Technol. A. 6: 544-547.

Mizutani W, Shinego M, Sakakibara Y, Kajimura K, Ono M, Tanishima S, Ohno K and Toshima N (1990): Scanning tunneling spectroscopy study of adsorbeb molecules. J. Vac. Sci. Technol. A8(1): 675-678.

Mizutani W, Shinego M, Ono M and Kajimura K (1990): Voltage-dependent scanning tunneling microscopy images of liquid crystals on graphite. Appl. Phys. Lett. 56(20): 1974-1976.

Phillips MC and Chapman D (1968): Monolayer characteristics of saturated 1,2-diacylphosphatidylcholines and phosphatidylethanolamines at the air-water interface.

Biochim. Biophys. Acta. 163: 301-313.
Rohler H. (1990): Research report: Scanning tunneling micro-
    scopy - methods and variations. Solid State Physics.
Salmeron M, Beebe T, Odriozda J, Wilson T, Ogletree DF and
    Siekhaus W. (1990): Imaging of biomolecules with
    scanning tunneling microscope: Problems and prospects.
    J. Vac. Sci. Technol. A8(1): 635-641.
Smith DPE, Kirk MD and Quate CF (1987): Molecular images and
    vibration spectroscopy of sorbic acid with the scanning
    tunneling microscope. J. Chem. Phys. 86: 6034-6038.
Smith DPE, Briant A, Quate C, Rabe JP, Gerber C and Swalen
    JD (1987): Images of a lipid bilayer at molecular
    resolution by scanning microscopy. Proc. Natl. Acad.
    Sci. USA, Biophysics 84: 969-972.
Smith DPE, Horber JKH, Binnig G and Nejoh H (1990):
    Structure, registry and imaging mechanism of alkyl-
    cyanobiphenil molecules by tunneling microscopy. Nature
    344: 641-644.
Stark G, Strassle M and Takacz Z (1986): Temperature-
    jump and voltage-jump experiments at planar lipid
    membranes support an aggregationl (micellar) model of
    gramicidin A ion channel. J. Membrane Biol. 89: 23-37.
Stemmer A, Hefti A, Aebi U and Engel A (1988): Scanning
    tunneling and transmission electron microscopy on
    identical areas of biological specimens. Ultramicroscopy
    25: 111-112.
Urry DW (1972): A molecular theory of ion-conducting chan-
    nels: a field-dependent transition between conducting
    and nonconducting conformation. Proc. Natl. Acad. Sci.
    USA 69: 1610-1614.

STUDY OF THE INFLUENCE OF THE SIDE CHAIN DIPOLES ON THE CONDUCTANCE OF ION CHANNELS FORMED BY GRAMICIDIN ANALOGUES

Genoveva Martinez, Miguel Sancho
Departamento de Física Aplicada III
Universidad Complutense, 28040-Madrid, Spain

Victoria Fonseca
Departamento de Física Atómica, Molecular y Nuclear
Universidad Complutense, 28040-Madrid, Spain

INTRODUCTION

Gramicidin A (GA) is a small polypeptide able to form transmembrane channels. It consists of 15 amino acids with alternating L- and D- configurations:

HCO–L–Val–Gly–L–Ala–D–Leu–L–Ala–D–Val–L–Val–D–Val–L
–Trp–D–Leu–L–Trp–D–Leu–L–Trp–D–Leu–L–Trp–NHCH$_2$CH$_2$OH

Fluorescence and conductance measurements in black lipid membranes (Veatch and Stryer, 1977; Szabo and Urry, 1979) have shown that the conducting form of the molecule is a head to head dimer of π(L,D) helices in which the polar peptide groups line a central cavity, while the side chains point outside towards the lipid medium. An important question for understanding the ion exchange through the structure concerns to the role of these residues. Recent experimental studies have focused on different Gramicidin analogues obtained by synthetic procedures, in which certain amino acids have been substituted by others with different polarity (Barrett

Russell et al., 1986; Heitz et al., 1988; Fonseca et al., 1989). The measured conductances for the studied modified gramicidins are in general lower than that of GA, in some cases by more than one order of magnitude. The important modification of the ion transport is most likely not the result of changes in channel structure (Mazet et al., 1984; Andersen et al., 1987). This leaves the electrostatic interactions either direct or through inductive changes in the electron densities of the carbonyl oxygens which are close to the ion pathway, as the other possible mechanism by which the side chains influence channel function. From the study of gramicidins where the most polar amino acid (Trp) at different positions has been replaced by other residues, and from measurements of surface potential in monolayers, Heitz et al. (1988) conclude that the dipole moments of the side chains modulate directly the conductance and that not only the dipole moment but also its orientation are important.

Electrostatic models have proved very useful in estimating energy barriers in ion channels (Jordan, 1982; 1986). Our approach includes the influence of the dipole moments of the side chains on the ion permeation characteristics. We apply the model to simulate the electric behavior of both pure and hybrid channels formed by two different modified monomers. Our calculations explain the main experimental results and support the interpretation of the long range dipole–ion interaction as an important component in channel functioning.

ELECTROSTATIC MODEL OF GRAMICIDIN AND ANALOGUES

The geometrical and dielectric characteristics of the model have been discussed in detail in a previous paper (Sancho and Martínez, 1991). The main parameters are represented in Figure 1. We assume rotational symmetry and neglect the effect of ion strength in the bulk medium on the electrical potential profile.

In this two–dielectric model the bulk water and the pore interior are represented by the same constant ε_1; the membrane and pore former are characterized by ε_2. A half–width h = 12.5 Å and an effective radius of 2.8 Å are taken for the channel. Two electrodes, situated at $\mp$ 6h and polarized at $\pm$ V_0, simulate the applied potential.

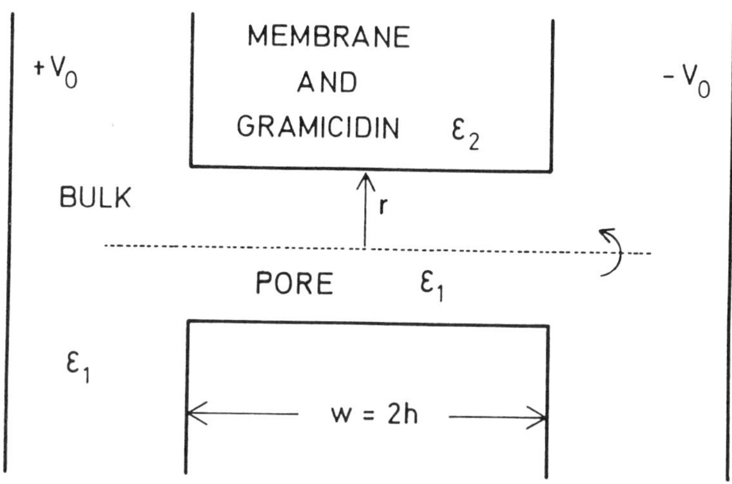

FIGURE 1. Cross-section of the cylindrical geometry used in the modeling of a gramicidin channel.

Dipolar rings with radius of 5 Å, total moment p and coaxial with the pore, are superposed to this geometry to account for the influence of polar side chains on the ion when it traverses the channel. The dipole moment is assumed to be distributed with uniform density along the circumference. The dipolar rings are situated at axial mean positions $z = \mp 8.5$ Å and have variable orientations with components p_r and p_z. The effect of the backbone charge distribution is represented by two rings of radius 3.5 Å and $p_z = 3.8$ D, each of them located at the midpoint of their respective monomers, with the moment pointing toward the nearest end of the channel.

SOLUTION OF THE ELECTROSTATIC PROBLEM

For the geometry of Figure 1, we have the following set of integral equations (Martínez and Sancho, 1991)

$$\phi(\vec{r}) = \frac{2\varepsilon_1}{\varepsilon_1 + \varepsilon_2}\, \phi_s(\vec{r}) + \frac{1}{2\pi(\varepsilon_1 + \varepsilon_2)} \int_{S_c} \frac{\sigma(\vec{r}')}{R}\, ds'$$

$$+ \frac{\varepsilon_1 - \varepsilon_2}{2\pi(\varepsilon_1 + \varepsilon_2)} \int_{S_{12}} \phi(\vec{r}')\, \frac{\partial}{\partial n'}\left(\frac{1}{R}\right) ds', \quad \vec{r} \in S_{12}$$

$$\phi(\vec{r}) = \text{const.} = \phi_s(\vec{r}) + \frac{1}{4\pi\varepsilon_1} \int_{S_c} \frac{\sigma(\vec{r}')}{R}\, ds'$$

$$+ \frac{\varepsilon_1 - \varepsilon_2}{4\pi\varepsilon_1} \int_{S_{12}} \phi(\vec{r}')\, \frac{\partial}{\partial n'}\left(\frac{1}{R}\right) ds', \quad \vec{r} \in S_c$$

where $R = |\vec{r} - \vec{r}'|$; ϕ_s is the source term, including the potential produced by the ion and the dipoles, in the medium of dielectric constant ε_1. S_c represents all the conducting surfaces and S_{12} is the interface between the dielectrics.

To obtain the solution of the integral equations, we have applied a moment method which allows the obtainment of the charge densities on the conductors and the potentials at the interface (Sancho and Martínez, 1991). Then the total potential energy of the ion at any position along the axis is calculated as

$$U = e\, (1/2\, \phi_i + \phi_v + \phi_g + \phi_d)$$

where $1/2\ \phi_i$ represents the image self-potential, ϕ_v the contribution of the applied voltage, and ϕ_g, ϕ_d those of the dipole moments associated with the backbone and side groups, respectively.

RESULTS

We have calculated energy profiles for monovalent ions taking different moments and orientations of the dipole rings representing the side chains. To investigate the ion transport, we have used the Nernst–Planck continuum approach. The basic equation describing the ion electro-diffusion across the channel is (Levitt, 1986)

$$I = eDA\ \frac{C_1 e^{v_1} - C_2 e^{v_2}}{\int e^{v}dz}$$

where $v = U/kT$; D is the diffusion coefficient and A the area available for the ion; the integral is calculated between the positions of both electrodes where reduced potentials are v_1 and v_2, using a trapezoidal rule.

In this model the short-range forces on the ion resulting from interactions with water and peptide atoms are not explicitly considered but roughly included through the diffusion coefficient. The width of the energy barriers that we have found, much larger than the estimated diffusion length of the ion, is consistent with this approach.

Pure Channels

To simulate the behavior of those channels formed by two monomers of the same type, we place dipole moments at both sides, symmetrically oriented with respect to $z = 0$. The variation of the energy profiles with magnitude and orientation of the dipole moment associated to the side chains is illustrated in Figure 2. In assigning values to p we have taken into account that the most polar amino acids in GA are the four tryptophanes at positions 9, 11,

13 and 15, with approximate dipole moment of 2.1 D each. Some analogues of interest are: GT in which all the four Trp residues have been substituted by Tyr with estimated p $\simeq$ 1.6 D, GT′ which contains four Tyr(Bzl) with p $\simeq$ 1.2 D, GM⁻ (GM enantiomer), with four Phe residues and p $\simeq$ 0.4 D, and GN with four Nap, approximately neutral.

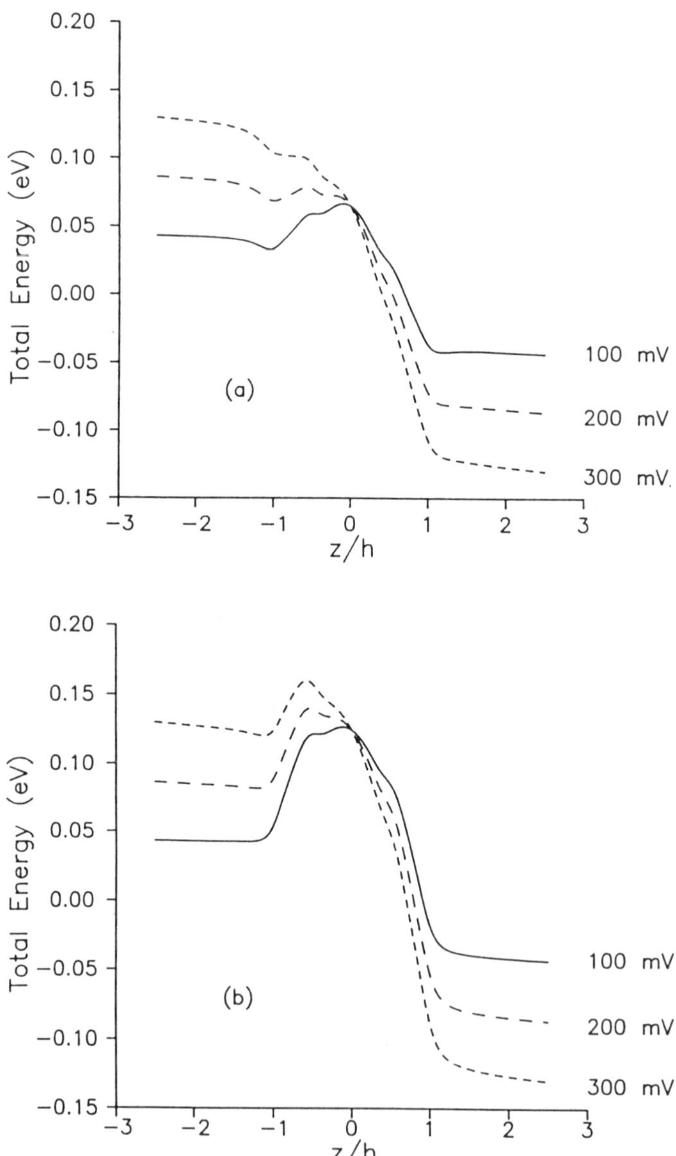

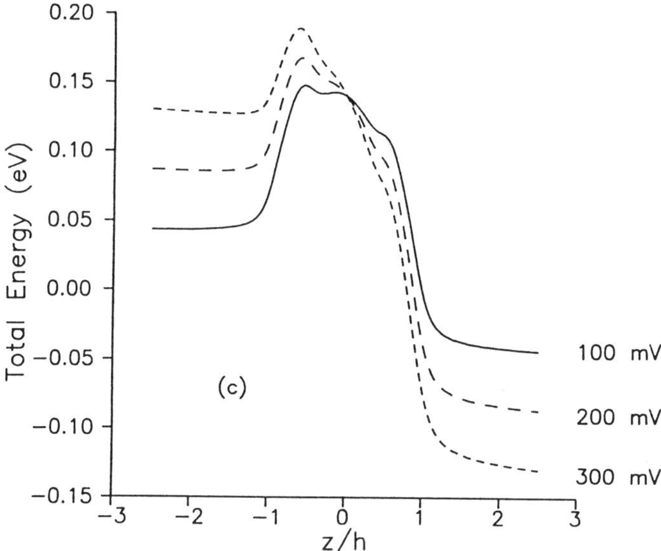

FIGURE 2. Calculated total energy profile of the permeating ion for three polarizations of the channel and different magnitude and angle with respect to the z-axis of the dipole moment: (a) $p = 8.5$ D, $\vartheta = 90°$; (b) $p = 5.0$ D, $\vartheta = 116°$; (c) $p = 0.6$ D, $\vartheta = 90°$.

We have estimated total values of p for these analogues and tried to fit experimental results changing the orientation. Figure 3 shows the calculated conductance curves. Names in brackets refer to the gramicidin analogues giving similar behavior in experiments. In the case of GT' two curves are drawn; radial orientation (GT'1) gives conductance values higher than those measured while the assumption of a negative axial component (GT'2) allows to fit the experimental data. For comparison, experimental values of the conductance of GA and analogues in DPhPC membranes (Fonseca et al., 1989), are shown in Figure 4.

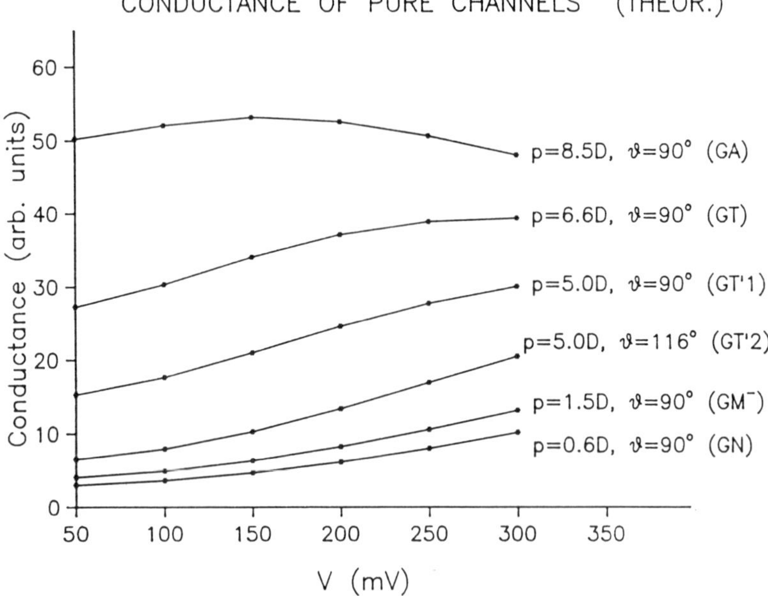

FIGURE 3. Calculated conductance vs. applied voltage for different values of the dipole moment and orientation.

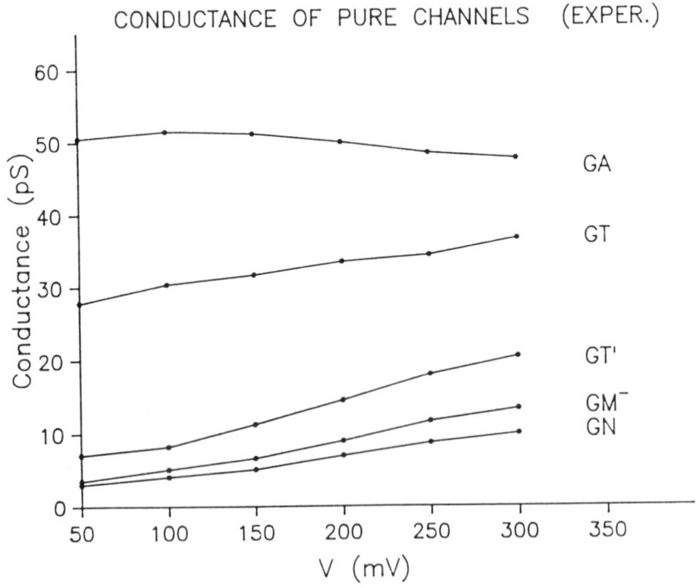

FIGURE 4. Measured conductance of pure channels formed by GA and analogues, as a function of the applied voltage.

Hybrid Channels

Taking different moments for the dipole rings of each monomer, we can simulate the contribution of the side chains in hybrid channels. Figures 5 and 6 illustrate the energy profiles and calculated conductances for different combinations of dipolar contributions. In computing these curves we use the same configuration and dipole moments assigned to each monomer in Figure 3. There is an asymmetry between the behavior of GA–GX and GX–GA hybrids, the first one having a higher and steeper conductance vs. applied potential curve.

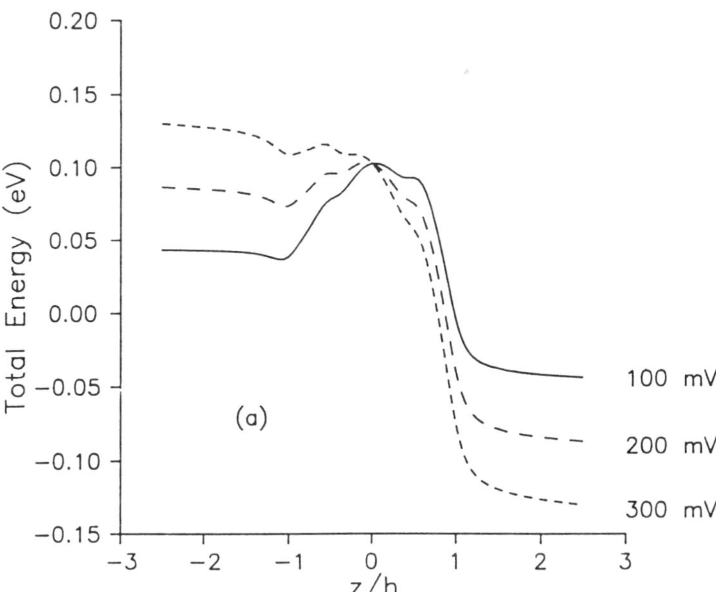

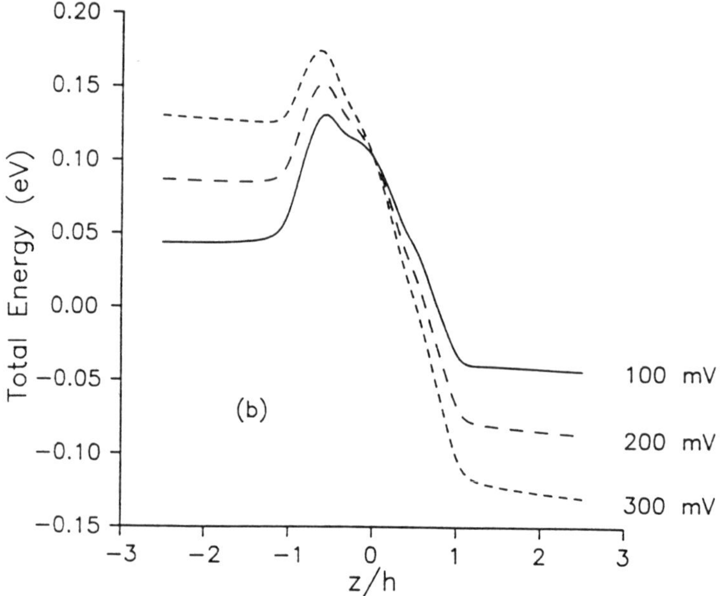

FIGURE 5. Total ion energy profile in hybrid channels.
(a) GA–GN hybrid; (b) GN–GA hybrid.

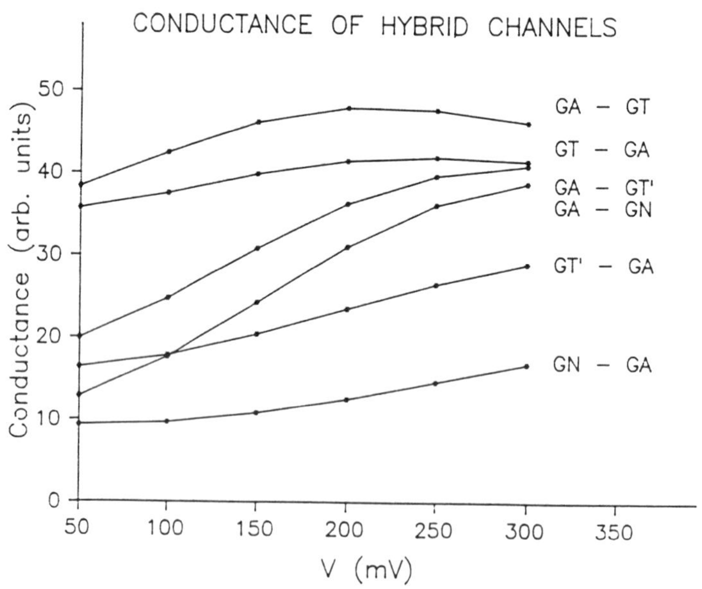

FIGURE 6. Calculated conductance vs. applied voltage for
different combinations of dipolar contributions simulating
hybrid channels.

DISCUSSION

We have described an electrostatic model which can explain the important modulation of the transport characteristics of Gramicidin channels in which certain amino acids are replaced by others with different polarity. The model contains several oversimplifications besides those implicit in a macroscopic treatment. We have considered only two dielectric media to simulate the pore structure, assumed cylindrical symmetry and neglected ionic strength effects. However, it is relatively simple and has the advantage of including the important influence of the electrical polarization of the membrane which is difficult to incorporate in microscopic approaches.

The image energy barrier of the ion in the cylindrical pore is about 200 meV for an electrical radius of 2.8 Å. A dipole ring of radius 5 Å, $p = 1$ D and with its center at $z = 5$ Å, can reduce this barrier in about 12 meV. But this effect depends on the orientation, being in general stronger for radial dipoles than for axial ones (Sancho and Martínez, 1991).

The comparison between our calculated conductance curves and the experimental ones for pure channels indicates that, contrary to the Urry's standard conformation (Venkatachalam and Urry, 1983), the orientation of all the four Trp in GA favors the ion passage. Besides, the relative magnitude of conductances suggests a predominantly radial orientation of p. For all the simulated analogues, the conductance is smaller and steeper with voltage, as the radial moment decreases. In one case (GT') the fit gives a different orientation for the total dipole moment.

In the study of hybrid dimers, the model predicts a larger and steeper conductance for the configuration GA–GX than for the reverse one GX–GA, in agreement with experiments (Fonseca et al., unpublished results). The reason for this behavior can be found by examining the energy profiles of Figure 5. The shifting of the barrier toward the less polar side produces higher but narrower profiles in the case of GX–GA heterodimers than those found for GA–GX hybrids. The resultant G–V curves are determined by the variation of height and –specially– width of these profiles with the applied voltage.

We conclude that an electrostatic model incorporating the long range dipole–ion interaction can be very useful for interpreting the structure–function relation in ion channels. The good qualitative agreement found with measurements on pure gramicidin channels and also on hybrids ones in which case no additional fitting parameters have been introduced, gives confidence to the model predictions.

REFERENCES

Andersen OS, Koeppe II RE, Durkin JT and Mazet JL (1987): Structure-function studies on linear gramicidins: site-specific modifications in a membrane channel. In: *Ion Transport through Membranes*, New York: Academic Press.

Barrett Russell EW, Weiss LB, Navetta FI, Koeppe II RE, and Andersen OS (1986): Single channel studies on linear gramicidins with altered amino acid side chains. *Biophys J* 49: 673–686.

Fonseca V, Daumas P, Ranjalahy-Rasoloarijao L, Heitz F, Lazaro R, Trudelle Y and Andersen OS (1989): Gramicidin Channels that have no Tryptophan residues. *Biophys J* 55: 502a.

Heitz F, Daumas P, Van Mau N, Lazaro R, Trudelle Y, Etchebest C and Pullman A (1988): Linear Gramicidins: Influence of the Nature of the Aromatic Side Chains on the Channel Conductance. In: *Transport Through Membranes: Carriers, Channels and Pumps*, Pullman A et al, eds. Dordrecht, The Netherlands: Kluwer Academic Publishers.

Jordan PC (1982): Electrostatic modeling of ion pores. *Biophys J* 39: 157–164.

Jordan PC (1986): Ion Channel Electrostatics and the Shapes of Channel Proteins. In: *Ion Channel Reconstitution*, Christopher Miller, ed. New York: Plenum Press.

Levitt DG (1986): Interpretation of biological ion channel flux data. *Ann Rev Biophys Biophys Chem* 15: 29–57.

Martínez G and Sancho M (1991): *Application of the integral equation method to the analysis of electrostatic potentials and electron trajectories.* In: Advances in Electronics and Electron Physics. Hawkes P, ed. New York: Academic Press.

Mazet JL, Andersen OS and Koeppe II RE (1984): Single-channel studies on linear gramicidins with altered amino acids sequences. *Biophys J* 45: 263–276.

Sancho M and Martínez G (1991): Electrostatic modeling of dipole-ion interactions in gramicidinlike channels. *Biophys J* 60: 81–88.

Szabo G and Urry DW (1979): N-acetyl gramicidin: single channel properties and implications for channel structure. *Science* 203: 55–57.

Veatch W and Stryer L (1977): The dimeric nature of the Gramicidin A transmembrane channel: conductance and fluorescence energy transfer studies of hybrid channels. *J Mol Biol* 113:89–102.

Venkatachalam CM and Urry DW (1983): Theoretical conformational analysis of the gramicidin A trans-membrane channel. I. Helix sense and energetics of head-to-head dimerization. *J Comput Chem* 4: 461–469.

IDEALIZED MODEL OF COUPLED PROCESSES IN MITOCHONDRIAL PROTON TRANSFER

Tofik M. Nagiev

Nagiev Institute of Theoretical Problems
of Chemical Technology, Baku City,
Azerbaijan Republic, USSR

INTRODUCTION

To solve the theoretical problem of constructing a chemical
system in which catalysis, chemical coupling, and the transport of
matter through a membrane should be combined in the optimum manner,
enhancing the effect of their mutual influence, the author has used
as model system the mitochondrial energy process, in which enzymatic
catalysis, the coupling of reactions, and the transport of ingredients through a membrane interact in an ideal manner.

The conclusions and recommendations that are given below cannot
be considered as yet one more key to the interpretation of the
coupled oxidative phosphorylation that takes place in cell mitochondria. Rather, biochemists must evaluate the results of this investigation critically and on no account regard the chemical model of an
idealized coupled process as a complete imitation of the biochemical
process.

Interpretation and Discussion

Let us consider the respiratory process (Lehninger, 1970)

$$CH_3COOH + 2O_2 = 2CO_2 + H_2O \qquad\qquad (1)$$

as the sum of two components. The Krebs cycle

$$CH_3COOH_{(1)} + 2H_2O_{(1)} \rightarrow 2CO_{2(g)} + 8H_{(g)} \qquad (2)$$

$$\Delta G^0_{298} = 406.6 \quad kcal/mole.$$

The respiratory chain

$$4H_{(g)} + O_{2(g)} = 2H_2O_{(1)} \qquad (3)$$
$$\Delta G^0_{298} = -307,73 \quad kcal/mole$$
$$(4H^+ + 4e + O_2 = 2H_2O)$$

In the functional respect, the respiratory process (1) consists of a section in which the enzymatic activation of substrate by the dehydrogenases of the Krebs cycle (2) (the endergonic component of the respiration process) and the binding of hydrogen atoms by the oxygen of the enzymes of the respiratory chain (3) in the form of the final product water (the exergonic component) take place.

According to biochemists, there are three material particles possessing an excess of chemical energy with the aid of which a linking (coupling) channel between respiration and phosphorylation can be established - the hydrogen atom, the electron, and the proton. They are all present in the bound state, forming the corresponding complexes with electron-transferring enzymes.

The electron and the hydrogen atom cannot be intermediary substances between the oxidation of CH_3COOH and phosphorylation; the former is strongly bound to the enzyme system of the respiratory chain, and the latter is always coordinated with the corresponding enzymes (Lehninger, 1964). These considerations enable us to arrive at the working hypothesis that it is in fact H^+ that is the intermediate substance that readily leaves the catalytic system of the respiratory process and may thereby be involved in other coupled reactions or again take place in the formation of H_2O, now in the final state of respiration.

We may note that oxidative phosphorylation sets a severe demand on the structure of the membrane (it must remain undamaged).

It appears to us that in the well-known equation for the formation of ATP (Lehninger, 1970) H^+ should also be included as an initial reactant, and then we obtain a summary equation with a complete balance of charges:

$$H^+ + \text{Adenosine-}\overset{O^-}{P}\text{-O}\sim\overset{O^-}{P}\text{-O}^- + \text{HO-}\overset{O^-}{P}\text{-O}^- \rightarrow \text{Adenosine-}\overset{O^-}{P}\text{-O}\sim\overset{O^-}{P}\text{-O}\sim\overset{O^-}{P}\text{-O}^- + H_2O$$

Thus, as it were, for the synthesis of ATP, the respiratory chain is generating H^+ all the time and this is also necessary for binding oxygen molecules in the form of water in the respiratory process.

These considerations lead us to the opinion that H^+ is nothing other than a highly active intermediate particle of two coupled reactions: on the one hand, the respiratory process and, on the other hand, oxidative phosphorylation.

Since it was clear to biochemists that the H^+ ion is involved in the process of oxidative phosphorylation, an alternative hypothesis of the formation of ATP has appeared as a counterbalance to chemical coupling, and this has acquired the name of the chemiosmotic hypothesis of the mechanism of oxidative phosphorylation. In the creation of this theory an enormous contribution has been made by Mitchell (Lehninger, 1970), who turned his attention just to the weak aspects of the chemical coupling hypothesis. In his theory, he started from the fact that in the process of transferring an electron along the respiratory chain, H^+ ions are "ejected" into the cytoplasm, as a result of which a gradient of H^+ ions arises which is determined by their concentration in the intramitochondrial space and the surrounding medium.

Mitchell also suggested that ATPase is oriented in the plane of the mitochondrial membrane in such a way that a molecule of water formed in dissociation form (H^+ and OH^-) leaves the surface of the membrane, the H^+ passing into the internal volume of the mitochondrion, lowering its pH, and OH^-, liberated into its external medium, conversely, raising the pH.

Let us attempt to give the scheme in the form of a combination of individual chemical reactions:

process of respiration

$$CH_3COOH + 2O_2 = 2CO_2 + 2H_2O,$$

phosphorylation (oxidative)

$$ADP^{3-} + P_I^{3-} + 2H^+ = ATP^{4-} + H_2O$$

dissociation
ATP-asa

$$H_2O = H^+ + OH^-$$

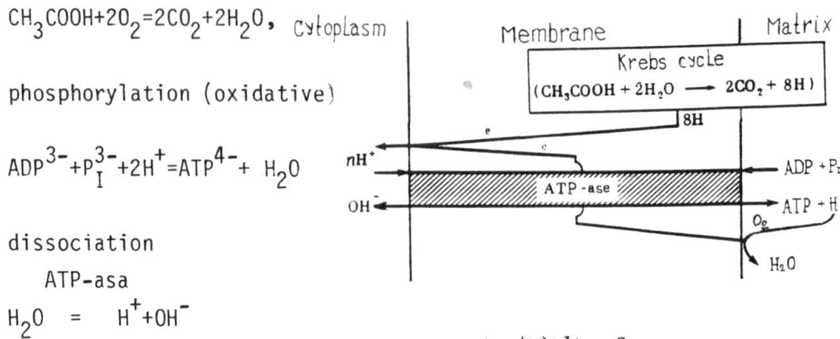

Fig. 1. Approximate scheme of the structural organization of chemical coupling in mitochondrial processes.

As is known, the first two reactions are coupled, and the intermediate substance - the carrier of the inducing action of the first reaction (respiration) to the second (phosphorylation) is (presumably), as mentioned above, the H^+ ion. As can be seen from the reaction for the synthesis of ATP, the H^+ ions is the oxidant the energy of which is sufficient to cause reaction.

For the sake of clarity, let us make use of Fig. 1, from an inspection of which follow a number of important consequences that are necessary for understanding the coupling mechanism. For the formalization of the subject under discussion we shall assume (and this is fully substantiated) that the cytoplasm and the matrix form part of one (single) reaction system but are simply separated from the scheme that the membrane fulfills another, and possibly the main, function enzymatic catalysis: the coupled process of respiration and phosphorylation are catalyzed by the membrane on one side of it - namely, in the matrix.

It is quite natural for an analogy to suggest itself between such a simplified representation of the mitochondrial processes and the phenomenon of the coupling of chemical processes on membrane catalysts that is known in chemistry (Gryaznov and Smirnov, 1972). The transport of a highly active intermediate product through a membrane is a necessary condition for coupling on membrane catalysts in any system.

On membrane catalysts, however, the coupled processes usually take place in the parts of the reactor separated by the membrane. From this follows an important consequence: for chemical coupling on a membrane catalyst the question of the occurrence of the processes in a single part or in various parts of the reactor is not of fundamental importance. In all probability, depending on the nature of the reactions being coupled, one of these possibilities may prove to be preferred; for example, when severe demands are set on the regulation of the rate of chemical coupling.

More attractive appears to be the case where the two coupled reactions are realized on one side of the membrane, as is the case, for example, in mitochondria. This conceals a profound truth, which consists in the fact that the presence of the substrates and products of both the reactions being coupled in the matrix (i.e.,) in one section of the reaction system) permits a finer harmonization of the regulation of the rates of their occurrence - a rise in the concentration of ADP in the matrix intensifies the process of respiration and, conversely, an accumulation of considerable amounts of ATP retards respiration.

In actual fact, it is difficult to imagine how reaction substrates and products present in different "halves" of a reaction system can participate in processes of regulation and inhibition, since the enzyme systems of both reactions will be located on different sides of the membrane.

However, as it is not difficult to note, when we have completely exhausted the active centers for the formation of the products of the induces reaction, we deprive the first (inducing) reaction of the possibility of forming the final products. Thus, in aiming at the maximum accumlation in the system of the products of the secondary reaction the yield of the products of the primary reaction is unavoidably minimized - in the limit, to zero. How is it possible to find a compromise solution in which the products of both reactions in the system are synthesized at the maximum rate? Probably it is that in which each active center must participate in the formation of the products of the coupled reaction in turn.

The mechanism upon which the chemiosmotic hypothesis of the synthesis of ATP is based permits the use of individual specimens of a highly active intermedicate particle - the H^+ ion - with the maximum efficacy in each of the coupled reactions.

In making the work of ATPase concrete, let us assume that the H^+ ions, the so-called active centers, arrive at the membrane from the cytoplasm, while the substrates -ADP and inorganic phosphate - and the phosphorylation reaction itself related to the inner surface of the mitochondrial membrane.

A water molecule is formed as one of the products of oxidative phosphorylation, and this, not being liberated into the bulk, dissociates into H^+ and OH^- there and then in the membrane. And it is just here that the chemiosmotic mechanism works with great elegance: the anion OH^- is desorbed into the cytoplasm and the H^+ ion into the matrix, where its appearance as an active center is connected with he formation of water in the final stage of the process of respiration.

Of course, it is thanks to the process of respiration that the electrical potential difference and the H^+ concentration gradient that play such an important role in the process of transporting H^+ ions appear on different sides of the membrane, without which, let us repeat yet again, there would be no chemical coupling.

Thus, within the framework of the concept of the chemical coupling of mitochondrial processes that has been put forward here, the nature of the active coupling center has been given concrete form (H^+) and the exceptional role of the membrane in the structural organization of coupling has been shown (Nagiev, 1987).

It is possible to convince oneself, according to the new treatment that these two theories are not alternatives but, rather, supplement one another and form, as it were, different sides of a single mechanism of the bioenergetic activity of the cell connected with the accumulation, transformation, and transport of chemical energy. It is precisely this "joint" mechanism of chemical coupling and the chemiosmotic hypothesis that has permitted an approach from purely chemical aspects to an explanation of the action of uncoupling agents of respiration and oxidative phosphorylation. The action of agents uncoupling respiration and oxidative phosphorylation and also a break in the mitochondrial membrane lead to a sharp fall in the electric potential difference on the two sides of the membrane and, as a consequence, to a cessation of the synthesis of ATP. According to this idea, when the mitochondrial surface is broken, H^+ ions must likewise be consumed in both reactions, but with the difference that they now become competing reactions, and, as it were, a "battle" is joined for the capture of H^+ ions. In actual fact, oxidative phosphorylation is incapable of competing with respiration for the capture of H^+ ions.

In view of this circumstance, chemical coupling, which in principle, could take place even with a broken membrane, is not realized in practice. It is possible to organize the work of a chemical system with the maximum indices for all the coupled reactions only with the use of membrane catalysts having a closed surface.

Application of Proton Transfer for creating inorganic enzymes mimics

For the last years there is intensively developing a new branch of catalysis connected with the creation of catalysts of new generation so called biomimics which model separate functions of enzymes in ordinary chemical systems. Such models of enzymes-enzyme mimics for oxidation reactions are observed and analyzed (Nagiev, 1989). One of the most likely models of oxidation biomimics of $PPFe^{3+}OH$ (ferrous protoporphyrine) supported on Aluminum oxide which is modeling catalyze and peroxidase reactions.

With this aim, a system modeling the remarkable property of cytochrome P-450 was developed and its catalytic possibilities have been investigated. The oxidation of propylene by hydrogen peroxide in the gas phase on $PPFe^{3+}OH/AL_2O_3$ has been performed. The optimum conditions for the oxidation of propylene have been established: the highest yield (60 wt %) corresponds to a temperature of 160°C and a C_3H_6 : H_2O_2 ration of 1:1 (molar) with a contact time of 1.9 sec. According to our ideas, the mechanism of the formation of allyl alcohol on the monoxygenation of propylene by hydrogen peroxide is describe by the following scheme:

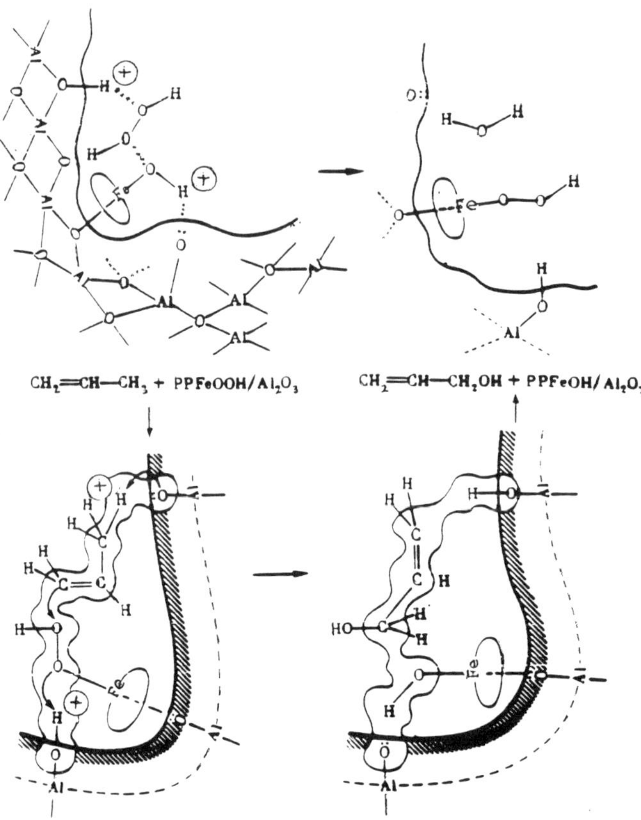

In stage 1 the transfer of a proton from a molecule of hydrogen peroxide to a basic center takes place and is accompanied by the cleavage of an O-H bond and by the transfer of an electron to the oxygen atom, and so on; in state II the proton bound to acid center is transferred to oxygen, and this is accompanied by the cleavage of an O-O bond, and so on. The whole sequence of electron and proton transfer takes place practically simultaneously without high consumptions of energy.

One of the possible ways of improving such system is modifying of organic ligand of oxidative-reduction site or its replace with more simple and available organic ligand of the analogous function. We have done this in such a way with PPFe^{3+} OH/AL$_2$O$_3$ using EDTA instead of protoporphyrin for Fe^{3+} ion coordination. According to such idea there were synthesized two types of biomimics distinguished only in valence state of Fe. They were tested in coupled oxidation of simple derivatives of thiophane with one basic goal - to prepare thyopen-1-monoxide and its derivatives. Thus, we have undertaken a detailed study of the oxidation reaction of dibromodimethylthiophane by hydrogen peroxide in the presence of $>Fe^{3+}$ OH/AL_2O_3). In result we could have succeeded in synthesis of 1-monoxide in a very simple way by the following reactions:

Yield, % mass:

and have characterized its PMR and mass-spectra.

According to the ideas developed in the activated intermediate $>Fe^{3+}OOH/AL_2O_3$, which as a result of assistance of acid-basic groups of AL_2O_3 support, forms stable products of the catalytic cycle with the outer substrate which then separate from mimics. This process may be described as follows:

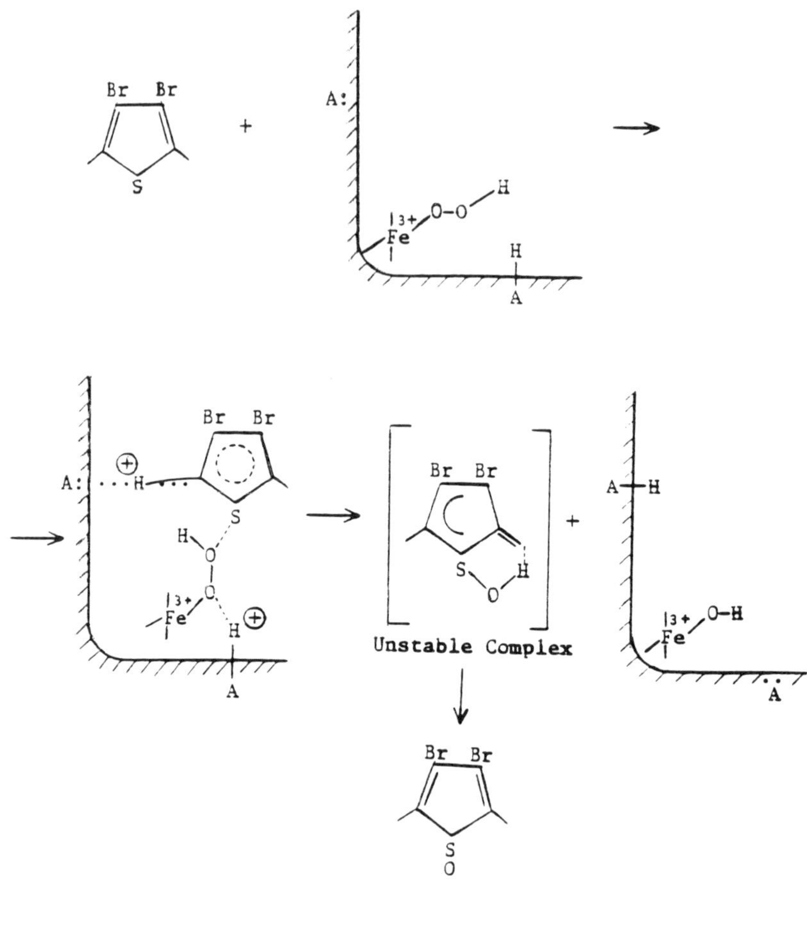

The reaction mechanism of monoxygenation of illustrated above means that an oxygen atom of hydrogen peroxide is transferred to the substrate molecule by addition of OH, at the same time a proton is transferred from the substrate to the matrix. By the end of conversion at the second state the catalyst Fe^{3+} OH/AL_2O_3 is regenerated. Thus the whole catalytic cycle is completed.

ACKNOWLEDGEMENTS

I thank Prof. James Terner for assistance in preparing the manuscript.

REFERENCES

A. L. Lehninger (1970): Biochmeistry, Worth, New York.

A. L. Lehninger (1964): The Mitochondrion, Benjamin, New   York.

V. M. Gryaznov and V. S. Smirnov (1972): Two Processes in One Reactor
    [in Russian], Moscow, P. 48.

T. M. Nagiev (1987): Idealized Model of Coupled Processes Constructed
    on the Principles of the Functioning of Mitochondria. Vestnik
    Moskovskogo Universiteta. Khimiya, vol. 42, N5, pp 447-451.

CHANNEL GATING BY DIVALENT CATIONS AND PROTONS

C A Pasternak
Department of Cellular and Molecular Sciences
St George's Medical School
Cranmer Terrace
London SW17 ORE
UK

INTRODUCTION

Calcium is known to protect cells against many kinds of membrane damage, in particular that resulting in an increased leakage of ions and low molecule weight metabolites across the plasma membrane. Nearly a century ago True (1914) showed that the loss of ions from lupin roots, caused by cutting them with a knife, was prevented by incubation in Washington, DC tap water: on analysis, the predominant ingredient in the water was found to be calcium. Half a century later Frankenhauser and Hodgkin (1957) showed that the electrical properties of squid axons deteriorated if external calcium was reduced five-fold; magnesium was less effective than calcium at preventing such membrane "leakiness".

Since then, extracellular (Bashford et al 1989) calcium as well as zinc and in certain cases protons, have been shown to prevent leakage induced by a number of membrane - damaging agents, such as haemolytic viruses (Pasternak and Micklem 1973), bacterial and animal toxins (Bashford et al 1986; Menestrina et al 1990), immune molecules (Micklem et al 1988; Bashford et al 1988) and synthetic agents such as polylysine or triton X 100 (Bashford et al 1986). In every case, the order of efficacy is Zn^{2+} > Ca^{2+} > Mg^{2+} (Pasternak 1987). In order to explore the mechanism of the effect, we have studied the increase in conductivity that ensues when such 'pore-forming' agents are added to synthetic planar lipid bilayers.

ELECTRICAL CONDUCTIVITY ACROSS PLANAR LIPID BILAYERS

Using the technique of Montal and Mueller (1972), it is found that haemolytic agents such as S. aureus α toxin (Menestrina 1986), A. hydrophila haemolysin (Wilmsen et al 1990), C perfringens ϕ toxin (Menestrina et al 1990) or the cytolysin from cytotoxic lymphocytes (Bashford et al 1988) induce the formation of single channels that, in the presence of divalent cations or protons, show voltage - dependent closure. The phenomenon is similar to that by which divalent cations and protons promote closure of endogenous ion channels (Hille 1984), that has also been studied in planar lipid bilayers (Cukierman et al 1988). Non haemolytic agents like diphtheria toxin (Alder et al 1990 a), heat shock proteins (Alder et al 1990 b), triton (T K Rostovtseva and A A Lev, unpublished experiments) or other agent are also closed by divalent cations (and H^+), but in this case closure is not voltage-dependent. While there is some evidence that in the case of voltage - dependent closure, divalent cations and protons interact with the agent itself, in the case of voltage - independent closure this is unlikely to be the case, especially as an agent like triton lacks any credible binding sites. Hence divalent cations and protons probably interact with the lipids themselves, even when these are neutral molecules like glycerol monoolein (GMO).

This result suggests that the mechanism by which divalent cations and protons gate channels induced across lipid bilayers may resemble that by which 'surface conductance' (Lev 1990) along pure lipids in the absence of pore-forming agent is affected by divalent cations and protons (Y E Korchev, V Osipov and A A Lev, unpublished experiments), and opens up a novel approach to this problem.

ACKNOWLEDGEMENTS

I am grateful to many colleagues for stimulating discussion and for permission to cite unpublished material. The work reported in this article was supported by the Cell Surface Research Fund.

REFERENCES

Alder, G. M., Bashford, C. L., & Pasternak, C.A. (1990a) Action of diphtheria toxin does not depend on the induction of large, stable pores across biological membranes. J. Membr. Biol. 113: 67-74

Alder, G.M., Austen, B.M., Bashford, C.L. Mehlert, A and Pasternak, C.A. (1990b) Heat Shock Proteins Induce Pores in Membranes. Biosci Rep 10: 509-518

Bashford, C. L., Alder, G. M., Menestrina, G., Micklem, K. J., Murphy, J. J., & Pasternak, C. A. (1986) Membrane damage by hemolytic viruses, toxins, complement and other cytotoxic agents: a common mechanism blocked by divalent cations. J. Biol. Chem. 261 9300 -9308

Bashford, C. L., Menestrina, G. Henkart, P. A., & Pasternak, C. A. (1988) Cell damage by cytolysin. Spontaneous recovery and reversible inhibition by divalent cations. J. Immunol. 141: 3965-3974

Bashford, C. L., Rodrigues, L. & Pasternak, C. A. (1989) Protection of cells against membrane damage by haemolytic agents: divalent cations and protons act at the extracellular side of the plasma membrane. Biochim. Biophys. Acta 983: 56-64

Cukierman, S., Zinkand, W. C., French, R. J., & Krueger, B. K. (1988) Effects of membrane surface charge and calcium on the gating of rat brain sodium channels in planar bilayers. J. Gen. Physiol. 92: 431-447

Frankenhauser, F. B. & Hodgkin, A. L. (1957) The action of calcium on the electrical properties of squid axons. J. Physiol. 137: 218-244

Hille, B. (1984) Ionic Channels of Excitable Membranes. Sinauer Associates, Sunderland, Massachusetts pp. 303-353

Lev, A. A. (1990) Sterol-dependent inactivation of gramicidin-A induced ionic channels in the cell and artificial lipid bilayer membranes In: Tenth School on Biophysics of Membrane Transport, Wroclaw p 231

Menestrina, G. (1986) Ionic channels formed by staphylococcus aureus alpha toxin: voltage-dependent inhibition by divalent and trivalent cations. J. Membr. Biol. **90:** 177-190

Menestrina, G., Bashford, C. L., & Pasternak, C. A. (1990) Toxicon **28:** 477-491

Micklem, K. J., Alder, G. M., Buckley, G. D., Murphy, J. & Pasternak, C. A. (1988) Protection against complement-mediated cell damage by Ca^{2+} and Zn^{2+}. Complement **5:** 141-152

Pasternak C.A. (1987) Virus, toxin, complement: common actions and their prevention by Ca^{2+} or Zn^{2+}. BioEssays. **6:** 14-18

Pasternak, C. A., & Micklem, K. J. (1973) Permeability changes during cell fusion. J. Membr. Biol. **14:** 293-303

True, R. H. (1914) The harmful action of distilled water. Am. J. Bot. **1:** 255-273

Wilmsen, H.U., Pattus, F. & Buckley, J. T. (1990) Aerolysin, a haemolysin from <u>Aeromonas hydrophila</u>, forms voltage-gated channels in planar lipid bilayers. J. Membr. Biol. **115:** 71-81

CHARGE TRANSFER EFFECT ON COELOMIC CELLS
IN EXALTED BIOLUMINESCENCE OF <u>LAMPITO MAURITII</u>

K.S.V. Santhanam and N.M. Limaye
Tata Institute of Fundamental Research
Colaba, Bombay 400 005, India

INTRODUCTION

The most intriguing aspect of bioluminescence of coelomic cells of <u>Lampito mauritii</u> lies in understanding the production of the active ingredients, luciferin, luciferase and peroxide; a recent report (Santhanam and Limaye, 1989) described this in terms of a synthetic machine operating within the cell.The aerobic cells produce bioluminescence for long periods (atleast 48 hours) upon continuous supply of oxygen. This long lasting bioluminescence is characteristic of several other earthworm species such as <u>D.Longa</u> (Wampler, 1978; Edwards and Lofty, 1972).It has been shown (Santhanam and Limaye, 1989; Limaye and Santhanam, 1988, Bellisario et al., 1972) that the coelomic cells do not produce bioluminescence in the absence of oxygen in the medium; a hydrogen atmosphere quenches the bioluminescence. With a view to understand more on the mechanism of cellular bioluminescence, we conducted electrochemical experiments to identify the electroactive components of the cell that are linked to the production of the active ingredients.As the synthesis of luciferin and luciferase are genetic (we do know that out of the 33 genera of earthworms only 16 of them are bioluminescent), specific directions for the amino acid sequence of a polypeptide are expected to be transcribed from DNA (deoxy ribonucleotide)-a gene for the polypeptide which transfers from the nucleus to the cytoplasm as a messenger RNA. The ribosomes are directing the synthesis of polypeptide bonds (McGilvery and Goldstein, 1979) and the luciferin. The entire mechanism is complex and requires a detailed study. Herein we use differential pulse voltammetry (dpv) to identify the component of DNA that is linked to the production of bioluminescence. The base pairs of DNA adenine-thymine, guanine-cytosine have been examined previously (Berg, 1978) and (Dryhurst and Pace, 1970) by electrochemical techniques; they are characterised by specific oxidation potentials. We wish to report here that coelomic cells of <u>Lampito mauritii</u> contain an identifiable base guanine/guanosine in dpv and a charge transfer effect of guanine/guanosine oxidation results in the exalted bioluminescence.

The role of guanosine triphosphate (GTP) in the conversion of adenosine diphosphate to adenosine triphosphate (ADP to ATP) has been illustrated using firefly luciferin assay (Santhanam and Limaye, 1989, Anderson et al., 1978); however, in this assay there is no direct reaction of GTP with luciferin and luciferase. We wish to demonstrate in the present study that guanosine and much less GTP reacts with peroxide to produce photons which links guanine in cellular biuminscence of <u>Lampito mauritii</u>.

EXPERIMENTAL

Chemicals:- The chemicals were obtained from the following sources listed here; guanine hydrochloride (SRL Chemicals), guanine (BDH), guanosine (SRL Chemicals), guanosine phosphate (Sigma Chemicals), tris-hydroxy methyl aminomethane (tris) (SRL Chemicals), NaOH (BDH), Na_2HPO_4 (Sarabhai Chemicals), KH_2PO_4 (Glaxo Laboratories), boric acid (S.D.Fine Chemicals), K phthalate (Glaxo Laboratories), and KCl (Merck). Hydrogen peroxide (SRL) was kept in the refrigerator until use. Cytosine (Aldrich), adenine (aldrich) and thymine (Aldrich) were stored in the dark to avoid any photodegradation. Argon and nitrogen gases were of high purity (99.99%) and was equilibrated with water.

Apparatus: The differential pulse voltammetry (dpv) of coelomic cells was performed by using model 264 polarographic analyser (EG&G PAR) with either a platinum wire working electrode (A= 0.12 cm^2) or carbon fibre electrode (A= 0.0021cm^2) and a platinum gauze counter electrode ($A = 0.90$ cm^2). A saturated calomel electrode was used as the reference electrode. The working electrode was cycled between 0 and -0.70 V in background solution to remove the oxide layers. The initial and the final potentials of the experiments were set at 0 V and 0.90 V. The scan rate was adjusted at 5 mv/s. The pulse amplitude was maintained at 100 mv. The current range was adjusted depending on the concentrations employed in the experiments; it is usually in the range of 500 μA to 5 mA. The current-voltage and luminescence-voltage curves were recorded using dual pen $X_1 - X_2$ -t recorder on a Houston Omnigraph 2000. The electrochemical cell was placed in the Turner luminometer and was covered adequately to protect it from stray light. The background experiments performed without the potential sweep were used to judge this criterion. The Turner luminometer was set on auto range mode for measuring the light output. The total light integration window was set to 600 s, which is longer than the dpv experiment. The luminometer output was connected to the recorder for the simultaneous recording of the dpv and bioluminescence.

Preparation of Coelomic cells: The coelomic cells were obtained from <u>Lampito mauritii</u>; the worms were removed from the soil and left in double distilled water for 15 minutes to remove soil sticking to the surface. This procedure was repeated atleast three times to ensure that no soil contaminants entered into the cell extracts. Thus this procedure is an improved one over the previously described method (Limaye and Santhanam, 1989). The worms were dried between the folds of Whatman paper no.41 and the coelomic cells contained in the post-clitellar segments were extracted into either borax buffer or tris-buffer or phthalate buffer or phosphate buffer or 0.1M KCl or 0.1M K_2SO_4

The different buffers provided an advantage of examining the coelomic cells at widely different pH values. The cell activity was examined by uv-vis spectrophotometry. All experiments were performed at room temperature (22.5°C).

Concentration of Coelomic cells: The concentration of

the coelomic cells was determined by using haematocytometer and counting the number of cells under a sensitive microscope. The average number of cells estimated in the present experiments are in the range of 1×10^7 cells/ml. The u.v.absorption was recorded to obtain the O.D. of the coelomic cells in solution.

Preparation of solutions: The buffer solutions were prepared by the methods described in the Handbook of Chemistry and Handbook of Biochemistry (CRC Handbook of Chemistry, 1985 ; Handbook of Biochemistry, 1984).

Spectral measurements: UV-visible spectra of the coelomic cell extracts and the base pair solutions were recorded using a JASCO UV-160 with a microcomputer controlled double beam recording spectrophotometer. The data is stored in the memory of the computer for further analysis.

Differential pulse volt-bioluminescence procedure: 3 ml of coelomic cells in appropriate buffer was taken in a 10 ml electrochemical cell. Two platinum gauze electrodes were placed in the cell which are separated by atleast 1 cm. A saturated calomel reference electrode (SCE) was also placed in the cell; the electrode was located very close to the working electrode. The cell was placed inside the Turner luminometer; the opening of the luminometer was fitted with a rectangular black box having a sliding door. The entire box was covered with four layers of black cloth to prevent any light leaks into the luminometer. The electrode connections were given through the side of the box which was fitted with three bannana jacks having o-rings; this portion was also covered with black cloth after the leads from the polarographic analyser was connected. The luminometer which is microprocessor controlled, is set in the auto position with an integration time set to 600 s.At the end of the set time the luminometer printed out the integrated photons. The luminometer output is connected to the y-input of the X-Y recorder. The polarographic analyser voltage output is connected to the X-input. It is set to differential pulse mode. The sweep rate of the analyser is set at 5 mv. The initial voltage is set at 0 V in most of the experiments. The final voltage is adjusted to 0.90 V or 1.0 V. In this experiment the voltage vs. bioluminescence intensity is recorded.

Procedure for differential pulse voltammetry: The cell described above containing the coelomic cells is used. In these experiments during the voltage sweep, the differential current output of the analyser is connected to the X-Y recorder in the place of the luminometer output.The amplitude of the pulse in these experiments are generally in the range of 25 mv to 100 mv.

For recording differential pulse voltammogramms of DNA base pairs, a solution in the μM range is used. To confirm that these current-voltage curves are originating from the selected base pair, the concentrations are successively increased and the peak current values are measured. The u.v. absorbance of the solutions are used to compute the concentrations.

Macroscale electrolysis: Macroscale electrolysis of guanine was performed with a platinum gauze as the working electrode (15 cm^2) and a bigger platinum gauze electrode as the counter electrode. A 'H' type cell fitted with a fritted glass disc to separate the anolyte and catholyte was used. The potential of the working electrode was controlled at 1.00 V vs.SCE. A stream of argon gas was bubbled through the solution during the electrolysis. The u.v. absorbance of the solution was recorded before and at the end of the electrolysis.

Chemiluminescence procedure: A 10 ml chemiluminescent cell was placed inside the Turner luminometer. About 0.2 ml of guanine or guanosine or guanosine phosphate was taken with Fischer micro pipette. An equal volume of 0.7% hydrogen peroxide was injected into the solution. The luminometer integration time was set at 600 s with an auto scaling. The experiment was performed in the complete absence of external light. The luminescence intensity is recorded on an X-Y recorder. The luminometer is calibrated using luminol-peroxide reaction (Anderson et al., 1978; Limaye and Santhanam, 1987).

RESULTS

The differential pulse volt-bioluminescence curves of coelomic cells of <u>Lampito mauritii</u> at different pH values are depicted in Figure 1. In these experiments a voltage ramp is applied to the

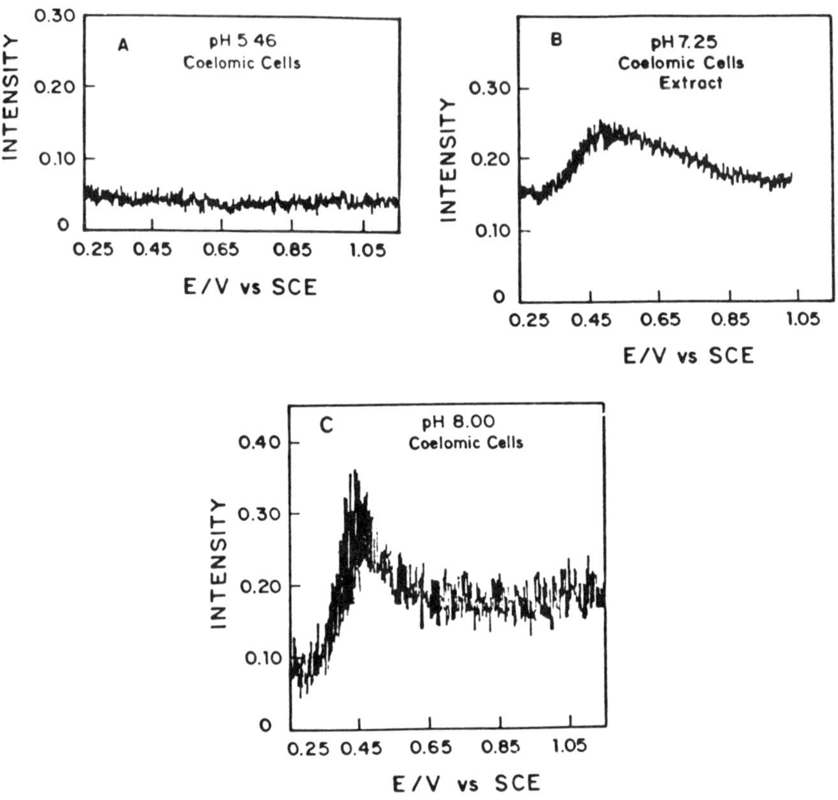

FIGURE 1. Differential volt-bioluminescence of coelomic cells of <u>Lampito mauritii</u> in different buffers in the pH range of 5 to 9

electrochemical cell containing a platinum mesh electrode and a saturated calomel reference electrode with a fixed height modulation pulse superimposed on it. The bioluminescence intensity is measured during these modulations continuously without sampling before the application and at the termination of the pulse. The appearance of an intensity maximum in the differential pulse volt-intensity curves suggests an

exalted bioluminescence occurring at well defined
potentials; the bioluminescence reaches a peak which
is a pH dependent process. Figure 2 shows the

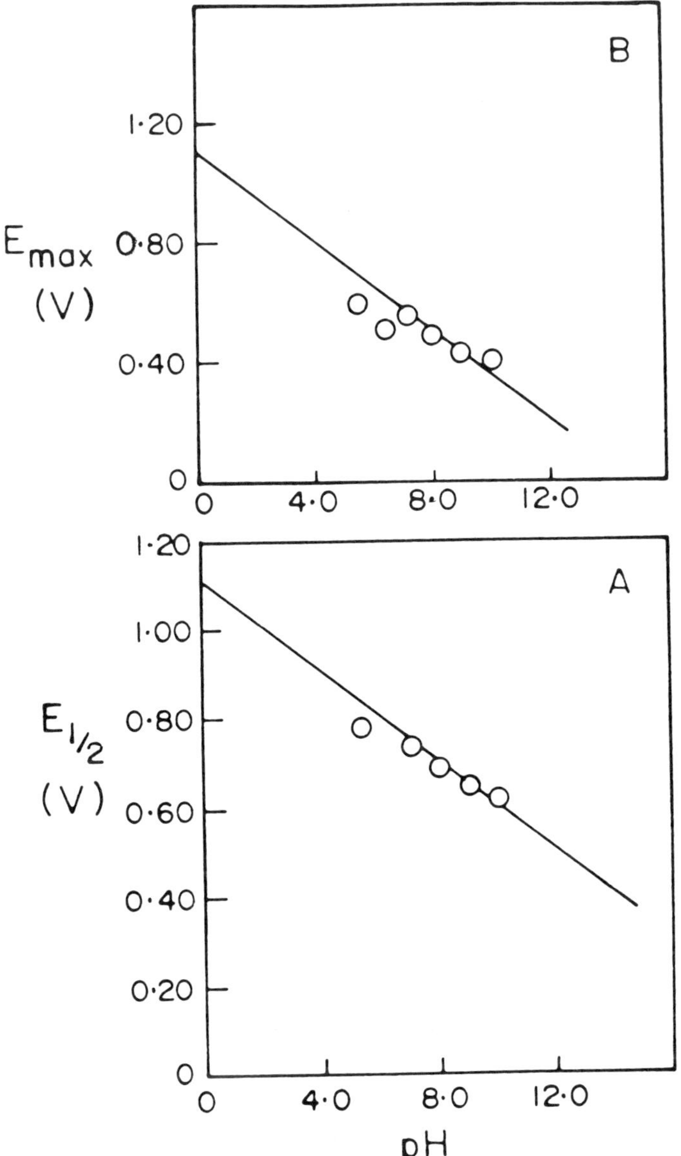

FIGURE 2. The plot of A) E_{max} vs. pH of differential
pulse volt-bioluminescence of coleomic cells B) $E_{1/2}$
vs.pH of coelomic cells.

bioluminescence peak shift with pH. This bioluminescence peak shift may be represented as

$$E_{max} = k - 0.054 \text{ pH}$$

(1)

where k is the expected potential at zero pH and has a value of k = 1.113. The data could not be collected at

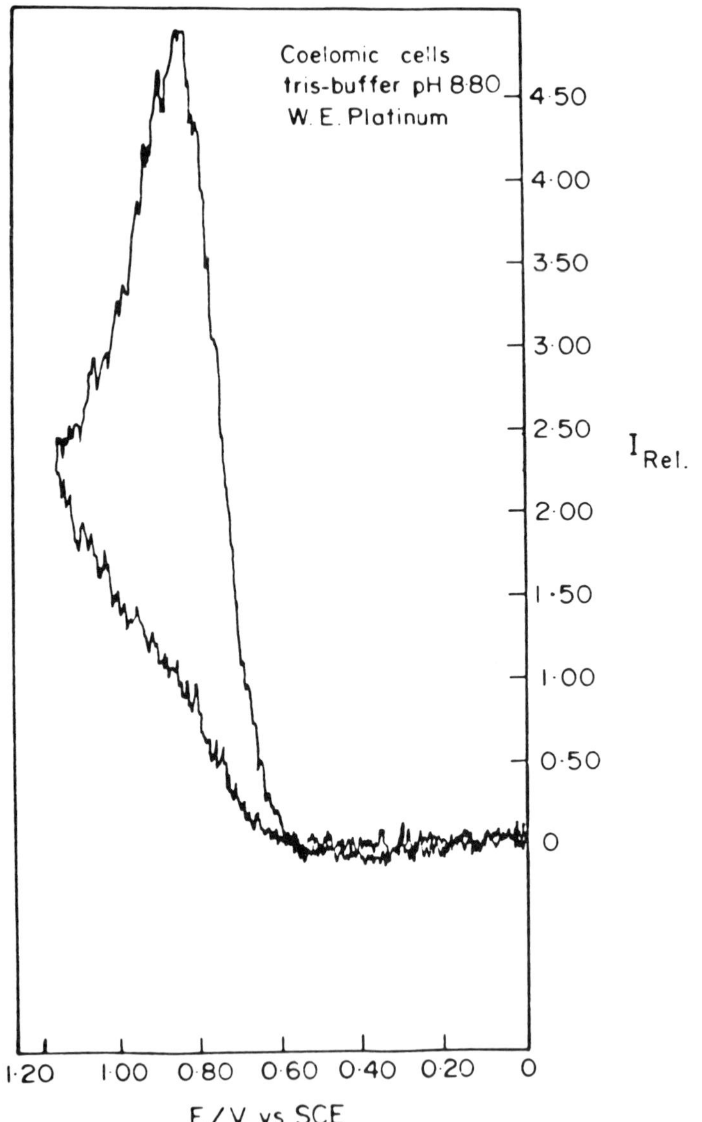

FIGURE 3. Bioluminescence of coelomic cells during linear voltage sweep conducted at 20 mv/s at a platinum working electrode.

lower pH values the bioluminescence is quenched; this is to be expected as the cells are inactivated at this pH (DeLuca and McElroy, 1986)

When the experiments are conducted by applying a linear voltage sweep to the platinum electrode, the bioluminescence intensity increases at a well defined voltage and reaches a maximum; following this maximum the bioluminescence follows the pattern shown in figure 3. The peak positions shift unlike in the case of differential pulse volt-bioluminescence. A common feature of figure 1 and figure 3 is that the bioluminescence intensity not reaching the background level at the positive potentials past the peak due to a continuing process. However, if the potential is returned to the starting voltage, the bioluminescence reaches the initial value. These results suggest that the coelomic cells are electroactive or discharging an electroactive component into the suspended medium. The electroactivity was masked in the earlier experiments in cyclic voltammetry due to the very low concentrations of it ($< \mu$M); the faradaic current is masked by the background.

Differential pulse voltammetry of coelomic cells:
Figure 4 shows the differential pulse voltammetry of coelomic cells at pH 9.85 The appearance of the differential current maximum suggests the onset of an electrochemical process during these experiments. From the plot of peak potential vs. pH (Figure 2B) the peak position at any pH can be defined by

$$E_p = 1.113 - 0.0523 \, pH \qquad (2)$$

The slope of the curve which is 0.052 V is indicative of one electron and one proton transfer oxidation at the electrode. The expected value of the slope for this process is 54 mv based on the temperature used in the experiments. At pH=0, the expected value of the differential pulse voltammetric peak is estimated at 1.11 V. As the cells are deactivated at pH=0, this experiment was not successful.

The differential pulse voltammetric curves always exhibit a broad adsorption wave starting from 0.10 V; this prewave reaches a maximum at 0.32 V.

This isattributed to the adsorption of the electroactive species in the coelomic cells.

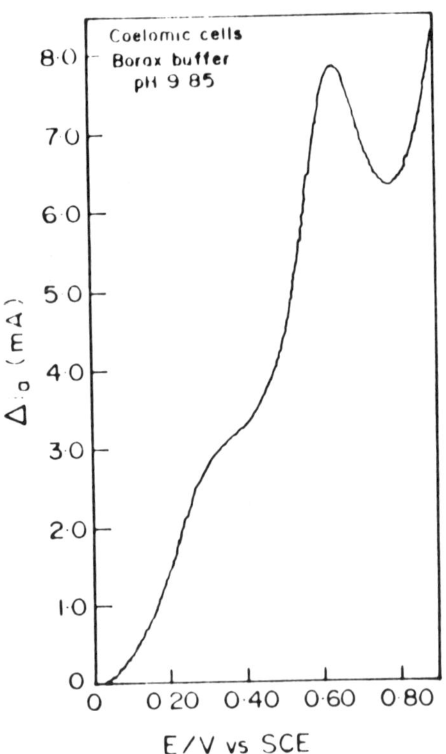

FIGURE 4. Differential pulse voltammetry of coelomic cells   at pH 9.85. Sweep rate 5 mv/s; pulse amplitude 100 mv.

<u>Differential Pulse Volt-Bioluminescence of Coelomic cells Obtained From Lampito mauritii in Ice-Cold Tris-buffer of pH 8.95</u>

An exalted bioluminescence is observed at a platinum mesh electrode when the potential reaches 0.37 V vs.SCE. The bioluminescence maximum is reached at about 0.50 V. At positive potentials, the bioluminescence reaches a steady value, which is about two times the initial bioluminsecence. When the pulse amplitude was varied from 25 mv to 100 mv.,the bioluminescence intensity also showed an increase in the same manner as previously reported for the coelomic cells obtained by dissecting the worm.

The differential pulse voltammetry of the coelomic cells showed current maxima at 0.32 V vs SCE and at 0.63 V vs SCE. Both bioluminescence and current maxia peaks show shifts with pH of the medium (Refer earlier section).

Differential pulse voltammetry of Base pairs:

In an attempt to understand the exalted bioluminescence of coelomic cells, the differential pulse voltammetry of guanine, adenine,thymine and cytosine are examined under the same experimental conditions.

Guanine: The differential pulse voltammetry of guanine at different pH values, in the positive potential region shows a characteristic broad adsorption peak followed by a diffusion controlled oxidation peak; the position of these peaks are pH dependent. Figure 5 shows the guanine electrochemical oxidation at different pH values. The peak shift can be correlated to pH as (see Figure 6)

$$E_{1/2} = 1.19 - 0.058 \text{ pH} \tag{3}$$

In the pH range studied here the electrochemical oxidation of guanine occurs in two stages (Dryhurst and Pace, 1970); the first stage involes removal of two electrons and two protons resulting in the formation of 2-amino-6,8-dioxypurine which upon further oxidation produces 2-amino-6,8-dioxypurine-4,5-diamine.The diamine is unstable and decomposes with time resulting in electroactive products.

Guanosine:The differential pulse voltammetry of

guanosine shows a pattern very similar to guanine. It exhibits a strong adsorption peak at 0.37 V vs. SCE in tris-buffer of pH 8.80. This is followed by a small peak at 0.75 V vs. SCE. The first peak increases with increasing concentration of guanosine; the second peak sluggishly changes with concentration due to the adsorption wave. Guanosine undergoes cyclic voltammetric irreversible oxidation at graphite electrode with very little adsorption unlike in the

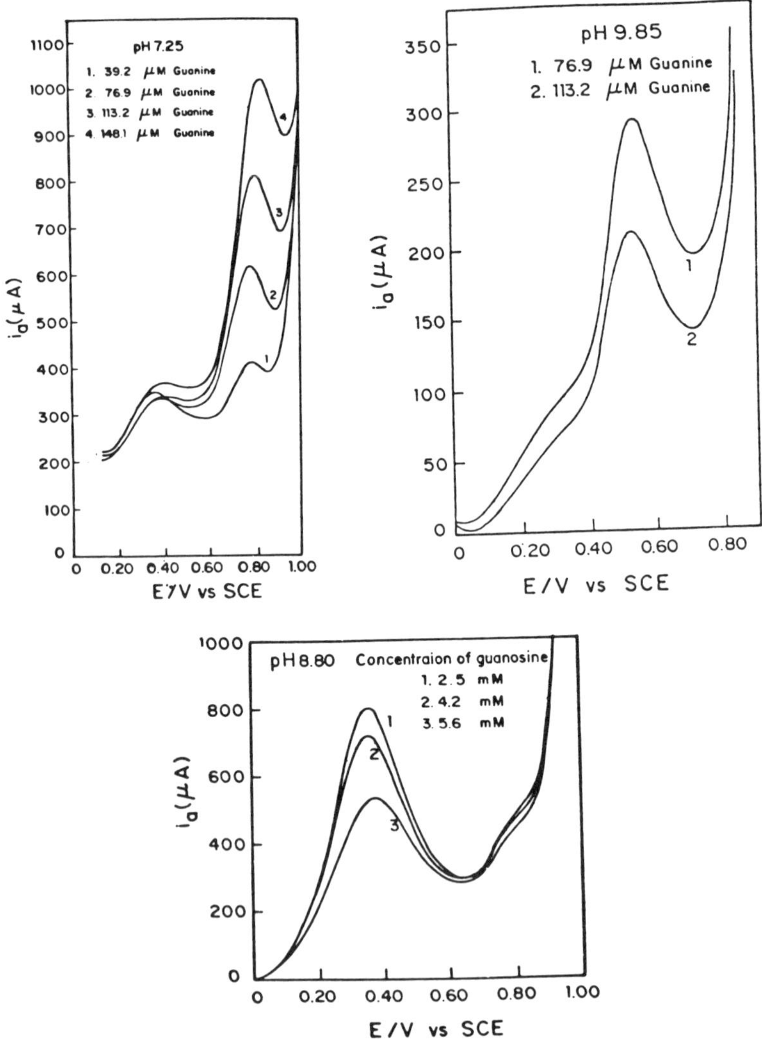

FIGURE 5. Differential pulse voltammetry of guanine in buffer solutions of different pH values. Sweep rate 5 mv/s; pulse amplitude 100 mv.

case of platinum working electrode.

Guanosine mono phosphate (GMP): The differential pulse voltammetry of GMP in tris-buffer of pH 8.95 shows a pattern of an adsorption wave at 0.30 V vs. SCE followed by a sharp peak at 0.70 V vs. SCE. The peak positions shift with pH in the manner shown for guanine.

Adenine: The electrochemical oxidation of adenine as a function of pH has been examined at the pyrolytic graphite electrode (Dryhurst and Elving, 1968); in the pH range of 0 to 11 the oxidation is very close to the background discharge. In alkaline medium the voltammetric peak is not visible. In acid medium the oxidation at pH 2.3 consumes 6 electrons per adenine molecule and produces a dicarbonium ion intermediate. The reaction products of parabanic acid, oxaluric acid, urea, carbon dioxide and ammonia have been isolated (Dryhurst and Elving, 1968). In tris-buffer of pH 8.83, the differential pulse voltammetric oxidation peak merges with the background discharge at a platinum electrode.

Cytosine: The anodic oxidation potential of cytosine has been reported (Berg, 1978) in phosphate buffer at pH 7.2 at graphite electrode; the peak potential is estimated at 1.20 V vs. mercury/mercurous sulphate electrode. It's oxidation relative to guanine and adenine is far positive. The measurements carried out in tris-buffer of pH 8.95 at platinum electrode show a very faint wave in the background discharge due to cytosine. But no quantitative measurements are possible.

Thymine: The electrochemical oxidation of thymine in phosphate buffer of pH 7.2 occurs at 1.05 V vs. mercury/mercurous sulphate (Berg et al., 1978). The oxidation in tris buffer at pH 8.95 shows a faint wave near the background discharge. Hence no quantitative measurements were undertaken.

Chemiluminescence of Guanine: Guanine reacts with hydrogen peroxide to produce chemiluminescence; Figure 7 shows the intensity- time curve during this reaction. The intensity of this reaction varies with the concentration of guanine and peroxide. At low

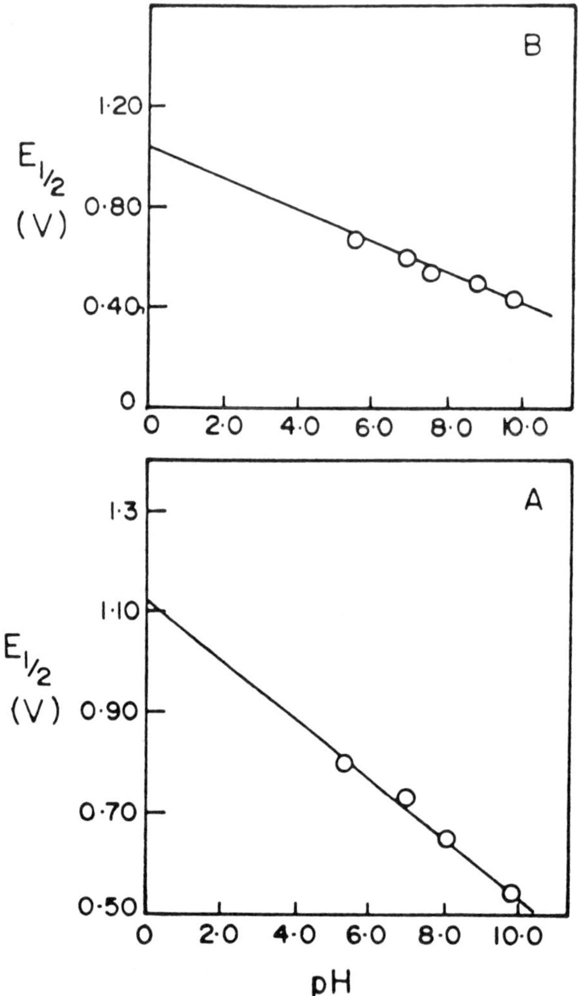

FIGURE 6. Effect of pH on the differential pulse
voltammetry of A) Guanine B) Guanosine

concentrations of guanine (0.20 mM) and 0.7% peroxide,
the emission intensity reaches a   maximum. At higher
concentrations of peroxide (30%) the chemiluminescence
is reduced and reaches a background value. Figure 8
shows the plot   of guanine concentration   vs.
integrated   photonic output. The quenching of the
chemiluminescence at higher peroxide concentration is

attributed to hydroperoxide of guanine; a similar situation is observed with other systems (Bellisario et al., 1972)

The chemiluminescence reaction was also monitored through spectrophotometry; guanine in tris-buffer shows absorption maxima at

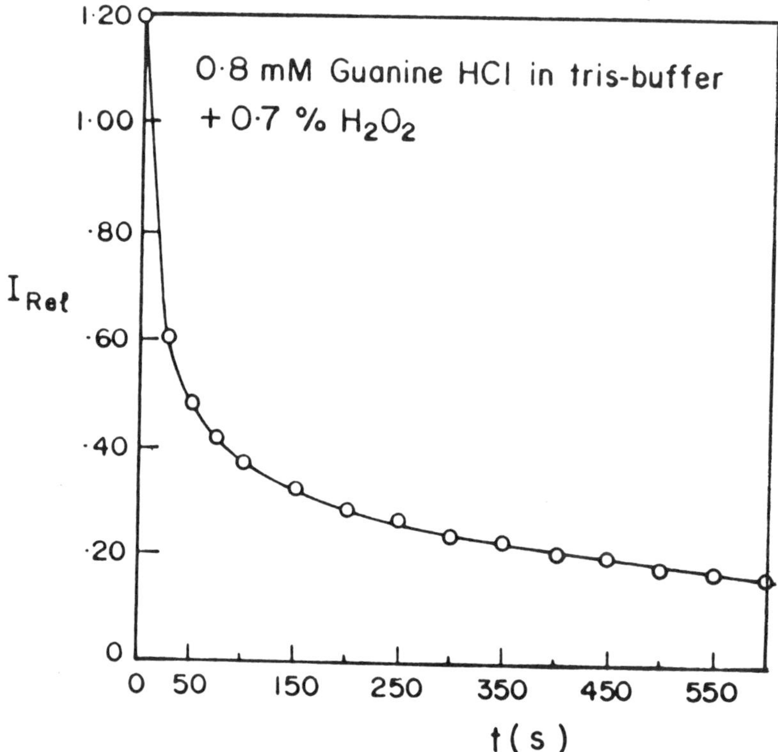

FIGURE 7 Chemiluminescence of guanine in tris-buffer of pH 8.95. Concentration of guanine: 0.80 mM; Hydrogen peroxide   concentration:

244 nm and 272 nm. Addition of hydrogen peroxide removes these peaks during chemiluminescence and generates a product absorbing at 220 nm.
The loss of conjugated system is indicated during chemiluminescence.

Guanosine: The chemiluminescence of guanosine by reaction with peroxide shows a pattern similar to guanine. The linearity in chemiluminescence output

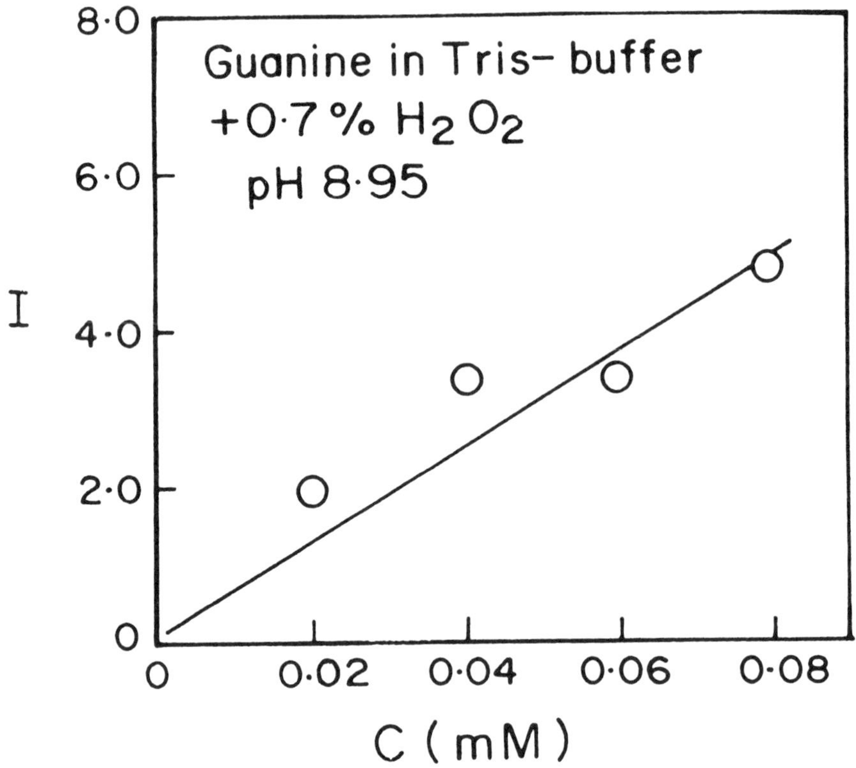

FIGURE 8   Integrated photonic output vs concentration of   guanine.

with concentration of guanosine is noticed here; however, for any selected concentration of guanosine the chemiluminescence output is lower than with guanine. For example, the integrated photonic output for 0.1 mM guanine is 23.6 in comparison to guanosine yielding 10.4. With 0.5% hydrogen peroxide, the chemiluminescence output for guanine and guanosine yielded values 34.2 and 8.6. The absorption spectral pattern follows the trend of disappearing conjugation during the reaction.

Guanosine Monophosphate: The chemiluminescence of guanosine phosphate could not be observed by reaction

with hydrogen peroxide under the same conditions as
described earlier.

DISCUSSION

The results obtained here suggest   that the
coelomic   cells of Lampito mauriti are undergoing
three types of reactions

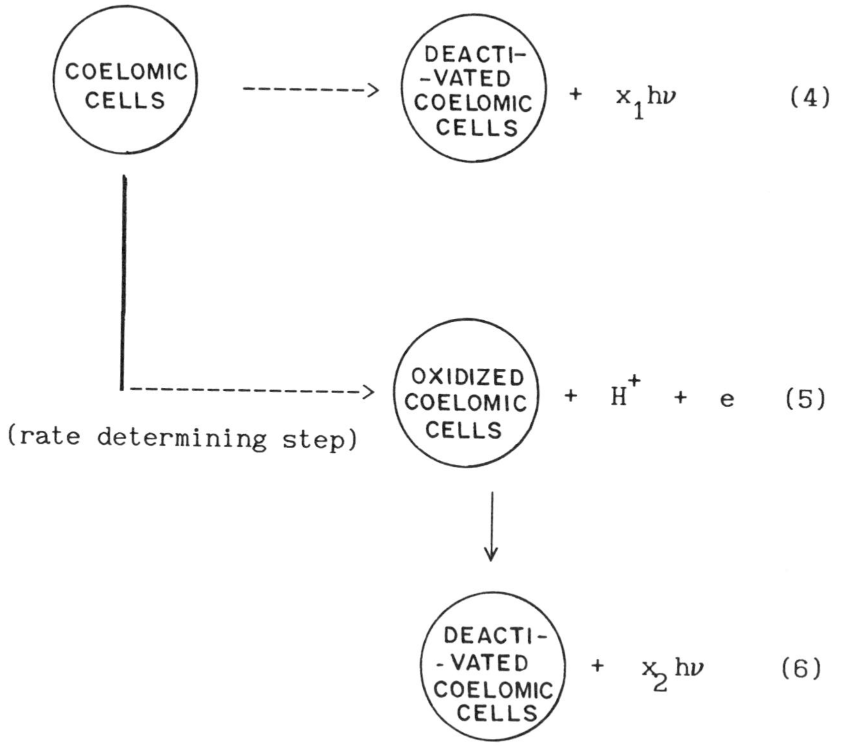

$$\text{COELOMIC CELLS} \dashrightarrow \text{DEACTI--VATED COELOMIC CELLS} + x_1 h\nu \qquad (4)$$

(rate determining step)

$$\text{COELOMIC CELLS} \dashrightarrow \text{OXIDIZED COELOMIC CELLS} + H^+ + e \qquad (5)$$

$$\text{OXIDIZED COELOMIC CELLS} \rightarrow \text{DEACTI--VATED COELOMIC CELLS} + x_2 h\nu \qquad (6)$$

where $x_2 > x_1$ i.e. the number of photons x_2 is greater

than the first step. This scheme implies that there is
an electroactive component in the cell getting
oriented during the differential pulse experiments;
this hypothesis derives support from two experiments.
The magnitude of the bioluminescence output increases
with increasing modulating pulse of 25 mv to 100 mv.
Table 1 shows the relative   intensities in the three
experiments. A linear sweeping of potential of the
electrode does not show the exalted bioluminescence at
the potential where it is observed in dpv experiment.
The pH shift of the

TABLE 1. Bioluminescence output as a function of the modulating amplitude in tris-buffer of pH 8.95

Amplitude mv	E_{max} mv	I Photons/s
25	550	4.2×10^{10}
50	550	12.0×10^{10}
100	550	14.8×10^{10}

bioluminescence maximum indicates that the rate determining step as depicted by reaction (5) involves one electron. The identification of the nature of the electroactive species that is undergoing the charge transfer reaction is difficult to conclude; however, based on the data presented here it appears to be guanine. Compare the pH shift observed with the cells (equation 1) and with that of guanine (equation 2); the shifts are very similar. Guanosine also shows a similar shift in half-wave potential with pH; however, its intensity is lesser than guanine. We also examined the absorption of the cells by spectrophotometry; it shows a maximum at 280 nm and 420 nm. It is not possible to discern the guanine absorption peak appearing at 240nm (Rao, 1967) as the absorption by the solvent is very strong at this wavelength. Upon subjecting the solution to a modulating pulses of amplitude of 100 mv and duration of a msec. for a period of half an hour the absorption in this cut off region increases, while 280 nm peak absorption remains the same. The peak appearing at 280 nm is attributed to luciferase (Ohtsuka et al., 1976) and this concentration does not appear to increase during the differential pulse experiments. During these experiments a substance which has an absorption in the 240 nm region is released into the medium.

The effect of dissolved oxygen on the differential pulse volt-bioluminescence and differential pulse voltammetry has been examined. When inert gas is bubbled through the solution, the bioluminescence reaches a negligible level. The differential pulse voltammetric current at 0.67 V increases in the absence of dissolved oxygen. If 30 μm guanine or guanosine solution is examined by differental pulse voltammetry in the presence and absence of oxygen, the peak currents are not altered indicating oxygen is not directly involved in affecting the peak current of the coelomic cells. When the coelomic cells react with oxygen, bioluminescence is produced; this is considered to proceed via the production of peroxide. The peroxide reacts with guanine or guanosine to produce bioluminescence (see the results section). When this reaction takes place, the guanine or guanosine concentration decreases in the differential pulse voltammetric experiments.

The electrochemical oxidation of guanine or guanosine at 0.90 V vs. SCE does not produce chemiluminescence; the product of this electrolysis when added to coelomic cell in tris-buffer of pH 8.95 does not produce exalted bioluminescence. This suggests that the exalted bioluminescence arises from the charge transfer process at the electrode; during the differential pulse experiments there appears to be guanine or guanosine released in the cell which reacts with oxidase to produce bioluminescence. In other words small modulation of the voltage produces the exalted bioluminescence. If we assume that the DNA duplicating process requires the unwinding of two strands of DNA,and the synthesis of two new chains begins as soon as the two original chains start to unwind; DNA precursor units such as shown in figure 9 floating loosely in the cell appear to reach the (Crick, 1981) bioluminescence zone where oxidase is located.As shown by the experiments described here, the loose precursors such as adenine or cytosine or thymine or the nucleosides do not produce chemiluminescence upon reaction with hydrogen peroxide. The charge transfer effect on the coelomic cells appears to slow down the synthesis of new chains of DNA and as a result guanine is available for the

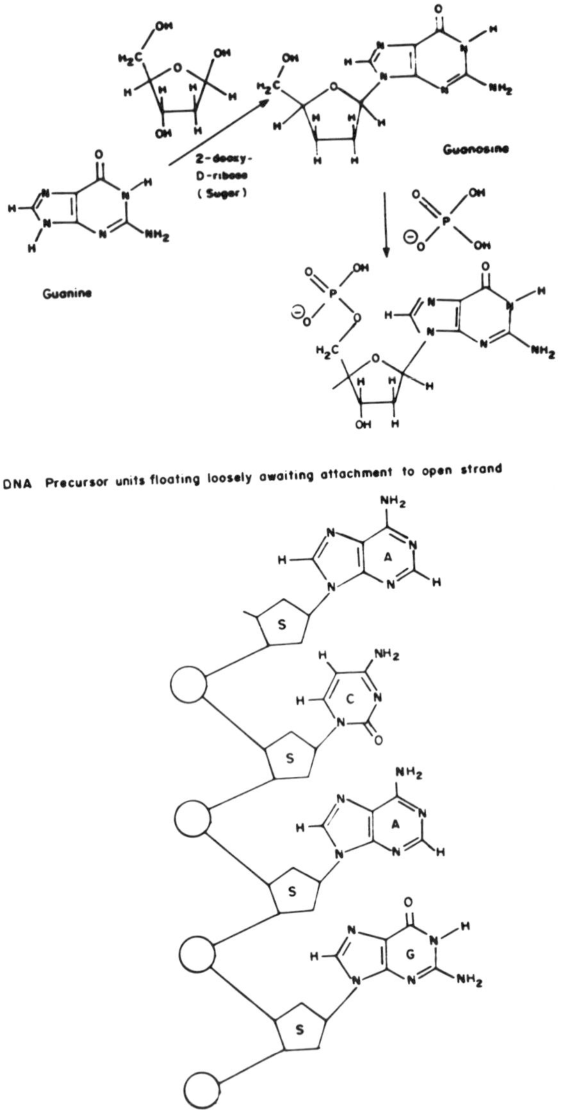

FIGURE 9 Proposed mechanism for exalted bioluminescence of <u>Lampito</u> <u>mauritii</u>

bioluminescence.

ACKNOWLEDGEMENT

The authors thank Miss.Poonam Narula for carrying out the chemiluminescence experiments during this study.

REFERENCES

ANDERSON JM., FAINI JG. and WAMPLER JE (1978): construction of instrumentation for bioluminescence and chemiluminescence assays. In:Methods in enzymology, DeLuca MA, ed. New York: Academic Press.

Bellisario R, Spencer TE and Cormier MJ (1972): Isolation and properties of luciferase, a non-Heme peroxidase from the bioluminescent earthworm, Diplocardia longa. Biochemistry, 11: 2256.

Berg H, Gollmick H, Bauer E, Horn G, Flemming J and Kittler L (1978): Redox processes during photodynamic damage of DNA I. Results obtained by several physico-chemical methods. Bioelectrochemistry and bioenergetics, 5: 335.

Crick FHC (1981): The structure and hereditary material. In: Genetics, Davern CI, ed. SanFrancisco: Freeman WH & Co.

DeLuca MA and McElroy WD (1986): Methods in Enzymology vol.133. New York: Academic Press

Dryhurst G and Elving PJ (1968): Electrochemical oxidation of adenine: Reaction products and mechanisms. J electrochemical soc , 115: 1014

Dryhurst G and Pace GF (1970): Electrochemical oxidation of guanine at the pyrolytic graphite electrode. J electroanal chem , 117: 1259

Edwards CA and Lofty JR (1972): Biology of earthworms. London: Chapman and Hall

Jamieson BGM and Wampler JE (1979): Bioluminescent Australian earthworm. Aust J Zool , 27:637

Limaye NM and Santhanam KSV (1987): Secretory differences in electrobioluminescence of Lampito mauritii upon segmentary electron input. Bioelectrochem and bioenergetics , 17: 105

Limaye NM and Santhanam KSV (1988): Evidence for two parallel mechanisms operating in the production of the excited state in electrobioluminescence of Lampito mauritii. Bioelectrochem and bioenergetics, 19: 9

McGilvery RW and Goldstein G (1979): Biochemistry: A

functional approach. Philadelphia: Saunders WB Company

Ohtsuka H, Rudie NG and Wampler JE (1976): Structural identification and synthesis of luciferin from the bioluminescent earthworm D.Longa. Biochemistry , 15: 1001

Rao CNR (1967): UV-VIS Absoprtion spectroscopy, London: Butterworths

Santhanam KSV and Limaye NM (1989): A proposed scheme for the activated bioluminescence of Lampito mauritii by ferrous ion injection. Bioelectrochem and bioenergetics ,22: 231

Wampler JE and Jamieson BGM (1980): Earthworm bioluminescence: Comparative physiology and biochemistry. Comp Biochem Physiol , 66B: 43

EMULSION BIOELECTROCHEMISTRY: BACTERIORHODOPSIN PHOTOTRANSFER OF PROTONS THROUGH THE INTERFACE WATER/LIPID IN OCTANE

Alexander G.Volkov, Maya I. Gugeshashvili,

Vladimir I. Portnov and Vladislav S. Markin
The A.N.Frumkin Institute of Electrochemistry, the USSR Academy of Sciences, 31 Leninsky Prospekt, Moscow 117071, U.S.S.R.

L.N. Chekulaeva
Institute of Biophysics, USSR Academy of Sciences, Pushchino, Moscow region, U.S.S.R.

INTRODUCTION

The interface of two immiscible liquids is widely used as a model biological membrane for investigation of fundamental processes of photosynthesis, biocatalysis, fusion and interaction of cells, elucidation of mechanisms of functioning of ionic pumps, and electron exchangers (Boguslavsky and Volkov, 1987). The bacteriorhodopsin sheets from *Halobacterium halobium* were one of the first membrane systems studied upon the octane/water interface (Boguslavsky et al., 1975, 1976; Boguslavsky and Volkov, 1987; Drachev et al., 1983; Hwang et al.,1977, Post et al.,1984). These studies were carried out in various laboratories, but the results itself as well as their interpretation significantly differed. The progress was achieved due to studies of (Post et al.,1984) who proposed the way to immobilize bacteriorhodopsin sheets in the system with extended surface, octane-in-water emulsion. In this case the control of phototransfer of protons was significantly simplified: the measurements of concentration of protons in aqueous medium of emulsion was carried out using the conventional pH-meter. In our opinion the emulsion system proposed in (Post et al., 1984) is quite promising, since it facilitates the quantitative examination of the transfer of various ions through the water/lipid interface during functioning of membrane ion pumps using the ion selective electrodes. The vast interfacial area makes it possible to obtain the products of heterogeneous reactions in macroscopic quantities, whereas the low dielectric constant of non-aqueous phase may result in a broad decrease of activation energy (Kharkats and Volkov, 1987). The emulsion enzymology enables one to study naturally immobilized membrane enzymes in conditions close to the native ones.

The paper (Post et al.,1984) explicitly showed that bacteriorhodopsin immobilized upon the water/lipid-in-octane interface is capable to phototransfer of protons from water into octane. The photoeffect was not inhibited by uncouplers of oxidative phosphorylation and depended upon the ionic strength of the solution.

The molecular mechanism of immobilized bacteriorhodopsin upon the octane/water interface remained, however, unclear. According to results of several studies [(Boguslavsky et al.,1975, 1976; Hwang et al., 1977; Post et al., 1984) the bacteriorhodopsin immobilized upon the interface was capable to phototransfer of protons through the interface. The other authors (Drachev et al., 1983) claimed that the bacteriorhodopsin   became denaturated upon the octane/water interface and in presence of phospholipids it formed the closed structures with the so-called "third water" upon the octane/water interface. Upon irradiation the bacteriorhodopsin was thought to pump protons from aqueous phase into the "third water". This process was not suggested to be accompanied by transfer of protons directly from water into octane. In order to register the photoeffect the method of Volta-potential measurement were used in (Boguslavsky et al., 1975, 1976; Drachev et al., 1983; Hwang et al., 1977). This method yielded information concerning the variation of the potential difference and the charge upon the interface but did not specify, whether the transfer of charges from one phase into another occurred since both the variation of the orientation of dipoles upon the interface and transfer of charges could affect the value of Volta-potential. A direct method of estimation of transfer of ions through interface was suggested in (Post et al., 1984) based upon the ion-selective electrodes. This method was used in present study.

In the present communication we consider the capacity of bacteriorhodopsin to photoinduced transfer of protons through the water/lipid interface in octane and analyzed the mechanism of fusion of purple membranes with monolayer of lipids upon the octane/water interface.

MATERIALS AND METHODS

We used the soybean phospholipids (asolectin) from Sigma (USA). The composition of asolectin was assessed analytically using the thin layer chromatography (37% phosphatidilcholine, 29% phosphatidilethanolamine, 3% phosphatidilserine, 7% cardiolipin, 6% neutral lipids).

The salts KCl and $NaNO_3$ of high purity grade were doubly recrystallised, the octane (chemical purity grade) was subjected to additional purification using oleum and distillation procedures. The pentachlorophenol (PCP) was recrystallised from alcohol three times. All the solutions were prepared with double-distilled water.

The purple membranes were isolated from *Halobacterium halobium* according to the method given in (Oesterhelt and Stoeckenius, 1974). The quantity and concentration of bacteriorhodopsin was estimated specrophotometrically assuming $\varepsilon_{570} = 63000$ $M^{-1}cm^{-1}$.

The emulsion was prepared sonifying the two phase system octane/water (the weight ratio being 1 (octane)/ 4 (water)) using the ultrasonic desintegrator UZDN-1 three times with duration of 15 seconds, the interval between sonification procedures being 45 seconds. The desintegrator was tuned to the maximum power upon the frequency of 22 kHz, the cells were cooled with water. The emulsion yielded the reproducible photoeffects during 48 hours. Usually we used the following emulsions: asolectin 60 mg, bacteriorhodopsin sheets 1.85 mg, octane 0.8 ml, aqueous phase 1.52 ml. In several experiments the water phase contained 2 M $NaNO_3$. The values of pH were registered using the glass and Ag/AgCl- electrodes connected through the electrometer designed by V.S.Sokolov (Institute of Electrochemistry, the USSR Academy of Sciences) on the basis of Keithley-301 instrument with the time resolution of 1 ms, the results being displayed by two-coordinate recorder H-306. The Ag/AgCl-electrode was brought into the contact with emulsion using the salt bridge filled with 50 mM KCl. The selective electrode with respect to ions H^+ ESL-11G-04 and Na^+ - ESL-51G-04 were used as the glass electrodes. The measurements of pH was carried out either in emulsions or in emulsions covering the layer of aqueous phase containing the concentrated salt solution in accord with results obtained in (Post et al., 1984). Having succeeded in reproduction of the main results of (Post et al., 1984) we somewhat modified the experimental procedures. The Australian scientists covered 3 ml of aqueous phase with 100 μl of emulsion. During illumination of emulsion the photoresponse appeared in 4-5 minutes, and during the switching off the light the value of pH attained the initial level in 5-6 minutes. We placed the glass electrode immediately into the emulsion in order to eliminate the diffusion restrictions in aqueous phase. In this case the photoeffect appeared in 2 s and during switching off the light it vanished in several seconds.

The measurements were carried out also in octane-in-water emulsions, the weight ratio being 1 (octane) / 8 (water). In this case 2M $NaNO_3$ were initially dissolved in water phase. The sonification resulted in obtaining octane-in-water emulsion covering the layer of aqueous electrolyte solution. The measurements of pH were carried out in the upper layer. According to the measurements of the size of emulsion particles using the method of dynamic of light scattering (Autosizer K7027, Malvern, England, the courtesy of Dr.Yu.A.Ermakov, Institute of Electrochemistry, USSR Academy of Sciences) the specific size of emulsion droplets was equal to 2 μm. These results were confirmed by electron microscopy data (Post et al., 1984).

We used the light source OI-28 with halogen lamp (75 W) as an radiation source. The light was focused upon the emulsion by additional lens. In control experiments the ultraviolet and blue part of spectrum were cut off using the filters VS-8 and ZhS 18 + 19 from light source spare parts kit. In order to increase the intensity of light the opaque mirror was replaced by a transparent one. The action spectrum was measured using the interferential filters with the transmission band $\lambda_{0.5} = 8$ nm. The power of light was measured by thermocouple. Absorption spectra were recorded on spectrophotometer "Specord-M40".

RESULTS

While irradiating emulsions with white light containing all necessary components we observed the variation of proton concentration in aqueous phase of emulsion. Fig.1 displays the reversible variation of pH in octane-in-water emulsion containing bacteriorhodopsin sheets as a function of radiation intensity. When the light was switched on, the alkalinization of aqueous phase was also observed. The aqueous solution was a concentrated solution of a strong electrolyte.When one of the components (either bacteriorhodopsin or non-organic salt) was excluded from emulsion, the variation of pH was not observed. If the prepared emulsion did not contain phospholipids we detected   denaturation of bacteriorhodopsin. When the light was switched off the photoresponse raised very quickly and attained the even level in approximately 2-3 s. During switching off the value of pH returned to the initial level significantly slower. The photoeffect was reversible and we repeated it several times.

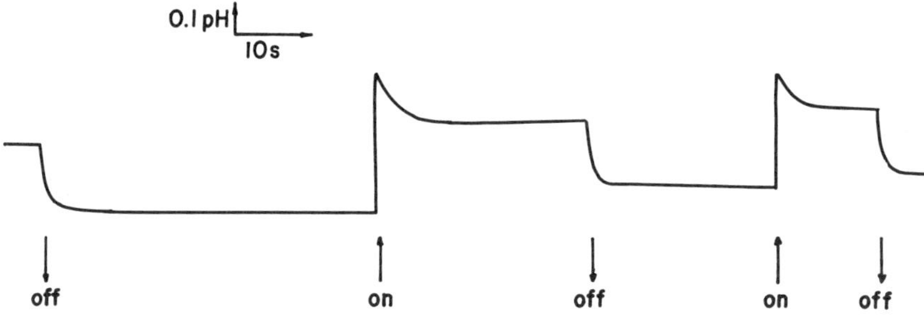

FIGURE 1.The reversible light-dependent pH shift in octane-in-water emulsion containing bacteriophodopsin sheets; pH of emulsion in dark condition being 6.0.

The observed ΔpH signal can be ascribed both to the variation of pH in aqueous phase of emulsion and to processes developing upon the film from bacteriorhodopsin and lipids which could cover the glass electrode. In order to verify the source of photoresponse we purposefully covered the glass pH-electrode by the film containing bacteriorhodopsin sheets+ asolectin + octane and then immersed it into the aqueous solution of 2M $NaNO_3$. This resulted in impulse photoresponse quickly relaxing to the dark level. When the light was switched off we also observed outbursts at the recorder output which quickly relaxed to the initial value. The effect was absent if the bacteriorhodopsin film was washed with ethanol from the   electrode surface. Thereby, the light-dependent variation of   pH given in Fig.1 cannot be ascribed to the bacterio-rhodopsin immobilized upon the surface of glass electrode. When bacteriorhodopsin sheets or $NaNO_3$ was excluded from the system under study there was no variation of pH during irradiation.

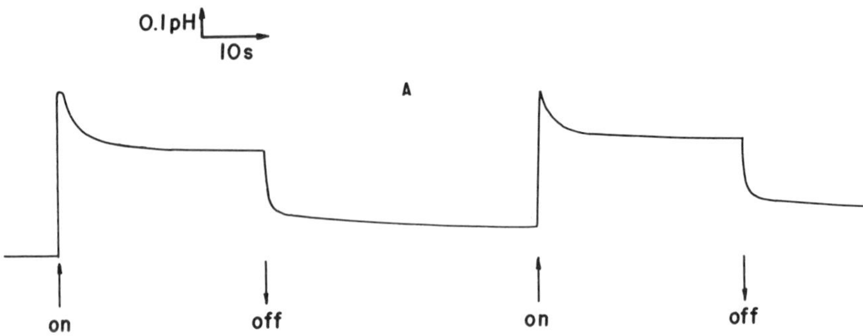

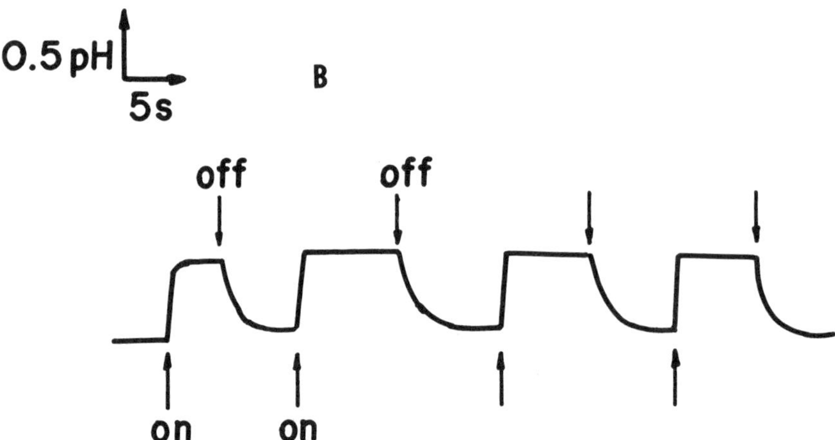

FIGURE 2. The variation of pH as the function of time after introduction of (a) - 2 mM of pentachlorophenol, (b) - 30 μM TTFB.

The system of two non-mixing liquids (oil-water) can form 6 types of structures:
   (a) two individual non-mixing liquids;
   (b) water-in-oil emulsion / aqueous phase;
   (c) oil/oil-in-water emulsion;

(d) two non-mixing liquids and emulsion upon the interface;

(e) oil-in-water emulsion or water-in-oil emulsion;

(f) water-in-oil emulsion / oil-in-water emulsion.

While studying the processes upon the oil/water interface one should specify with which of above six structures the experimentalist deals with. It is also necessary to point out that the transition from one structure to another can be carried out in respectively smooth conditions varying the concentrations of surface active compounds or that of electrolyte. This transition can be also caused by chemical reaction. Unfortunately, in majority of papers concerning the oil/water systems having plain or extended interface this problem was not analyzed. In present study we investigated the two types of structures: oil-in-water emulsion or emulsion/aqueous solution. As it was the case, the studies carried out in above two systems gave similar results.

By introducing the uncouplers of oxidative phosphorylation into the emulsion (pentachlorophenol or TTFB, Fig.2) we not only failed to observe the inhibition of proton transport, but even detected a slight increase in the photoeffect although not exceeding 0.05 of pH unit, i.e. lying within the experimental errors.

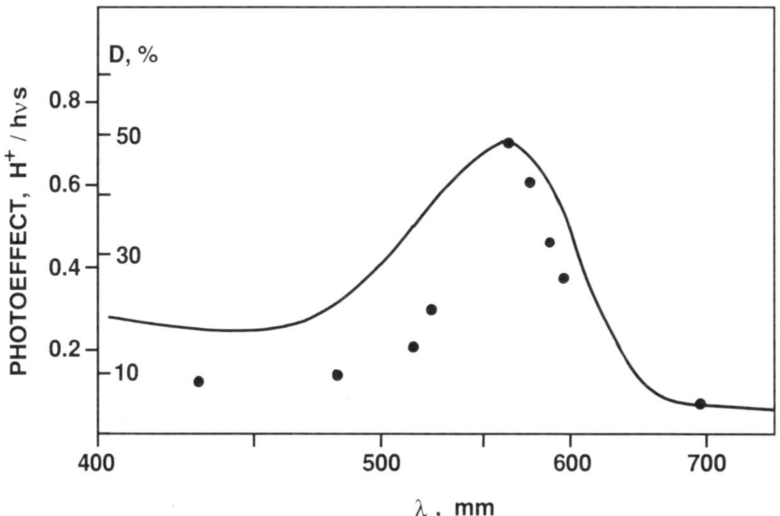

FIGURE 3. The absorption spectrum of bacteriorhodopsin sheets in water and the action spectrum of reaction. The photoeffect was calculated with respect to the energy of incident quant. The conditions are the same as those given in Fig.1.

Using Na$^+$-selective electrodes we staged the control experiment which indicated that the phototransfer of H$^+$-ions takes place in the studied emulsion, and there was no variation of the concentrations of ions of alkaline metals.

The absorption spectra of bacteriorhodopsin sheets in water and action spectrum are given in Fig.3. As this figure indicates, the spectrum of reaction follows the absorption spectrum of bacteriorhodopsin.

We also carried out special experiments investigating the emulsions without NaNO$_3$. In this case we failed to observe photoeffect. However, the addition of 1 mM PCP into emulsion results in reversible photoeffect attaining only 0.05 of pH units.

DISCUSSION

The bacteriorhodopsin is the sole protein of purple membranes (bacteriorhodopsin sheets), ordered lipoprotein complexes with diameter 0.5 μm. The bacteriorhodopsin plates are unclosed structures containing 25% of lipids.

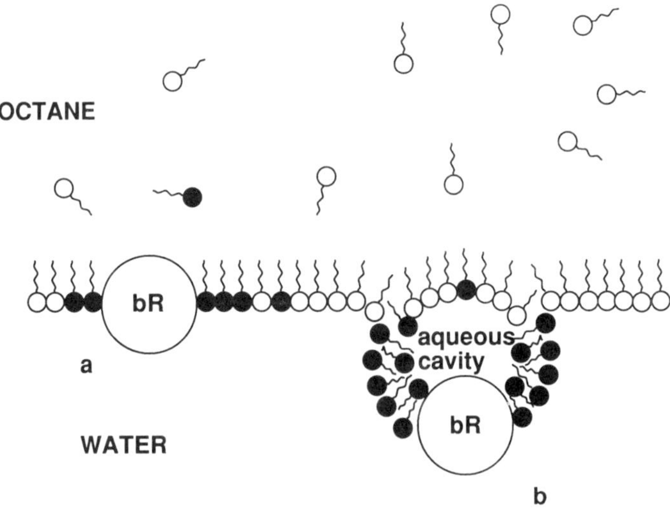

FIGURE 4. The shematic diagram of possible ways of inclusion of bacteriorhodopsin sheets upon the water/lipid interface: a- building of bacteriorhodopsin in monolayer, b- trilaminar structure (system with "third" water).

Two types of implantation of bacteriorhodopsin into the water/octane + lipid interface are possible (Fig.4): fusion with lipid monolayer (a) or creation of Plateau interface with aqueous cavity between the bacteriorhodopsin and octane (b) (the so-called "third" water). The proton transport in the systems with "third" water may be inhibited using uncouplers of oxidative phosphorylation, dinitrophenol, pentachlorphenol, TTPB. Since the introduction of uncouplers in emulsion does not decrease the photoeffect (Fig.2) one can conclude that in above systems the protonofores do not transfer protons in closed membrane structures with aqueous cavities, but rather can become acceptors of protons in interface area on the part of octane. The

pentachlorphenol is a weak acid dissociating in PCP^- and H^+. PCP^- is adsorbed upon the octane/water interface charging octane negative with respect to water (Volkov, 1984). Let us estimate the maximum value of the possible photoeffect if PCP^- operates as the surface acceptor of protons. The radius of the emulsion droplet is about 1 μm. The volume of octane is 0.8 ml. The total amount of droplets is 2×10^{11} with the total surface of about 2 m^2.

Assuming the surface concentration of PCP^- being equal to 10^{-11} mole/cm one obtains that the shift of pH from initial value of pH 6.1 does not exceed 0.1 pH units. The photoeffect observed in our experiment (ΔpH 0.05)

presumably suggests the functioning of PCP^- as a surface proton acceptor.

The switching on the light adds to the transfer of protons from water in octane the passive ionic co-transport. According to paper (Bell, 1931) the distribution coefficient of acid between water and octane permits non-organic acids to be present in octane in sufficient quantities. HCl dissolves in octane in concentrations up to 0.21 g/l, HNO_3 having even higher solubility (Bell, 1931). As it has been shown in paper (Post et al., 1984) the higher is the size of the salt anion and lower is the cation radius, the higher becomes the value of ΔpH during emulsion irradiation. This fact can be easily explained since according to theory of resolvatation of ions (Markin and Volkov, 1989) between two non-mixing liquids the higher is the particle radius, the more negative is the standard free energy of transport of ion from water into oil.

In present study we investigated the phototransfer of H^+ by the proton pump of bacteriorhodopsin through the water/lipid interface. In the similar way one can consider the transfer of ions and electrons through the water/oil interface during functioning of membrane enzyme systems and photosynthetic dyes. The methods of emulsion photo-electrochemistry can be applied to utilize the solar energy and create the highly effective systems of organic synthesis during development of coupled reactions.

REFERENCES

Bell RP (1931): The electrical energy of dipole molecules in solution and the solubilities of ammonia, hydrogen chloride, and hydrogen sulphite in various solvents. *J Chem Soc :* 1371 -1382.

Boguslavsky LI, Kondrashin AA, Kozlov IA, Metelsky ST, Skulachev VP and Volkov AG (1975): Charge transfer between wayer and octane phases by soluble mitochondrial ATPase, bacteriorhodopsin and respiratory chain enzymes.*FEBS Letters* 50: 223-226.

Boguslavsky LI, Boytsov VG, Volkov AG, Kozlov IA and Metelsky ST (1976): Light-dependent translocation of H+ from water to octane by bacteriorhodopsin. *Bioorg Khimiya* 2: 1125-1134.

Boguslavsky LI and Volkov AG (1987): Redox and photochemical reactions at the interface between immiscible liquids. In: *The interface structure and electrochemical processes at the boundary between two immiscible liquids*. Kazarinov VE, ed. Berlin: Springer Verlag.

Drachev LA, Kaulen MD, Skulachev VP and Vojtsitsky VM (1983): Bacteriorhodopsin-mediated photoelectric responses in lipid/water systems. *J Membr Biol* 65: 1 - 12.

Hwang SB, Korenbrot JI and Stoeckenius W (1977): Proton translocation by bacteriorhodopsin through an interface film. *J Membr Biol* 36: 137 - 158.

Kharkats YuI and Volkov AG (1987): Membrane catalysis: syncronous multielectron reactions at the liquid-liquid interface. *Biochim Biophys Acta* 891: 56 - 63.

Markin VS and Volkov AG (1989): The Gibbs free energy of ion transfer between two immiscible liquids. *Electrochim Acta* 34: 93 -107.

Oesterhelt D and Stoeckenius W (1974): Isolation of the cell membrane of *Halobacterium halobium* and its fraction into red and purple membrane. *Methods in Enzymology* 31: 667-678.

Post A, Young SE and Robertson RN (1984): Light-induced proton translocation by bacteriorhodopsin at the interface of an octane-in-water emulsion and inhibition by a retinotoxin. *Photobiochem Photobiophys* 8 :153 - 162.

Volkov AG (1984): A possible mechanism of the photooxidation of water sensitized by chlorophyll adsorbed at the interface. *Bioelectrochem Bioenerg* 12: 15 -24.

Effect of Electrochemical Processes and Electromagnetic Fields
on Biological Systems

CELLULAR EFFECTS OF EXTREMELY LOW FREQUENCY (ELF) ELECTROMAGNETIC FIELDS

Stephen F. Cleary, Li-Ming Liu and Guanghui Cao
Physiology Department
Medical College of Virginia
Virginia Commonwealth University
Richmond, Virginia

INTRODUCTION

Epidemiological studies have revealed a possible association of cancer risk (predominently leukemia and brain center) with environmental and occupational exposure to low intensity, extremely low frequency (ELF) electromagnetic fields (EMF) most commonly 50- or 60 Hz electric power distribution frequencies. Assessments of the significance of this association have been impeded by factors such as: 1) inadequate EMF exposure assessments, 2) uncertainty regarding confounding factors inherent in epidemiological studies, 3) limited amount of chronic in vivo exposure data derived from animal studies, and 4) absence of theory explaining how such fields affect living systems under conditions involving relatively weak interactions or activation energies.

These same factors have impeded progress in the beneficial application of ELF EMFs. Clinical usage of EMF for the treatment of connective tissue disorders including: bone nonunions (Bassett et al., 1981; 1982), fresh fractures (Wahlstrom, 1984) and tendinitis (Binder et al., 1984), has been reported. Whereas such applications provide additional evidence that low intensity ELF magnetic and electric fields are indeed biologically active, there is much uncertainty about optimum treatment methods and mechanisms of interaction.

The most obvious reason for the lack of understanding of the potentially harmful or beneficial effects of EMFs is limited ability to quantify field-tissue interactions. This is attributed to spatially complex induced fields resulting from the highly inhomogeneous dielectric properties of tissue. In addition to such densitometric and dosimetric problems, reported ELF EMF in vivo and in vitro effects have involved other nonclassical responses including: 1) multiple frequency specificities (i.e. frequency or modulation "windows") at ELF frequencies, 2) multiple intensity specificities (i.e. intensity or power density "windows"), and 3) interaction energies below thermal background energy (i.e. kT). Whereas theories explaining such phenomena have been advanced, none provide adequate quantitative or predictive bases for reported effects of ELF EMFs on living systems (Cleary, 1991).

An approach taken to better understand the effects of ELF EMFs has been to conduct studies using mammalian cells or tissue exposed in vitro under controlled conditions. This approach enables quantitative determination of exposure parameters and relatively simple and unambiguous assays of cellular end points. Such studies have revealed a number of cell physiological changes induced by ELF EMFs including alteration of: 1) DNA, RNA and protein synthesis, 2) cation fluxes and binding, 3) immune responses, 4) membrane signal transduction and 5) proliferation. The results of these studies have been reviewed by Cleary (1991).

Reported here are studies of the effects of ELF electric fields on two cell model systems. First, effects of clinically relevant ELF electric fields on the proliferation of tendon explants in vitro will be summarized. Secondly, effects of 60 Hz sinusoidal electric fields on the proliferation and mitochondrial activity of transformed mammalian cells in vitro will be reported.

TENDON EXPLANTS

A chicken tendon explant model system was developed to investigate the effects of extremely low frequency (ELF), low amplitude, unipolar, square wave pulsed electric fields on fibroplasia in vitro. An electric field parameter set consisting of 1Hz, 1ms duration pulses, with a time-averaged current density of 7 mA/m^2(peak current density 7 A/m^2) induced maximal (32%) increase in fibroblast proliferation in tendon explants exposed for 4d. Exposure to the same field but at an average current density of 1.8 mA/m^2 had no effect on fibroblast proliferation, whereas exposure to current densities of >10mA/m^2 inhibited proliferation and relative collagen synthesis, without affecting noncollagen protein synthesis. Fibroplasia was significantly increased in explants oriented parallel to applied electric fields having current densities of 3.5 or 7 mA/m^2 but there was no detectable effect on explants oriented perpendicular to the same electric field. Fibroblast proliferation and relative collagen synthesis were inversely proportional to donor age for chickens in the 3 to 16 week age group used in this study. For these dependent variables (proliferation and relative collagen synthesis), there was no interaction between donor age and ELF electric field exposure.

Tissue Age and Extracellular Ca^{+2}

This tendon explant model system has been used to test two hypotheses: 1) the effect of weak ELF electric fields on tendon fibroplasia is dependent upon extracellular Ca^{+2}, and 2) the effects of ELF electric fields and extracellular Ca^{+2} depend upon tissue age.

To test these hypotheses chicken tendon explants obtained from chickens aged 2- to 15 weeks were exposed for 3 days to a 1 ms duration unipolar square wave 1 Hz pulsed electric field at a time-averaged current density of 7 mA/m^2. Explants were exposed or sham-exposed in rectangular cross section plastic cell culture dishes containing 4 ml culture medium with 1% fetal calf serum (FCS). Agar salt-bridge electrodes (12 cm length of 8 mm diameter Tygon tubing) delivered the current from a Wavetek function generator. Tantalum electrodes electrically coupled the constant current source to the agar salt-bridge electrodes. The methods and procedures were as described by Cleary et al. (1988). Explants were aligned with the longitudinal axis parallel to the applied E field. The four treatment groups were: 1) sham-exposed, 2) ELF electric field exposed, 3) sham-exposed with 1 mM EGTA in the culture media (i.e. zero extracellular Ca^{+2} sham-exposure) and 4) ELF electric field exposed with 1 mM EGTA in the culture medium (i.e. zero extracellular Ca^{+2} ELF E-field exposure). Effects on fibroblast proliferation were determined by measuring the rate of incorporation of tritiated thymidine (^{3}H-TdR), as described by Cleary et al., (1988). To determine the effect of tissue age, tendon explants were grouped into 2 categories: 1) donor age 6 weeks or more (i.e. adolescent or mature chickens), or 2) donor age less than 6 weeks (i.e. pre-adolescent or immature chickens).

The effect of donor age and decreased extracellular Ca^{+2} on tendon fibroplasia in the 4 treatment groups are summarized in Table 1, which indicates the mean % change in ^{3}H-TdR incorporation in explants exposed (or sham-exposed) in the presence or absence of 1 mM EGTA in the culture media. Fibroplasia was suppressed in all treatment groups by the addition of EGTA. The mean reduction in ^{3}H-TdR uptake in explants obtained from chickens 6 weeks or older was 20%, whereas for explants from chickens 6 weeks or younger 1 mM EGTA caused a 13% mean reduction in fibroplasia. These results are consistent with the known Ca^{+2} dependence of cell proliferation. The data indicates an interaction of Ca^{+2}-dependent cell proliferation, tissue age, and ELF electric field exposure. In the absence of extracellular Ca^{+2} there was a 32% reduction in proliferation in explants from chickens aged 6 weeks or more that were exposed to the 1 Hz E-field. The same E-field exposure of explants from younger chickens (<6 weeks) resulted in only a 4.4% reduction in proliferation in the absence of extracellular Ca^{+2}.

Table 1. Effect of EGTA (1mM) on ^{3}H-TdR Uptake

I.   Chickens aged 6 weeks or more

Mean (± SEM) % change in ^{3}H-TdR uptake*

A. ELF-exposed	-32.1 ± 6.4
B. Sham-exposed	-7.6 ± 21.9

Grand Mean (± SEM) = -20   ± 11.5

II.   Chickens less than 6 weeks of age

A. ELF-exposed	-4.4 ± 8.7
B. Sham-exposed	-21.4 ± 26.4

Grand Mean (± SEM) % change = -12.9 ± 12.4

* Mean % change in ^{3}H-TdR uptake =
$$\frac{(\text{TdR uptake w EGTA}) - (\text{TdR uptake w/o EGTA})}{(\text{TdR uptake w/o EGTA})} \times 100$$

The results of a series of experiments to determine the relationship between ELF E-field exposure, extracellular Ca^{+2}, and tissue donor age are summarized in Table 2. In the presence of normal extracellular Ca^{+2} the 7 mA/m^2 pulsed 1 Hz field consistently resulted in increased fibroplasia which varied from a maximum 108% increase in explants derived from a 15 week old chicken, to a 2.2% increase in explants from an 11 week old donor. In contrast, there were 1- and 38.2% reductions in fibroplasia in explants from 4- or 2-week old chickens, in the presence of normal extracellular Ca^{+2} concentration. These data indicate that the effect of the ELF E-field _per se_ on chicken tendon fibroplasia depends upon age. The consistent and statistically significant enhancement of fibroplasia previously detected under these E-field exposure conditions (Cleary et al., 1988), only occurred when explants were obtained from adolescent or mature chickens. In the case of explants from immature chickens the E-field had the reverse effect.

Table 2. Effect of EGTA and Pulsed 1Hz 7mA/m^2
Electric Field on ^{3}H-TdR Uptake by
Chicken Tendon Explants In Vitro

Chicken Age (Weeks)	% ^{3}H-TdR Uptake w/o EGTA[*]	p-value[**]	% ^{3}H-TdR Uptake with 1mM EGTA[*]	p-value[**]
15	108.2	0.09	0.7	0.98
12	7.9	0.61	-1.8	0.85
11	2.2	0.90	0.05	0.99
7	10.3	0.28	6.4	0.63
4	-1.0	0.91	-1.7	0.91
2	-38.2	0.02	2.9	0.83

*% ^{3}H-TdR Uptake = $\dfrac{\text{ELF - Sham}}{\text{Sham}}$ x 100

** p-value for two-tailed student's t-test

Column 4 of Table 2 indicates the % increase in fibroplasia in chicken tendon explants from the same donors exposed (or sham-exposed) to the ELF E-field in the absence of extracellular Ca^{+2} (i.e. 1 mM EGTA in culture medium). Under this condition the ELF E-field did not affect fibroplasia.

<u>Statistical Analysis</u>

The results of an unbalanced randomized block analysis of variance (ANOVA) (type III sums of squares) are indicated in Table 3. When grouped according to age, at normal extracellular Ca^{+2} concentration, there was a statistically significant increase (p=0.05) in fibroplasia in

Table 3. Results of Unbalanced Randomized
Block Analysis of Variance (Type III SS)

	F-value	Two-tailed p-value
I. Chickens > 6 weeks old		
a. w/o EGTA	2.21 (4.08)*	0.15 (0.05)*
b. w EGTA	0.0 (0.02)*	0.98 (0.88)*
II. Chicken $\leq$ 6 weeks old		
a. w/o EGTA	28.84 (24.05)*	0.001 (0.001)*
b. w EGTA	0.01 (0.05)*	0.93 (0.83)*

*log (^{3}H-dpm)

E-field exposed explants from chickens 6 weeks or older, for logarithmic transformed data. In the presence of 1 mM EGTA there was no E-field effect on fibroplasia. In the case of explants from chickens less than 6 weeks of age, at normal extracellular Ca^{+2} concentrations, there was a statistically significant reduction in fibroplasia in response to exposure to the ELF E-field. Again, as in the case of explants from older chickens, removal of extracellular Ca^{+2} nullified the effect of the ELF E-field. These data indicate that the effect of a 1 Hz pulsed E-field, having a time-averaged current density of 7 mA/m^2, on tendon fibroplasia, is dependent upon extracellular Ca^{+2} and upon tissue donor age.

TRANSFORMED CELLS

Experiments were conducted to determine the effect of 60 Hz sinsusoidal E-fields on proliferation and mitochondrial activity of transformed mammalian cells in vitro. These experiments were designed to test the following hypotheses: 1) 60 Hz E-fields affect proliferation of transformed cells in vitro, and 2) E-field effects on proliferation and mitochondrial activity are dependent upon time-averaged induced current density or E-field strength.

Cell Culture

Chinese hamster ovary (CHO) or human breast cancer (MCF-7) cells were maintained in monolayer culture in a 37°C, 5% CO_2 incubator. Cells were cultured in DME/F12 (CHO) and DMEM (MCF-7) with 10% fetal calf serum (FCS), 100 units/ml penicillin and 100 μg/ml streptomycin. Cells were subcultured weekly.

In the 60 Hz ELF exposure experiment 2.5 x 10^5 log-phase (3-4 days after subculture) cells were plated as 1 cm diameter spots at the center of a rectangular (24 mm x 67 mm) plastic cell culture dish containing 4 ml culture medium. CHO cells were exposed to the 60 Hz E-field in the presence of 1- or 10% FCS, whereas MCF-7 cells were exposed in culture medium supplemented with 1% FCS.

60 Hz Electric Field Exposure System

The 60 Hz electric field exposure system was described in detail by Cleary et al., (1988). In this system 2 or more 4-well-plates were used with the rectangular culture wells connected in series by agar salt-bridges consisting of a 12-cm length of 8-mm inner diameter Tygon tubing filled with sterile DME/F12 and 2% agar. The electrodes were electrically coupled to the agar salt-bridges by electrode baths containing DME/F12 or DMEM. Continuous sine wave electric current was supplied by a constant current generator (Wavetek 182). Tantalum electrodes were used to connect the current source to the electrode baths. An oscilloscope (Tektronix 468) was used to continuously monitor the current by measuring the voltage drop across a precision resistor placed in series with the cultures. Cells were exposed, or sham-exposed, continuously for 24h. E-field uniformity, as determined by microelectrode measurements, was ± 2%.

^{3}H-Thymidine Uptake Assay

The tritiated thymidine (^{3}H-TdR) incorporation assay used in this study was described by Freshney (1987). Immediately after ELF exposure cells were pulse labeled for 1 hour with ^{3}H-TdR (5 μCi/ml in DME/F12 or DMEM medium). After labeling, cells were scraped off the plate, divided and replated into two dishes, one for ^{3}H-TdR activity counting, and the other for MTT reduction assay of mitochondrial activity. ^{3}H-TdR incorporation was determined by liquid scintillation counting. The radioactivity (dpm) of samples was normalized to equal cell concentrations and a Student's t-test or unbalanced randomized block analysis of variance (ANOVA) (type III sum of squares) was used to compare the ELF E field-exposed and sham-exposed data.

Mitochondrial Activity Determination by MTT Reduction Assay

As reported by Green et al., (1984), MTT (3(4,5-dimethyl-thiazoyl-2-yl)2,5 diphenyltetrazolium bromide) is reduced by mitochondrial dehydrogenases to a purple formazan. The optical density of formazan provides a quantitative assay of mitochondrial activity.

In our study 100 μl MTT stock solution (5 mg/ml in PBS) was added to 10^6 cells in 2 ml DME/F12 medium. Cells were cultured with MTT at 37°C, 5% CO_2 in an incubator for 3 hours. The medium with MTT was then discarded. Cells were washed once with PBS; 1.5 ml/well acidified isopropyl alcohol (0.04 N HCl in isopropanol) was added to solubilize the MTT formazan. The optical density (OD) at 570 nm was measured spectrophotometrically. OD readings of ELF- and sham-exposed CHO were compared by a Student's t-test or ANOVA.

RESULTS

Chinese Hamster Ovary (CHO) Cells

The effect of 60 Hz E-fields having root mean square (RMS) current densities of 1-, 1.25- or 1.58 A/m² (0.6-, 0.75- or 0.95 V/m (RMS), respectively) on CHO ³H-TdR incorporation, is summarized in Table 4.

Table 4. Effect of 24 Hour 60 Hz E-field Exposure of
Log-phase CHO Cells on DNA Synthesis

Current* A/m²	FCS (%)	³H-TdR(dpm)±S.D. Exposed	Sham	%** Change	p-value
1	10	59,637 ±13,327 (11)***	85,533 ±23,414 (14)	-30.3	0.012
1.25	1	691±276 (6)	493±199 (6)	40	0.163
1.58	1	763±90 (6)	478±112 (6)	60	0.001

* Root mean square (RMS)   *** number of samples ( )
** % Difference = $\underline{\text{Exposed-Sham}}$ x 100
                                Sham

Log-phase CHO cells exposed to a 1 A/m^2 60 Hz E-field for 24 h at a FCS concentration of 10% had a statistically significant (p=0.012) 30% reduction in DNA synthesis relative to sham-exposed cells. The rate of DNA synthesis in cells exposed to either 1.25-, or 1.58 A/m^2, at a FCS concentration of 1%, on the other hand, was increased by 40- and 60% respectively. Whereas the 40% increase at 1.25 A/m^2 was not statistically significant (two-tailed T-test p-value 0.16), the increase induced by exposure at 1.58 A/m^2 was highly statistically significant (p=0.001). The data indicate that the 60 Hz sinusoidal electric fields used in this study modulate CHO proliferation and the field effect depends upon field intensity and FCS concentration.

The effect of these same field intensities on CHO mitochondrial activity is summarized in Table 5. E-field exposure at all current densities caused statistically significant increased CHO mitochondrial activity. The mean increase for all treatments was 48.5%.

Table 5. Effect of 24 Hour 60 Hz E-field Exposure of
Log-phase CHO Cells on Mitochondrial Activity

Current* A/m^2	FCS (%)	O.D. at 570nm (±SD) ± Exposed	Sham	%[**] Change	p-value
1	10	1.025 ± 0.141 (12)***	0.710 ± 0.178 (13)	44.4	0.0002
1.25	1	0.569 ± 0.057 (6)	0.439 ± 0.055 (6)	29	0.0033
1.58	1	0.560 ± 0.12 (6)	0.325 ± 0.077 (6)	72	0.0003

* Root mean square (RMS)

** % Difference = $\dfrac{\text{Exposed-Sham}}{\text{Sham}}$ x 100

*** number of samples ( )

<u>Human Breast Cancer (MCF-7) Cells</u>

Human breast cancer cells (MCF-7) were exposed to a 1 A/m^2, 60 Hz E-field for 24 h at 37°C at FCS concentration of 1%. Field exposure effects on ^{3}H-TdR and mitochondrial activity are summarized in Table 6.

Exposure resulted in a 22% increase in DNA synthesis, which was consistent with the results of exposure of CHO at this same FCS concentration. However, in the case of MCF-7, the increased rate of ^{3}H-TdR incorporation was not statistically significant. In agreement with the effect of 60 Hz E-fields on CHO mitochondrial activity, exposure of MCF-7 cells resulted in a statistically significant (p=0.034) 12.2% increase in mitochondrial activity. These data indicate that in the range of current densities used in these studies (i.e. 1- to 1.58 A/m^2), 60 Hz E-field exposure increased cell mitochondrial activity. Dose-response relationships are being determined for 60 Hz E-field effects on CHO and MCF-7 DNA synthesis and mitochondrial activity over a wider current density range, as is the dependency of E-field effects on FCS concentration.

Table 6. Effect of 24 Hour 60 Hz E-field Exposure (1A/m^2) RMS Current Density on MCF-7 DNA Synthesis and Mitochondrial Activity (1% FCS)

A. ^{3}H-TdR Uptake	mean dpm ± S.D.	
E-field Exposed Sham-Exposed	431,087 ± 62,402 352,434 ± 78,817	% Change = 22.3%
	T-test p-value = 0.45	
B. Mitochondrial Activity	mean opitcal density ± S.D.	
E-field Exposed Sham-Exposed	1.62 ± 0.049 1.44 ± 0.06	% Change = 12.2%
	T-test p-value = 0.034	

DISCUSSION

The results of these studies of the effects of ELF electric fields indicate that E-fields having distinctly different waveforms affect cell proliferation in vitro. Clinically relevant square wave unipolar 1 ms duration pulsed E-fields, with time-averaged current densities on the order of 1 mA/m^2, stimulated tendon fibroblast proliferation. E-field stimulation required the presence of extracellar Ca^{+2}, suggesting that field-induced transient alterations in Ca^{+2} concentration at fibroblast cell membrane surface were involved in the proliferative response. The Ca^{+2} dependence of the E-field effect occurred regardless of the age of the tendon explant donor chicken. However, in the presence of extracellular Ca^{+2}, the E-field effect was tissue-age dependent. In contrast to the stimulatory E-field effect on tendon explants from chickens 6 weeks or older, E-field treatment suppressed fibroplasia in explants from chickens less than 6 weeks old. This provides evidence of age-dependent differences in the basic mechanisms for the induction of fibroblast proliferation, which is consistent with well-known developmental responses.

Results of exposure of CHO and MCF-7 cells indicate that 60 Hz sinusoidal fields modulate the rate of proliferation of transformed cells in vitro. There was evidence that the qualitative nature of the E-field effect depended upon current density, as well as concentration of FCS in the culture medium, suggesting an interaction between these variables. This interaction is currently being investigated.

Since sinusoidal E-fields modulated CHO and MCF-7 cell proliferation, the effect did not depend upon time-averaged current density which obviously was zero in this case. Assuming the effect to be dependent upon instantaneous current density (or field strength), the results of 1 Hz pulsed field and 60 Hz sinusoidal fields may be compared. For the 1 Hz E-field, maximum 32% increase in cell proliferation occurred at an instantaneous current density of 7 A/m^2 (instantaneous field strength 4.2 V/m). CHO or MCF-7 cells, at the same FCS concentration (i.e. 1%), exhibited similar magnitude increases in proliferation when exposed to a 60 Hz sinusoidal E-field with a peak current density of 1.41 A/m^2 (peak field strength 0.85 V/m). Although the limited extent of the data precludes conclusions to be drawn, these data suggest that E-fields at current densities on the order of 1- to 10 A/m^2 (E-field strengths in the range of 1- to 5 V/m) affect cell proliferation in vitro.

These data also provide limited insight into the kinetics of E-field induced effect on proliferation. Assuming, on the basis of the Ca^{+2} dependence of the E-field effect on fibroplasia, that the effect involves transient E-field induced redistribution of ions at or near the cell surface, the E-field time structures provide estimates of the redistribution time constants. In the case of the 1 Hz square wave, cell proliferation was modulated by a 1 ms duration pulse. The half-period of the 60 Hz E-field used to modulate CHO or MCF-7 cell proliferation was 8.3 ms. Obviously, data for E-fields of shorter pulse-duration, or half-periods, are needed to determine the time constant threshold for E-field- induced modulation of cell proliferation.

Data on the effect of 60 Hz E-fields on the activity of CHO or MCF-7 mitochondria indicate a general stimulatory effect that does not appear to be directly correlated with the modulatory effect on cell proliferation. The mechanism responsible for E-field mitochondrial activation is unknown. Based upon reported effects of ELF electric or magnetic fields on cell membrane cation transport and binding, it is possible that mitochondrial activation may be related to the ATP requirement of membrane active transport processes. This possibility is being investigated.

ACKNOWLEDGEMENTS

This research was supported in part by the National Institute of Occupational Safety and Health Grant No. R01 OH02148. We thank Dr. David Gewirtz, Pharmacology and Toxicology Dept., Virginia Commonwealth University for providing the MCF cells.

REFERENCES

Bassett CAL, Mitchell SN, Gaston SR (1981): Treatment of ununited tibial diaphyseal fractures with pulsed electromagnetic fields, J. Bone Joint Surg [Am], 63A:511-523.

Bassett CAL, Mitchell SN, Gaston SR (1982): Pulsing electromagnetic field treatment in ununited fractures and failed arthrodoses, J. Am. Med. Assoc., 247:623-628.

Binderman I, Parr G, Hazelman B, Fitton-Jackson S (1984): Pulsed electromagnetic field therapy of persisten rotator cuff tendinitis, Lancet, 8379:694-698.

Cleary SF, Liu LM, Graham R and Diegelmann RF (1988): Modulation of Tendon Fibroplasia by Exogenous Electric Currents, Bioelectromagnetics, 9:183-194.

Cleary SF (1991): In vitro studies: Low frequency electromagnetic fields. In: Proceedings of Scientific Workshop on the Health Effects of Electromagnetic Radiation on Workers, USDHHS, Public Health Service, NIOSH Cincinnati, Ohio (in press).

Freshney RI (1987): Culture of Animal Cells, Alan R. Liss Inc., New York.

Green LM, Reade JL and Ware CF (1984): Rapid Assay for Cell Viability: Application to the Quantitation of Cytotoxic and Growth Inhibitory Lymphokines, J. Immun. Meth., 70:257-268.

Wahlstrom O (1984): Stimulation of fracture healing with electromagnetic fields of extremely low frequency (EMF of ELF), Clin. Orthop., 186-293-301.

ELECTROPERMEABILIZATION OF HUMAN CULTURED CELLS GROWN IN MONOLAYERS
II. Control of cell proliferation and DNA–replication

S. Kwee, B. Gesser and J. Celis
Institute of Medical Biochemistry and Danish centre for Human Genome
Research, University of Aarhus, Build.170, DK–8000 Aarhus C (Denmark)

INTRODUCTION

A technique was developed to introduce monoclonal antibodies into cultured
cells by electroporation, while retaining cell viability at the same time (Kwee et
al., 1990). This enabled us to study cell growth and DNA synthesis in response
to electroporation and the following uptake of specific antibodies.

Preliminary results were reported about these effects in transformed human
amnion cell cultures (AMA) and primary cell cultures of human fetal
fibroblasts (Kwee and Celis, 1991). It was found that electroporation itself, but
particularly the uptake of an antibody to the intermediate filament protein
vimentin, had distinct effects on cell proliferation. Especially in the primary cell
cultures the uptake of anti–vimentin resulted in a noticeable increase in the
number of cells synthesizing DNA. However, the effect seemed to be dependent on the age of the cell culture in question i.e. the older the primary cell
culture, the less effect on growth control. On the other hand the introduction of
an antibody to the nuclear protein, known as factor IV or primatin, caused extensive cell death.

Since the cell cultures in question belonged to two different cell types, it
was not quite possible to compare the effects. Therefore a primary cell line had
to be found, comparable to the immortal AMA cells. To this purpose we chose
human amnion cells and here results are reported of similar experiments performed on this primary cell culture.

EXPERIMENTAL

Materials
5–Bromo–2'–deoxyuridine and 5–fluoro–deoxyuridine were from Sigma.

Antibodies
Monoclonal antibody Dako–BrdUrd from Dakopatts, monoclonal antibody
mAB 1C4C10 (Celis et al., 1987), anti–vimentin (a gift from S.Blose).

Polyclonal antibody R22 was prepared by immunizing a rabbit with cyclin,
a nuclear protein, isolated from human lymphoblasts (MOLT–4 cells) and purified by chromatography and 2– dimensional gelelectrophoresis. The cyclin

fraction, cut out from the 2–dimensional gel was used to immunize the rabbit. The isolated antibody R22 recognized cyclin on a Western blot. The blank serum sample from a non–immunized rabbit was tested on a 2–D blot (B. Gesser).

All antibodies introduced into the electroporated cells were exhaustively dialized against steril HANKS buffer to remove azide and other preserving agents before use.

Secondary antibody Rhodamine conjugated rabbit antimouse and swine antirabbit FITC–conjugated from Dakopatts.

Cell lines

Transformed human epithelial amnion cells (AMA) and human amnion cells. All cells were grown as monolayer cultures in Dulbecco's modified Eagle's medium containing 10% calf serum, pennicillin and streptomycin.

Procedure

The apparatus and standard procedure for electroporation have been described before (Kwee et al., 1990). Effects of electropermeabilization and the uptake of antibody on cell growth and DNA synthesis were studied using the same technique as reported before (Kwee and Celis, 1991).

RESULTS AND DISCUSSION

Effect of electroporation on growth and DNA synthesis in human amnion cells

Contrary to transformed cells (AMA) the primary cells were much more sensitive to electroporation e.g. regeneration took longer time. In case of AMA cells pore closure, measured by the exclusion of the dye Trypan blue, was already completed after 30 min. In 90% of the primary cells resealing first took place after 2 h and in the last 10% no pore closure had taken place. Moreover growth was largely impaired at the electric field strength that had to be applied in order to introduce the antibodies. As previously observed (Kwee et al., 1990) in spite of the fact that cells appeared normal after 1–2 h, according to general criteria, they nevertheless could have suffered invisible damage, which first showed up after a longer period i.e. after incubating the cells in growth medium for another 24–48 h at 37°C. First then the resulting damage could be evaluated by the detection of reduced viability, diminished growth and a low number of cells in the S–phase.

Effect of anti–Vimentin in human amnion cells

The electroporated cells were incubated immediately with the antibody against vimentin, according to the procedure described before. In this cell line permeabilization of the outer membrane only took place on one side of the cell. This could be seen from the immunofluorescence stain of the antibody that was taken up into the cell (Fig.1). Further transport into the plasma of the cell seemed also to be considerably slower than in the case of AMA cells.

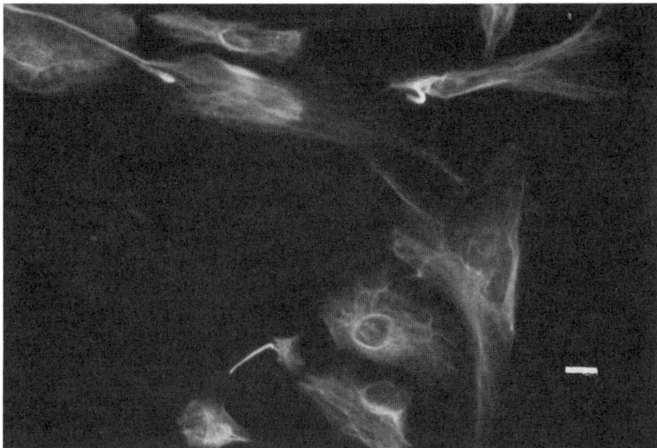

FIG. 1. Immunofluorescence staining of vimentin in human amnion cells shortly after uptake of anti–vimentin by electroporation. Bar: 10 μm.

After 24–48 h re–incubation there was a considerable increase in the number of cells in the S– phase (Fig.2). So in this cell line the uptake of anti-vimentin seemed to have the same effect as described before in other cell lines resulting in an increasing percentage of cells synthesizing DNA (Table I). We proposed previously that this could be due to the fact that the vimentin gene is associated with cell proliferation and that cell shape, which is controlled by the cytoskeleton, is tightly connected to growth control.

The effect of mAB 1C4C10 in human amnion cells
This antibody binds to primatin/nuclear factor IV, which are proteins that might be involved in DNA replication, repair and recombination (de Vries et al., 1989). Uptake of this antibody by electroporation resulted in lysis and extensive cell death. Even in low concentration (200 times dilution) the effect was the same i.e. cell growth was completely inhibited. So this shows that no cell proliferation is possible if the DNA repair mechanism is blocked in this way.

The effect of polyclonal antibody rabbit R22 in human amnion cells
This antibody is supposed to bind to PCNA/cyclin, the proliferating cell nuclear antigen, which is an auxiliary protein of the enzyme complex DNA polymerase–δ (Tan et al., 1986) and was also found to play a part in DNA repair. The uptake of this antibody into electroporated cells (Fig.3) resulted in a considerable increase of the number of cells synthesizing DNA (Fig.2), while a blank serum sample from non–immunized rabbits had no effect. Apparently cells are now forced to synthesize new DNA polymerase in response to the condition that the original enzyme is not functional any more, since part of it is inhibited by the antibody in question. This increased protein synthesis then initiates a chain of reactions resulting in DNA replication and cell proliferation (So and Downey, 1988).

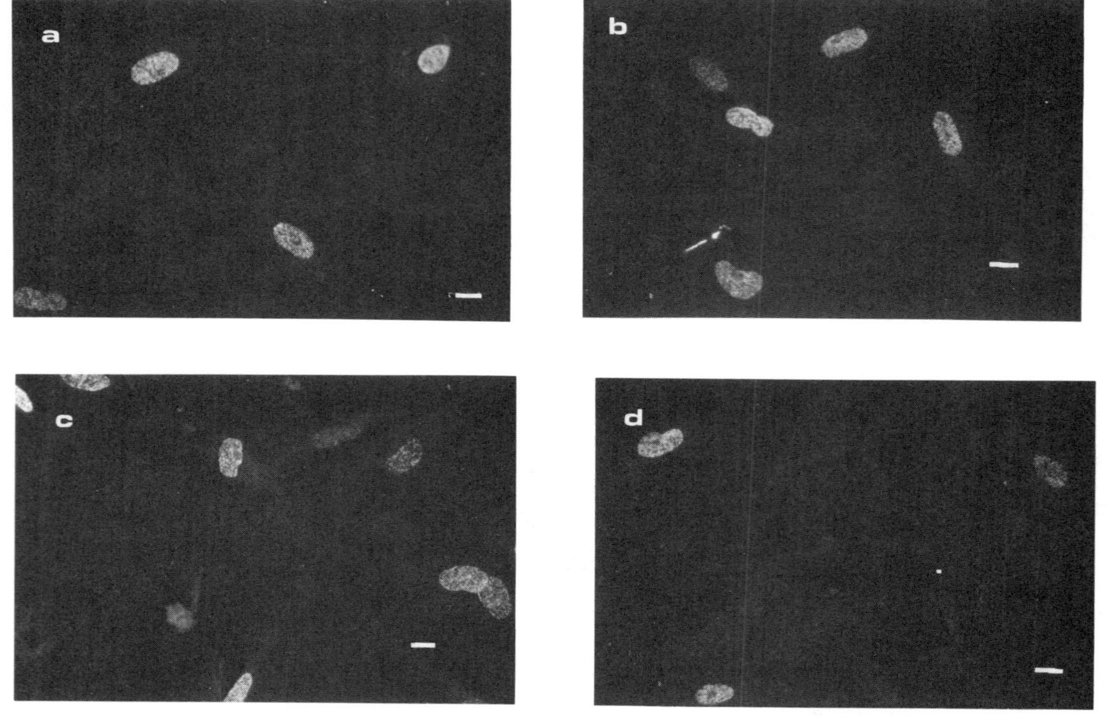

FIG.2. Immunofluorescence staining of BrdU-labelled human amnion cells after electroporation and 24 h re-incubation in growth medium at 37°C. Conditions stated as in Table I. (a) Non-electroporated cells. (b) Electroporation followed by pre-incubation with buffer for 30 min. (c) Electroporation followed by pre-incubation with anti-vimentin for 30 min. (d) Electroporation followed by pre-incubation with pAB R22. Bar: 10μm.

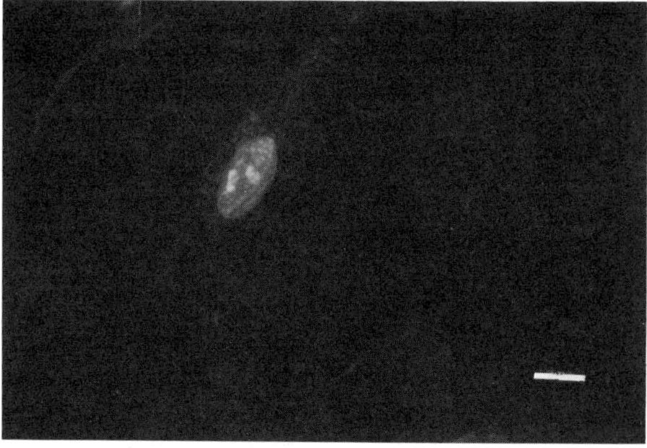

FIG. 3. Immunofluorescence staining of PCNA/cyclin in human amnion cells after uptake of pAB R22 by electroporation. Bar: 10 μm.

TABLE I. Growth and DNA synthesis in human amnion cells after electroporation. Applied field 400 V cm^{-1}, transient pulse length 0.4 ms, pulsing period 1 s, capacity o.1 μF

Passage number	Age /h	Treatment before re-incubation	Growth and DNA synthesis after 24 h re-incubation	/% cells in S-phase
1–4	24	none	good	25 + 5
1–4	24	e.p.[a] and buffer	slow	15 + 5
1–4	24	e.p. and anti–vim	good	60 + 20
1–4	24	e.p. and R 22	good	50 + 20
1–4	24	e.p. and 1C4C10	lysis	0
5–6	48	none	slow	10 + 5
5–6	48	e.p. and buffer	none	8 + 5
5–6	48	e.p. and anti–vim	normal	50 + 20
5–6	48	e.p. and R 22	good	55 + 5
7–9	48–72	none	none	8 + 2
7–9	48–72	e.p. and buffer	none	3 + 1
7–9	48–72	e.p. and anti–vim	none	5 + 2

a   e.p.: electroporation

The effect of age in human amnion cells

Primary amnion cell cultures are growing much slower than their trans-formed counterparts, the AMA cells.There was also a large variation in the life span of the individual cell cultures. In average vigorous growth diminished greatly after passage 4–5 and the maximum life time of a culture was 7–11 passages. The growth promoting effect of the antibodies against vimentin and against PCNA/cyclin was greatest in the younger and vigorously growing cells (Table I). As soon as growth started to decline in a cell culture, there was hardly any effect. This is in agreement with the results from our previous experiments with other primary cell cultures, i.e. external factors seem to influence cell proliferation only in younger and fast growing cell cultures. Studies on other phenomena in various cell cultures, such as transformation (Katenkamp et al., 1991), fusion of erythrocyte ghosts (Sowers, 1990) and the effect of extremely–low–frequency (ELF) magnetic fields (Goodman et al., 1989) reported the same observations i.e. that young cells are most sensitive to externally induced changes.

ACKNOWLEDGEMENTS

Research support was received from NOVO, Lægevidenskabens Fremme and P. Carl Petersen's Fond. The secretarial assistance from Ms. Lone Larsen is gratefully acknowledged.

REFERENCES

Kwee S, Nielsen HV and Celis JE (1990): Electropermeabilization of human cultured cells grown in monolayers. *Bioelectrochem Bioenerg* 23: 65–80.

de Vries E, van Driel W, Bergsma WG, Arnberg AC and van der Vliet PC (1989): HeLa nuclear protein recognizing DNA termini and translocating on DNA forming a regular DNA–multimeric protein complex. *J Mol Biol* 208: 65–78.

Gesser B: to be published.

Goodman R, Wei L–X, Xu JC and Henderson A (1989): Exposure of human cells to low–frequency electromagnetic fields results in quantitative changes in transcripts. *Biochim Biophys Acta* 1009: 216–220.

Katenkamp U, Atrat P and Hüller E (1991): Influence of the physiological state on the electric field mediated transformation efficiency of intact mycobacterial cells. *Bioelectrochem Bioenerg* 25: 285–294.

Kwee S and Celis JE (1991): Electroporation as a tool for studying cell proliferation and DNA synthesis in human cultured cells grown in monolayers. *Bioelectrochem Bioenerg* 25: 325–332.

Madsen P, Nielsen S and Celis JE (1986): Monoclonal antibody specific for human nuclear proteins IEF 8Z30 and 8Z31 accumulates in the nucleus a few hours after cytoplasmic microinjection of cells expressing these proteins. *J Cell Biol* 103: 2083–2089.

So AG and Downey KM (1988): Mammalian DNA polymerases α and δ : Current status in DNA replication. *Biochemistry* 27: 4591–4595.

Sowers AE (1990): The mechanism of electroporation and electrofusion. In: Electromagnetic field effects on molecules and biological, cells, *Proceedings electroporation symposium* 1990, Bielefeld University press p.56–57.

Tan CK, Castello C, So AG and Downey KM (1986): An auxiliary protein for DNA polymerase-δ from fetal calf thymus. *J Biol Chem* 261: 12310–12316.

EXTREMELY WEAK AC AND DC MAGNETIC FIELDS SIGNIFICANTLY AFFECT MYOSIN PHOSPHORYLATION

MS Markov, JT Ryaby, JJ Kaufman, AA Pilla

Department of Biophysics, Sofia University, Sofia, Bulgaria
Bioelectrochemistry Laboratory, Department of Orthopaedics,
Mount Sinai School of Medicine, New York, NY, USA

Recent studies show that extremely low level electromagnetic fields (EMF) are capable of producing significant bioeffects. These effects have most often been related to the induced electric field, while the role of the magnetic component remains ambiguous. Most of the effects of weak magnetic fields have been studied by assuming that the cell membrane is the EMF target (Markov and Blank, 1988; Marino, 1988). In fact, the actual biophysical pathways involved in the coupling of weak electromagnetic fields to biological systems still remains unclear. The physical mechanisms which allow extremely low frequency and DC magnetic fields of milli- and microtesla intensity to interact with cells and to produce physiological and/or biochemical reactions is difficult to model. Nevertheless, a number of studies have recently been reported concerning the effects of weak magnetic and electromagnetic fields on calcium dependent processes and reactions. Several resonance mechanisms have been proposed by Chiabrera et al., 1985; Liboff,1985; Lednev, 1991; and Chiabrera et al., 1991. They are related to the Lorentz force effects on ion transport (cyclotron resonance) or on ion/ligand binding kinetics, Recently, a quantum approach was taken which suggests that the ion movement in a binding site is quantitized, leading to parametric resonance. It has been reported that weak AC/DC fields tuned to calcium frequencies can significantly affect Ca^{2+}/calmodulin dependent myosin phosphorylation (Shuvalova et al., 1991). Myosin light chains are known to be capable of binding divalent cations and are phosphorylated by myosin light-chain kinase, which requires calmodulin, a calcium binding protein, to function.

This study was designed to critically examine the effects of extremely weak DC and low frequency AC magnetic fields on Ca^{2+}/calmodulin dependent myosin light chain phosphorylation in a cell free preparation.

MATERIALS AND METHODS

Myosin with completely dephosphorylated light chains was isolated from smooth and fast skeletal rabbit muscles. Phosphorylation of myosin light chains was performed by incubation of the protein mixture (myosin light chains, calmodulin, myosin light chain kinase) in plastic containers. Myosin light chains kinase was assayed in a volume of 0.08 ml of 40 mM Hepes buffer pH 7.0, 0.5 mM magnesium acetate, 1 mg/ml bovine serum albumin, 0.1% (w/v) Tween 80, 1 μM calmodulin. The optimum calcium concentration was found to be 30 μM. The enzyme was diluted in 25 mM Tris-buffer containing 0.2% (w/v) Tween 80 to prevent loss of enzyme through binding to plastic and to ensure an accurate measure of the enzyme-specific activity. The amount of reaction medium was prepared as a stock solution and then aliquoted to plastic containers. The reaction was initiated by adding 5 μlγ^{32}P-ATP to the reaction medium and was stopped by adding LSB stopping solution. All samples were incubated in a water bath at 37°C for 6 and 15 min. These reaction times correspond to early and middle points in phosphorylation kinetics.

The electromagnetic field exposure system consisted of a pair of 8 in. diameter circular AC coils and two orthogonal pairs of 8 in. square coils by which the vertical and horizontal components of the DC could be adjusted. This exposure system allows AC only, DC only and combined AC/DC magnetic fields to be applied. In all experiments, the horizontal component of the earth's magnetic field (HDC) was set to zero. The vertical component (VDC) was either set to zero or reduced to 20.9 μT. Sine wave AC magnetic fields in the 8-20 Hz frequency range were adjusted to 20.9 μT. All magnetic field values were measured to $\pm$ 0.1μT using Bartington MAG-01 and MAG-03 fluxgate magnetometers. All control samples were exposed to ambient geomagnetic fields (55 $\pm$ 3 μT vertical and 25 $\pm$ 3 μT horizontal) at the site of the experiment.

The extent of phosphorylation was evaluated by γ^{32}P-ATP incorporation into purified mixed light chains using Cherenkov emission. Aliquots (25 μl) were pipetted onto 3MM paper which was washed in 30% acidic acid at room temperature for 15 min. with continuous stirring. Three subsequent washes for 15 min. each were performed using 15% acidic acid solution. A Student's paired t-test was performed for each time and exposure condition.

RESULTS AND DISCUSSION

A summary of the results to date is shown in Figures 1 to 3. The first remarkable point is that nearly all of the magnetic field conditions employed significantly alter myosin phosphorylation. As seen in Figure 1, which concerns AC=16 Hz ("Ca^{2+} tuning" for DC=20.9 μT); the AC field alone; the DC field alone; and the AC/DC combination all decreased phosphorylation by $\geq$ 50 %. There was no significant difference between the results obtained at 6 and 15 min exposures.

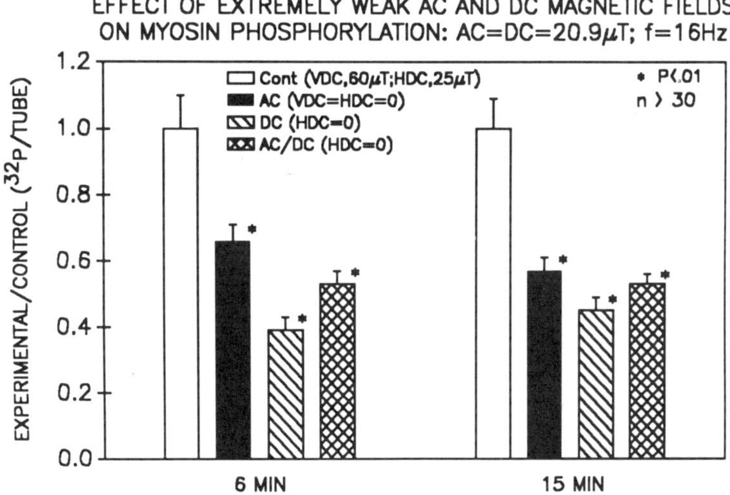

FIGURE 1. Effects of weak AC and DC magnetic fields on myosin phosphorylation. The AC/DC condition is "Ca^{2+} tuned" for 16 Hz (AC = VDC = 20.9 μT, HDC = 0). The AC only exposure is 20.9 μT at 16 Hz and HDC = VD = 0. For the DC only exposure, the vertical component is reduced by approximately 3X, and the horizontal component is set to zero. Note that for all exposure conditions, the DC components of the earth's magnetic field have been lowered. Myosin phosphorylation is significantly (p<0.01) reduced by $\geq$50% for all exposures, however interexposure differences are not significant.

The preliminary results in Figure 2 show that, for the combined AC/DC exposure, there appears to be some frequency dependence. However, this is somewhat complicated by the results in Figure 3, which show that, for the AC signal alone (VDC = HDC = 0), 8, 12, and 16 Hz all appear to decrease phosphorylation, while 20 Hz increases phosphorylation at 6 min., with no effect at 15 min.

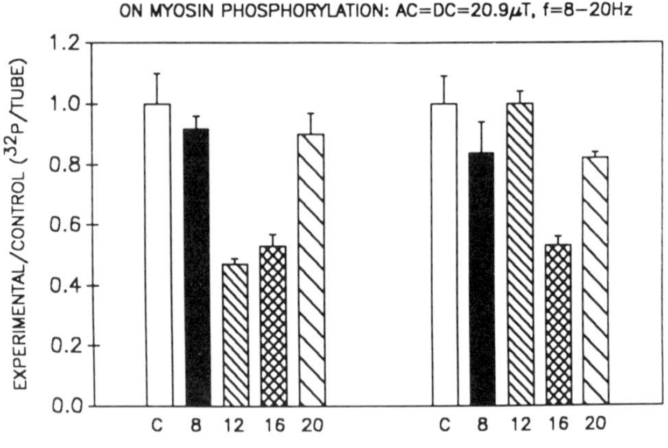

FIGURE 2. Frequency dependence of combined AC/DC magnetic fields effects on myosin phosphorylation. The AC frequency varied from 8-20 Hz. Amplitude was set to 20.9 μT which corresponds to "Ca^{2+} tuning" at 16 Hz. The vertical DC field was 20.9 μT (earth component lowered by 3x) and the horizontal component was set to zero (horizontal earth DC 25 μT). The effect appears to be maximal at 12-16 Hz.

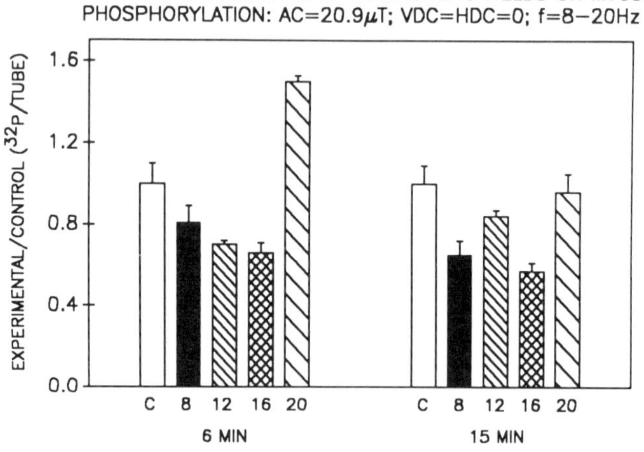

FIGURE 3. Frequency dependence of AC magnetic fields effects on myosin phosphorylation. The AC amplitude wa set to 20.9 μT ("Ca^{2+} tuning" at 16 Hz) and its frequency varied from 8-20 Hz. Both horizontal and vertical components of the earth's magnetic field were set to zero. Under these conditions, the rate of myosin phosphorylation was significantly decreased at all frequencies.

In summary, all of the results obtained thus far appear to show that combined DC/AC and DC and AC alone at 20.9μT ("Ca^{2+} tuning") significantly decreased phosphorylation in the 16 Hz experiment. Surprisingly, in each experiment and in every sample, 20.9μT DC significantly decreased phosphorylation. The frequency effects of AC alone and AC/DC combinations are complex, but tend to result in altered phosphorylation. However, no clear evidence for resonance was observed. It is also not clear that any of these results fit the quantum or classical theories. Both Chiabrera et al. (1991) and Adair (1991) suggest that the quantum approach predicts DC magnetic field effects on the probability of ion transitions among allowed energy levels.

ACKNOWLEDGEMENT

This work was supported by American Medical Electronics, Richardson, TX.

REFERENCES

Adair RK (1991): Criticism of Lednev's Mechanism for the Influence of Weak Magnetic Fields on Biosystems. EPA Scientific Advisory Board Hearings Record, Jan. 14-16, 1991.

Chiabrera A, Bianko B, Caratozzollo F, Gianetti G, Grattarola M and Viviani R (1985): Electric and Magnetic Fields Effects on Ligand Binding to Cell Membrane. In: *Interactions between Electromagnetic Fields and Cells*, Chiabrera A, Nicolini C and Schwan HP, eds.: 253-280.

Chiabrera A, Bianko B, Kaufman JJ and Pilla AA (1991): Quantum Dynamics of Ions in Molecular Crevices Under Electromagnetic Exposure. In: *Electromagnetics in Biology and Medicine*, Brighton CT and Pollack SR, eds. San Francisco Press Inc.: 21-26.

Chiabrera A, Bianko B, Kaufman JJ and Pilla AA (1991): Quantum Analysis of Ion Binding Kinetics in Electromagnetic Bioeffects. In: *Electromagnetics in Biology and Medicine*, Brighton CT and Pollak SR, eds. San Francisco Press Inc.: 27-34.

Lednev VV (1991): Possible Mechanism for the Influence of Weak Magnetic Fields on Biological Systems. *Bioelectromagnetics*, 12: 71-75.

Liboff AR (1985): Cyclotron Resonance in Membrane Transport. In: *Interactions between Electromagnetic Fields and Cells*, Chiabrera A, Nicolini C and Schwan HP, eds.: 281-296.

Marino A (ed.) (1988): *Modern Bioelectrochemistry*, Marcel Dekker.
Markov M and Blank M (eds.) (1988): *Electromagnetic Fields and Biomembranes*, Plenum Press Inc.
Shuvalova LA, Ostrovskaja MV, Sosunov EA and Lednev VV (1991): Weak Magnetic Field Influence on the Speed of Calmodulin Dependent Phosphorylation of Myosin in Solution. *Dokladi Academy of Science USSR*, 217: 227-231 (in Russian).

THE SENSITIVITY OF CELLS AND TISSUES TO WEAK ELECTROMAGNETIC FIELDS

A.A. Pilla, P.R. Nasser and J.J. Kaufman

Bioelectrochemistry Laboratory, Department of Orthopaedics, Mount Sinai School of Medicine, New York, NY 10029

INTRODUCTION

The question of whether or not weak environmental or therapeutic electromagnetic fields (EMF) can affect the behavior of living cells and tissues remains controversial. The biophysical community often maintains that basic physical principles cannot explain EMF bioeffects (Adair, 1991). Some clinicians, epidemiologists and biological scientists, on the other hand, are thoroughly convinced that there are real bioeffects caused by specific weak EMF (Savitz, 1988; Blackman, 1985). The physical argument against the possibility of an EMF bioeffect is usually based upon the ratio of the induced transmembrane voltage signal to the root mean square (RMS) thermal noise voltage. This signal to noise ratio (SNR) is usually calculated by assuming that the EMF target is a spherical cell of 10 μm radius. These calculations often lead to SNR<<1 for low frequency environmental EMF in the mG amplitude range. In this study, it is shown that, by evaluating SNR for connected cells in real tissues, sensitivity to exogenous EMF may be several orders of magnitude higher than previously thought.

CELL ARRAY IMPEDANCE MODEL

Organized tissue is developed and maintained, not solely by isolated 10 μm spherical cells, but also by an ensemble of complex geometry cells which have coordinated activity (Caveney, 1985). To this end, gap junctions provide pathways for ionic and molecular intercellular communication (Sheridan, 1985). They are present in all tissues including bone (Doty, 1981). These junctions increase the effective electrical "size" of cell arrays which changes EMF sensitivity, and therefore SNR. To calculate SNR for a cell array, a useful model is a distributed parameter linear electrical analog (transmission line) allowing the induced transmembrane voltage, V_M, to be evaluated as a function of frequency and position. This is similar to the electrophysiological models which have been proposed for current spread in electrotonically coupled tissues (Shiba, 1971) and the DC model proposed to account for tissue sensitivity to the weak electric currents commonly found in developing and regenerating tissues (Cooper, 1984). A first order electrical model for a linear cell array in gap junction contact is shown in figure 1. The propagation of V_M in this array is described by:

$$d^2V_M(x,\omega)/dx^2 = [(R_e + R_i)/Z_M(\omega)]\, V_M(x,\omega) \qquad (1)$$

where R_e and R_i are, respectively, the extracellular and intracellular resistances per unit length along the x axis and $Z_M(\omega)$ is the membrane impedance per unit length.

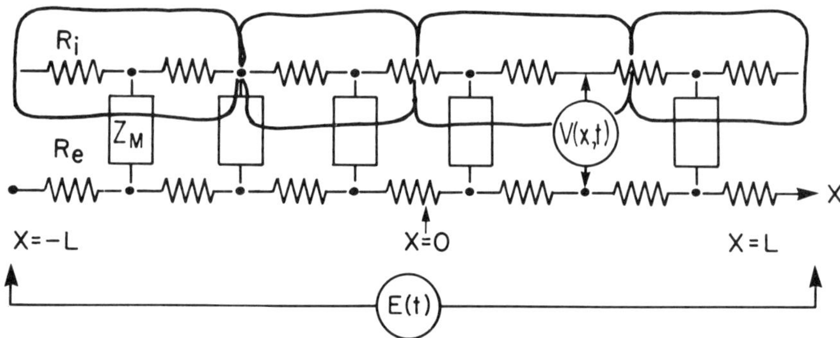

FIGURE 1. First order linear electric model for a cell array in gap junction contact. This is a distributed parameter system which can be used to describe the propagation of the applied signal, $E(t)$, along the array. The induced transmembrane potential, $V(x,t)$, is the quantity of interest. Its value at any position, x, and at any time, t, is determined by the extracellular, R_e, and intracellular, R_i, resistances and the particular membrane impedance, Z_M, per unit length.

Eq. 1 may be solved realizing that $(dV_M(x,\omega)/dx)_{x=\pm L} = -E(\omega)$, i.e. the electric field must be equal to the value of applied EMF across an array of length 2L, and $V_M(0,\omega)=0$. The result is:

$$V_M(x,\omega) = -[E(\omega)/\gamma]\,[\sinh(\gamma x)/\cosh(\gamma L)] \tag{2}$$

where $\gamma = [(R_e + R_i)\,Y_M(\omega)]^{1/2}$ and $Y_M(\omega)=1/Z_M(\omega)$.

The calculation of spatial amplification ($|V_M(x,\omega)/xE(\omega)|$) via Eq. 2 can be performed for specific parameters related to the cell array. A typical value for both R_e and R_i is $10^{10}\Omega/m$ since R_e and R_i would be expected to be of the same order for the cell volume percentage ($\approx 50\%$) in a typical tissue. The exact form of Z_M depends upon the model assumed. Three cases are shown in figure 2. The simplest and most classic case is a capacitance C_d in parallel with an ionic leak pathway, R_M. The admittance, Y_M, for this case is given by:

$$Y^I_M(\omega) = 1/R_M + C_d j\omega \tag{3}$$

Typical values for R_M range from 10^3 to $10^5\Omega$-m, and for C_d from 10^{-7} to 10^{-6} F/m.

The role of specific kinetics in the target pathway on EMF sensitivity can be examined for the ion binding model originally proposed in 1974 (Pilla, 1974). The linearized kinetics for this process are given by:

$$\Delta\Gamma(\omega) = [v/\Gamma j\omega]\,[-\Delta\Gamma(\omega) + aE(\omega)] \tag{4}$$

where $v=k_a-k_d$ ($k_a,k_d=$ adsorption, desorption rate constants respectively), Γ is the surface concentration of the adsorbing ion and a represents the potential dependence of adsorption ($\approx \partial\Gamma/\partial E=$const).

The membrane admittance per unit length now is:

$$Y^{II}_M(\omega) = Y^I_M(\omega) + 1/[R_A + 1/C_A j\omega] \tag{5}$$

where R_A is the equivalent adsorption (binding) resistance ($\propto 1/v$), ranging between 10 to 10^3 Ω-m), and C_A is the equivalent adsorption capacitance ($\propto \Gamma$), ranging from 10^{-6} to 10^{-5} F/m).

Ion binding can often regulate a follow-up biochemical reaction. The equivalent electric circuit for this is added in parallel across C_A, satisfying the model requirement that a change in surface concentration is necessary to trigger or modulate a biochemical process (Pilla, 1987). The membrane admittance is given by:

234

$$Y^{III}{}_M(\omega) = Y^{I}{}_M(\omega) + 1/\{R_A + 1/[C_A j\omega + 1/(R_B + 1/C_B j\omega)]\} \quad (6)$$

in which R_B is the equivalent resistance of the biochemical step (10^3 to 10^5 Ω-m) and C_B is its equivalent capacitance (10^{-5} to 10^{-3}F/m).

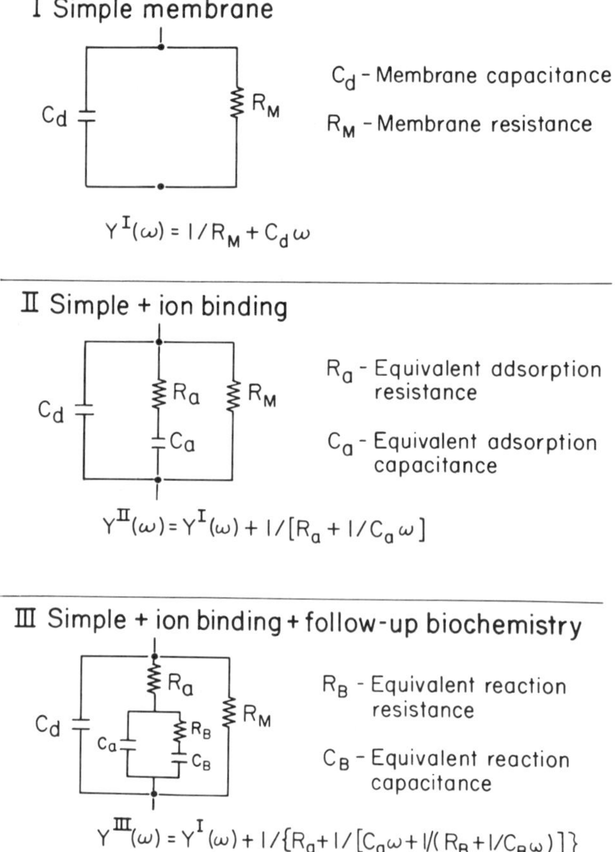

FIGURE 2. Electrical equivalent circuit models for membrane impedance. Case I accounts for the membrane capacitance, C_d, and transmembrane ionic leak pathway, R_M. Case II adds an ion binding pathway in parallel. R_a and C_a are the equivalent resistance and capacitance of adsorption respectively. Case III adds a follow-up biochemical pathway driven by a change in the surface concentration of the adsorbed ion, C_a. Any change in voltage across C_a can then affect current in the parallel biochemistry pathway, wherein R_B and C_B are its equivalent resistance and capacitance.

The effect of frequency and array length, L, on the spatial amplification of the transmembrane voltage ($|V_M/EL|$) is shown in figure 3. As can be observed, there is a substantial increase in $V_M(L,\omega)$ as L increases. The frequency response for a single cell ($L=10\mu m$) shows that V_M is maximum between 10^5 and 10^6 Hz. In contrast, for a 1mm cell array (e.g. Dipteran salivary gland) V_M is about 10^2 higher than for a single cell (for the same $E(t)$, see fig. 1), but only at frequencies below 100Hz.

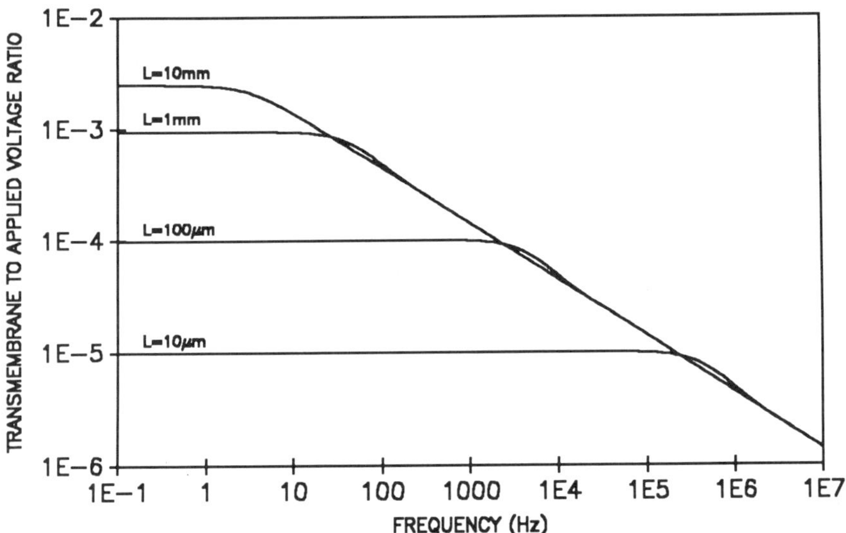

FIGURE 3. Frequency dependence of spatial amplification (V_M/EL) on the length of the cell array. A substantial increase is observed in the low frequency range, reflecting the increased propagation time as L becomes larger.

The use of other membrane models (cases II and III, fig. 2) results in more complex frequency behavior. A comparison of all three cases is shown in figure 4 for a 1mm cell array. As can be seen, successively lower frequencies are required for V_M to attain its maximum value. This means that if the EMF detector is an ion binding or a follow-up biochemical process, then the frequency band associated with either or both of these determines the frequency range over which maximum V_M is attained.

It is of interest to examine the effect of position along the cell array on V_M/EL. This is shown in figure 5 for $\omega=0$, which shows that maximum spatial amplification occurs at $x = \pm L$ (here L=2mm). The transmembrane voltage exhibits an approximate exponential dependence on x, being 0 at x=0, so that not all of the cell membranes in the array achieve maximum V_M. However, intercellular communication would be expected to allow all of the cells in the array to experience a similar bioeffect.

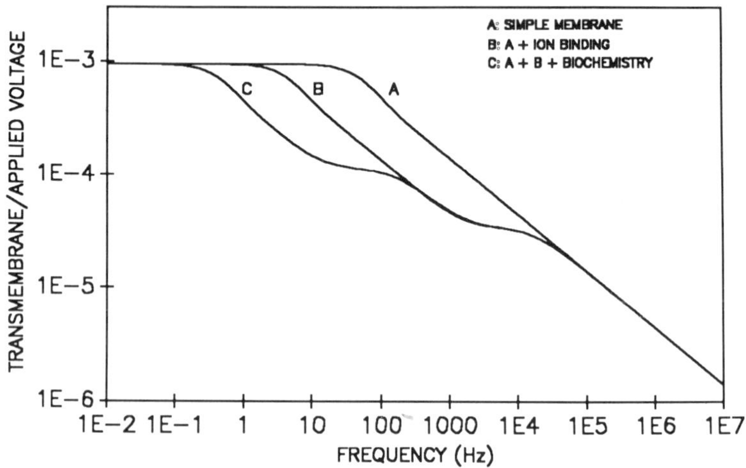

FIGURE 4. Effect of membrane impedance on the frequency response of spatial amplification. The required frequency range for maximum V_M/EL lowers as the kinetics of the membrane process becomes slower.

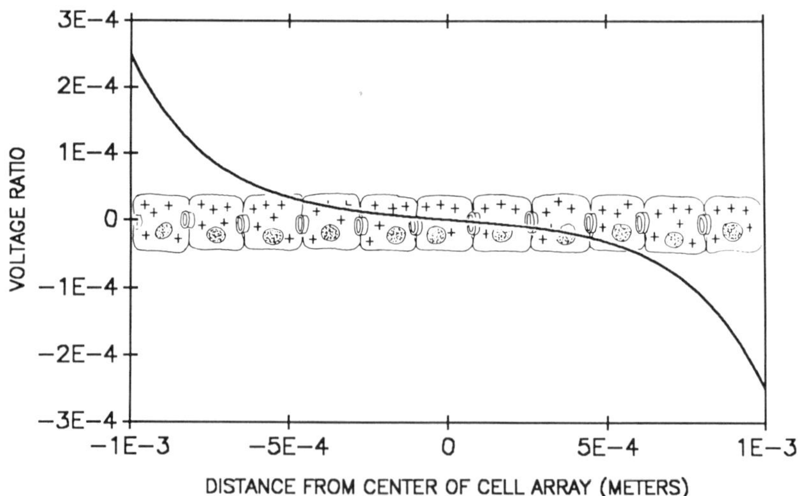

FIGURE 5. Variation of spatial amplification as a function of position along a 2mm cell array at ω=0. Although not all membranes experience the same V_M, cellular communication could allow an overall cellular response to be obtained.

SIGNAL TO NOISE CALCULATIONS

The noise sources in biological membranes are due to thermal, flicker (1/f), shot and conductance fluctuations (Stevens, 1972). The latter three usually relate to ion transport and their interpretation is model dependent. Thermal noise is present in all voltage dependent membrane processes and is the only noise source considered here for SNR calculations. The power spectral density, $S_n(\omega)$, of thermal noise is given by (DeFelice, 1981):

$$S_n(\omega) = 4kT\ Re[Z(x,\omega)] \tag{7}$$

where $Z(x,\omega)$ is the impedance of the cell array (fig. 1) and Re denotes its real part. $Z(x,\omega)$ is obtained by solving Eq. 1 and using the relation:

$$I(x,\omega) = -[1/(R_e + R_i)][dV(x,\omega)/dx] \tag{8}$$

which represents the propagation of current, $I(x,\omega)$, along the array.

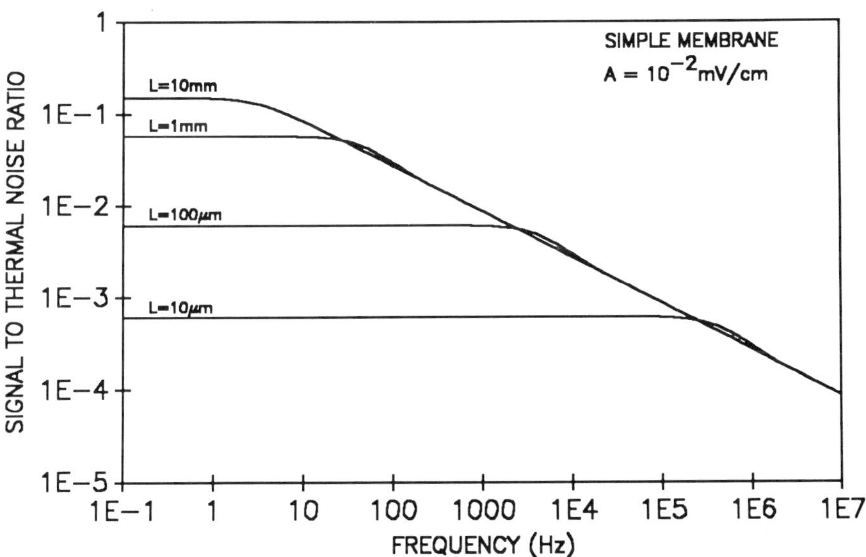

FIGURE 6. Frequency dependence of SNR on array length, L, integrating Eq. 7 to 10^7Hz. Maximum SNR is only achieved at low frequencies and only becomes useful for large L.

$Z(x,\omega)$ is obtained as:

$$Z(x,\omega) = [(R_e + R_i)/\gamma]\tanh\gamma x \tag{9}$$

This is to be contrasted with the normal procedure of employing the membrane impedance alone, neglecting the contribution from R_e and R_i which are electrically connected to the membrane (Adair, 1991; Weaver, 1990). As x becomes small enough (e.g. at the single cell limit) $Z(x,\omega)$ is given by (for the simple membrane):

$$Z(x,\omega) = 3(R_e + R_i)x/[3 + (R_e + R_i)(1/R_M + C_d j\omega)x^2] \tag{10}$$

for which the equivalent electric circuit looks like that for Case I (see fig. 2) except that R_e and R_i are in parallel with R_M, resulting in a new lower value for the transmembrane resistance. Note that when x becomes small enough for the cell membrane to be neglected, $Z(x,\omega)=(R_e + R_i)x$. This means that thermal noise across R_e and R_i could still be detected.

The most common approach to the evaluation of SNR uses the root mean square (RMS) noise voltage. This is calculated by taking the square root of the integration of Eq. 7 over all frequencies relevant to either the complete membrane response, or to the band width of

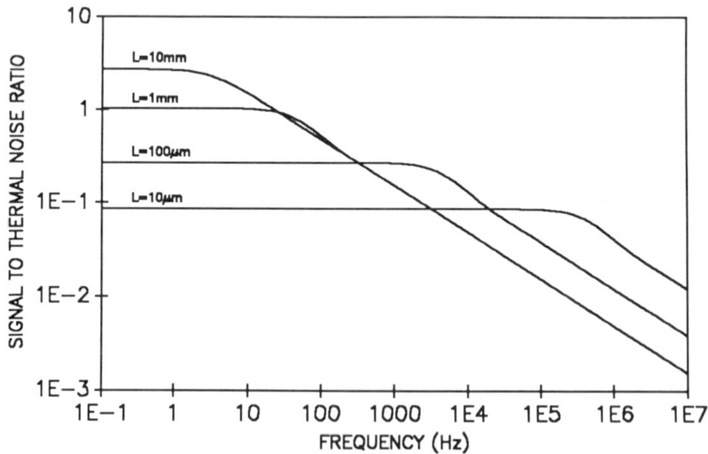

FIGURE 7. Frequency dependence of SNR on array length, integrating Eq. 7 to 10^2Hz. SNR for a single 10μm cell is still marginal, even for restricted noise and detector bandwidth.

the detector pathway. SNR is given by:

$$SNR = |V_M(\omega)| \, / \, RMS \tag{11}$$

Where $|V_M(\omega)|$ is the maximum amplitude of the transmembrane voltage at each sinusoidal frequency. In order to use Eq. 11 it is necessary to define the induced electric field $E(t)$. This quantity depends upon the EMF waveform, the geometry within which the cell array exists and its orientation with respect to the induced electric field lines (Pilla, 1983). Consider cylindrical geometry with the array placed at a radius, r = 2 cm. For $E(t)=A \sin\omega t$ where A is constant at all ω, the frequency characteristics of SNR vs. L are shown in figures 6 (Eq. 7 integrated to 10^7Hz) and 7 (integration to 10^2Hz) for the simple membrane model. For both cases, SNR is lowest in the highest frequency ranges. Restricting the noise bandpass to 100Hz significantly increases SNR at low frequencies, however it is still marginal for a $10\mu m$ cell. As L increases, SNR also increases as predicted by the model, but only at low frequencies.

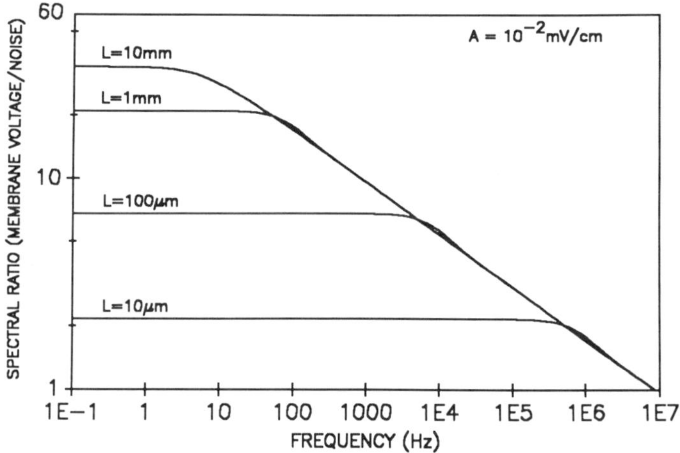

FIGURE 8. Frequency dependence of the spectral ratio of the induced transmembrane voltage and thermal noise voltage on array length. This reveals the frequency details which show that, at all frequencies (to 10^7Hz), $V_M(L,\omega)/S_n^{1/2} \geq 1$, even for a $10\mu m$ cell.

The preceding calculation of SNR does not allow the detail of the frequency response of both V_M and thermal noise to be considered. For example, noise components in the high frequency range may be irrelevant if the detector only has a bandpass of 10^2Hz. It is therefore of interest to examine the ratio of the frequency spectra of transmembrane voltage and thermal noise, $V_M(L,\omega)/[S_n(\omega)]^{\frac{1}{2}}$. This is shown in figure 8 for a sinusoidal input of constant amplitude (10^{-2} mV/cm) at each ω for the simple membrane model. As for SNR, the spectral ratio increases for larger L, but only at low frequencies. In the highest frequency ranges, increasing L provides no advantage. Of note is the fact that at <u>all</u> frequencies (to 10^7 Hz) the spectral ratio is >1, even for a 10μm cell.

DISCUSSION AND CONCLUSION

The results of this study indicate that significant spatial amplification occurs for cell arrays vs single cells, however both the propagation time and the particular membrane impedance pathway can considerably lower the frequency at which bioeffective SNR could be achieved. Calculation of SNR is most rigorously performed by considering all of the current pathways connected to the transmembrane point for which RMS noise is to be evaluated. In addition, the frequency response of the cell array appears to indicate that maximal spatial amplification may occur only over a specific frequency range. This reinforces the suggestion that tuning the input, $E(\omega)$, to the bandpass of the detector could lead to dose efficient and selective EMF bioeffect.

Much of the controversy concerning weak, non-thermal EMF bioeffects relates to the lack of reproducibility of *in vitro* cellular results. Part of this may be due to the restricted level of the induced electric field in culture plates compared with that in whole tissue. This is merely a scale-up problem, caused by the differing size and geometries of *in vitro* and *in vivo* EMF targets. The result is whole body, limb, or organ exposure to 1-10 mG, 60 Hz EMF may result in electric field levels equivalent to those obtained at 1 G in a culture dish.

In addition to purely geometric scale-up, this study suggests that spatial amplification of cells in gap junction contact vs that for isolated cells must be taken into account when assessing EMF sensitivity. While cell networks naturally occur in developing and repairing tissue, it is not clear that the formation of gap junctions between cells *in vitro* is sufficiently reproducible. This renders the problem of duplicating results *in vitro* exceedingly difficult, but does not change the increased EMF sensitivity gap junctions may provide.

Clearly, weak environmental EMF signals are often within sufficiently low frequency ranges for adequate SNR to be obtained in cell arrays of physiologically relevant size. All of the above lead to the conclusion that bioeffects, particularly in organized tissue, are indeed possible from exposure to remarkably low levels of electromagnetic fields.

ACKNOWLEDGEMENT

This work was partially funded by American Medical Electronics, Richardson, TX.

REFERENCES

Adair RK (1991): Constraints on biological effects of weak extremely low frequency electromagnetic fields. *Phys Rev A* 43: 1038-1049.

Blackman CF, et al. (1985): A role for the magnetic field in the radiation induced efflux of calcium ions from brain tissue *in vitro*. *Bioelectromagnetics* 6: 327-337.

Caveney S (1985): The role of gap junctions in development. *Ann Rev Physiol* 47: 319-355.

Cooper MS (1984): Gap junctions increase the sensitivity of tissue cells to exogenous electric fields. *J Theor Biol* 111: 131-148.

DeFelice LJ (1981): *Introduction to Membrane Noise*, New York: Plenum, 243-245.

Doty SB (1981): Morphological evidence of gap junctions between bone cells. *Calcif Tissue Intl* 33: 509-512.

Pilla AA (1974): Electrochemical information transfer at living cell membranes. *Ann NY Acad Sci* 238: 149-170.

Pilla AA, et al. (1987): Electrochemical kinetics at the cell membrane: A physicochemical link for electromagnetic bioeffects. In: *Mechanistic Approaches to Interactions of Electric and Electromagnetic Fields with Living Systems*, Blank, M and Findl, E, eds. New York: Plenum, 39-62.

Pilla AA, et al. (1983): Electrochemical and electrical aspects of low frequency electromagnetic current induction in biological systems. *J Biol Phys* 11: 51-58.

Savitz DK, et al. (1988): Case control study of childhood cancer and exposure to 60 Hertz magnetic fields. *Am J Epidemiol* 128: 21-38.

Sheridan JD et al. (1985): Physiological roles of permeable junctions: some possibilities. *Ann Rev Physiol* 47: 337-353.

Shiba H (1971): Heaviside's "Bessel Cable" as an electric model for flat simple epithelial cells with low resistive junctional membranes. *J Theor Biol* 30: 59-68.

Stevens CF (1972): Inferences about membrane properties from electrical noise measurements. *Biophys J* 12: 1028-1047.

Weaver JC, Astumian RD (1990): The response of living cells to very weak electric fields: the thermal noise limit. *Science 247: 459-461.*

SUCCESSFUL GENE TRANSFER IN PLANTS USING ELECTROPORATION AND ELECTROFUSION

James A. Saunders, Sally L. Van Wert, Camelia Rhodes Smith, Benjamin F. Matthews, and Stephen Sinden

Plant Sciences Institute
Beltsville Agricultural Research Center
USDA, ARS
Beltsville, MD 20705

ABSTRACT

Electroporation and electrofusion of plant tissues are rapid, reliable, techniques for genome modification. To accomplish these procedures two different wave forms have been developed and successfully used in plants during the last decade. We have shown that both square and exponential wave pulses are capable of incorporating genetic material into tobacco protoplasts. The production of an insect resistant potato hybrid, verified by phenotypic and genetic traits, has demonstrated the utility of electrofusion by the combination of two separate genomes. Specific genomic modifications are also possible as shown by the electroporation of marker genes into germinating tobacco pollen. The impact of these continuing investigations into the biological effects of electrical fields on plant cells, will enhance our ability to genetically manipulate biological systems.

INTRODUCTION

The use of brief high voltage pulses to affect modifications of plant cell membranes has become increasingly important in the field of biotechnology. Through these electrical membrane perturbations, plant protoplasts can be fused or foreign DNA can be inserted into plant cells, resulting in permanent modifications in the genome of the plant. These genomic modifications can be made with relatively simple, rapid procedures. These procedures exploit the ability of the plant cell membrane to respond to high voltage electrical pulses by reversibly forming areas of membrane instability. It is through

these unstable cell membrane regions that foreign DNA can be
introduced or cells can be fused providing for the effective use
of electroporation and electrofusion in gene transfer
experiments. This paper will describe several examples in which
electrofusion and electroporation have been used with plant
cells to demonstrate both the utility and importance of
electrical manipulations of plants.

ELECTROFUSION IN PLANTS

Plant cells have a unique role in somatic fusion experiments
by virtue of their ability to regenerate a mature organism from
a single cell. This characteristic of plant cells distinguishes
them from their animal cell counter-parts by offering the
researcher an opportunity to produce genetically altered stable
hybrids that can be vegetatively propagated. Several important
crop plants are amenable to this type of transformation
procedure and one of these, the potato, has yielded valuable
information on the potential use of somatic cell electrofusion
for crop development.

The development of protoplast fusion technology has aided in
the introduction of desirable breeding traits, such as disease
resistance from wild *Solanum* species, into the domestic
potato *(Solanum tuberosum)*, an important food crop. Wild
Solanum species are important sources for the improvement of
horticultural characteristics in the development of potato
cultivars, however, some potentially useful species are sexually
incompatible or difficult to cross. Agriculturally important
genetic traits that have been transferred from wild species of
potato by means of protoplast fusion include: 1) potato virus X
resistance, from *S. chacoense* (Butenko *et al.*, 1982);
2) atrazine resistance, from *S. nigrum* (Binding *et al.*,
1982); 3) potato leaf roll virus resistance, from *S.
brevidens* (Barsby *et al.*, 1984; Austin *et al.*, 1985;
Gibson *et al.*, 1988); 4) late blight resistance, from *S.
chacoense* (Butenko *et al.*, 1980) and from *S. brevidens*
(Helgeson *et al.*, 1988); 5) potato virus Y resistance,
from *S. brevidens* (Gibson *et al.*, 1988); and 6) male
sterility, from *S. berthaultii* (Perl *et al.*, 1990).
Some of these somatic hybrids and cybrids are fertile and may be
useful in potato breeding.

In addition to somatic hybrids and cybrids that are
potentially useful either directly as new cultivars or in potato
breeding, protoplast fusions using potatoes have produced a

number of genetically novel plants. In early pioneering protoplast fusion research Melchers *et al.* (1978) produced intergeneric somatic hybrids between a dihaploid potato *(S. tuberosum)* (2n = 2X = 24) and tomato *(Lycopersicon esculentum)* (2n = 2X =24), perhaps with the hope of producing french fries with ketchup. Although, some of the hybrids formed small tubers, none were fertile. Apparently the hybrids did not have the doubled chromosome number of a true amphitetraploid (2n = 4X = 48), which might be expected for fertile plants.

Male sterile cybrid potatoes that could serve as parents in the production of hybrid, true-potato seed have been obtained by the donor-recipient protoplast fusion procedure using irradiated donor protoplasts from a wild species, *S. berthaultii* (Perl *et al.*, 1990). These male-sterile cybrids have the nuclear genome of the potato cultivar Desiree and the plastomes of the wild species donor.

Protoplast fusion should also be useful in maintaining heterozygosity in modern potato breeding programs that use dihaploids (2n = 2X = 24) of *S. tuberosum* for more efficient selection and combination of agronomic traits (Wenzel *et al.*, 1979; Deimling *et al.*, 1988). The cultivated potato is tetraploid (2n = 4X = 48), heterozygotic, and genetically complex. Wenzel *et al.* (1979, 1982) proposed using anther derived or parthenogenically produced dihaploid and monoploid (2n = X = 12) lines for more efficient potato breeding, however, monoploid and dihaploid potato lines lack vigor. Doubling of dihaploid breeding clones with colchicine, or by other means, to produce tetraploid breeding lines and cultivars, does not maximize heterozygosity. Somatic hybridization of two desirable dihaploid breeding clones should result in both maximum heterozygosity by the complete combination of the two parental genomes. Deimling *et al.* (1988) demonstrated that somatic hybrids between dihaploid *S. tuberosum* parents can be efficiently obtained by mass fusion. Hybrids could be indentified by isozyme analysis without using a selection system or selectable markers.

Until recently, most potato somatic hybrids were obtained with chemical fusogens. Electrofusion techniques developed for tobacco protoplast fusion (Zimmerman and Scheurich, 1981; Bates, 1985) have been adapted for potato protoplasts. Early applications of electrical fusion to food crops demonstrated the somatic hybridization of potato and *S. phureja* (Puite *et al.*, 1986). Following developmental work on the technique of electrofusion with potato protoplasts (Tempelaar and Jones, 1985), somatic hybrids of potato and *S. brevidens* that have resistance to both potato virus Y and potato leaf roll virus were produced using electrofusion (Fish *et al.*, 1988; Gibson *et al.*, 1988). Our laboratory has produced a somatic electrofusion hybrid between *S. tuberosum* and *S. chaconse*. The hybrid contains the insect resistant alkaloid profiles of the wild potato and morphological traits of the domestic potato. Due to the potentially higher fusion efficiency, and higher protoplast survival rate in electrofusion experiments, electrofusion will likely replace chemically-induced fusion as the preferred technique for obtaining potato somatic hybrids. This is particularly true when large numbers of somatic hybrids are needed, as in potato breeding programs.

ELECTROPORATION IN PLANTS

Electroporation with plants cells has usually been applied to isolated protoplasts in which the cell wall has been removed by hemicellulase/cellulase digestion. The removal of the cell wall facilitates the uptake of DNA into the cell, however, subsequent cell division and regeneration of differentiated tissue requires that the cell wall be reformed. Although the cell wall of the plant cell is generally regarded as an imposing barrier, there are reports of electroporation of DNA through the cell wall into tissue sections (Dekeyser *et al.*, 1990). Our laboratory has also shown that DNA can be electroporated into plant cells, other than protoplasts, by using germinating pollen as the recipient tissue (Abdul-Baki et al., 1990; Matthews et al., 1990; see Pollen Electrotransformation discussion).

Evidence for the success of transformation after electroporation has been measured by radioactive labelling of DNA (Tsong and Kinosita, 1985), transient gene expression (Potter *et al.*, 1984; Smithies *et al.*, 1985), and the formation of stable transformants (Riggs and Bates, 1986; Stopper *et al.*, 1985). Successful and efficient introduction of DNA or RNA into cells by electroporation depend on several important variables. These parameters include the pulse wave shape, the pulse field strength, the pulse duration, resealing time of the pores induced in the cell membrane, the cell and nucleic acid concentrations in the electroporation medium, and the conditions under which the experiments are performed (Saunders *et al.*, 1989b). We have examined several of these parameters in protoplasts isolated from tobacco by following the uptake and expression of viral RNA as a model system. Particular attention was paid to the effect of the pulse wave shape on electroporation efficiency. The viability, number of protoplasts and, consequently, the efficiency of viral RNA uptake can be affected by the electroporation process. Therefore the amplitude, duration and shape of the electric pulse used to introduce the foreign genes into protoplasts should be optimized to create the best conditions for the maximum number of live protoplasts capable of receiving foreign genetic material.

There are two types of DC high voltage pulse wave forms, square and exponential. Each of these were examined to determine the optimum conditions each offered for the efficient introduction of viral RNA into tobacco protoplasts. Typically short pulses of less than 100 μsec are produced using the square wave generator while the exponential wave pulses are in the msec range. A constant pulse is obtained at varied field strengths with the square wave. At different voltages the shape of the exponential pulse wave may differ slightly due to the nature of the capacitors used in the pulse discharge generators and the change in the specific resistance of the electrolyte chamber. These factors add to the difficulty of measuring the duration of the exponential wave form. A number of conventions have been used to measure the duration of the exponential wave form, however, we have found that a convenient technique is to measure the time required to reach one half the peak voltage following the pulse.

The same field strength for both wave forms show a different effect on the protoplast survival number, viability, and RNA uptake and expression. At low field strength, the square wave pulse has a smaller effect on the viability of protoplasts than the exponental wave (Saunders *et al.*, Figure 1). RNA uptake and expression, using the square wave pulse, was less effective than the exponential wave for maximum yield of expression in viabile protoplasts. There is a narrow optimum field strength range, between 0.5 and 0.9 kV/cm, when the exponential wave form is used. Drastic decreases in protoplast viability occur above of this range. The square wave appears to be the more useful wave form over a broad field strength range, 2.0 and 4.0 kV/cm. This creates the opportunity to recover more viable protoplasts that have taken up RNA.

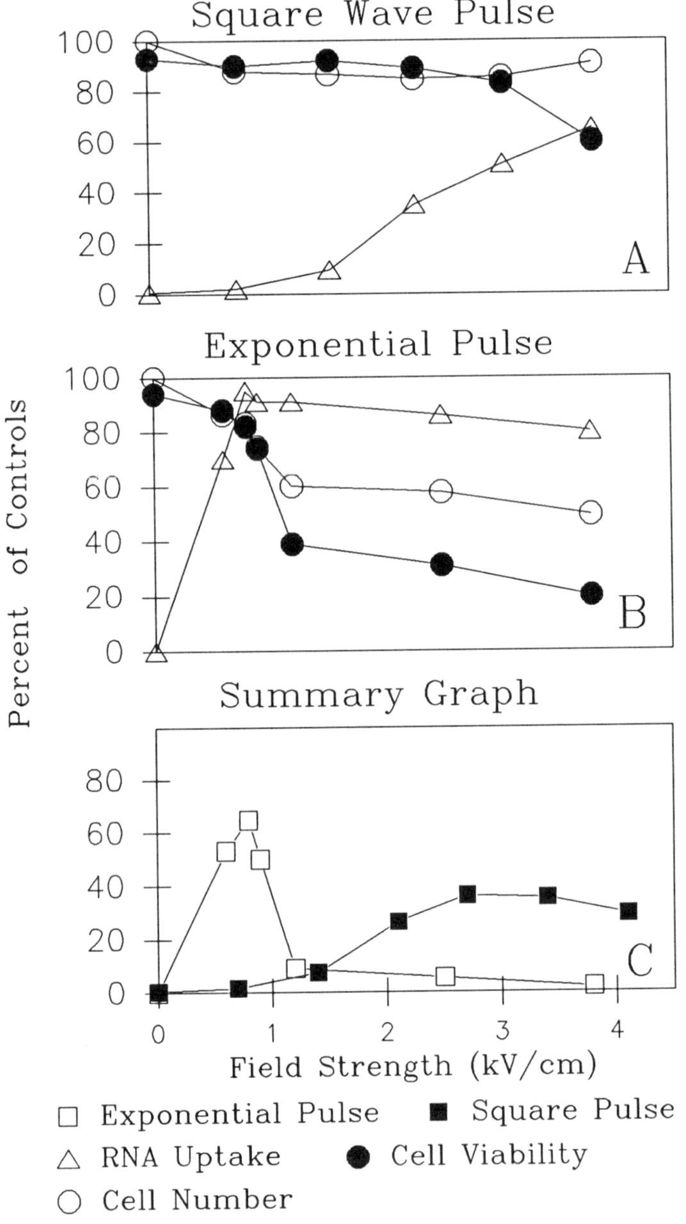

Figure 1. A) The effect of a square wave pulse on the RNA uptake and expression, protoplast viability, and number of protoplasts that survived the pulse. B) The effect of an exponential pulse on RNA uptake and expression, protoplasts viability, and the number of protoplasts that survived the pulse. C) A summary graph of A and B that shows the total number of surviving, viable, protoplasts that took up and expressed the electroporated RNA.

The pulse duration of the exponential wave form, when recorded in terms of $t_{1/2}$ or the interval of time it takes to decay to one half of the peak voltage after the pulse, shows that steady decreases in viability occur at $t_{1/2}$ values longer than 8 msec. The uptake of RNA when compared to the control protoplasts was over 80% at a pulse duration of 22 msec. Square wave pulses ranging from 20-100 μsec produced a gradual decline in the number of protoplasts due to pulse induced rupture. At the highest pulse duration 80% of the protoplasts remained when compared to the control protoplast's number and the viability of these protoplasts did not drop below 90% when compared to the controls. Following a 40 μsec pulse duration there was a steady increase in RNA uptake, although not as much as that seen after the exponential pulse. Overall the increase in square wave pulse duration will leave more viable protoplasts capable of taking up RNA.

When comparing the square wave and exponential wave pulses it can be seen that they both will introduce foreign genetic material into protoplasts in a rapid, efficient manner, if time is taken to optimize the electroporation parameters. However, in order to create a situation of reversible membrane breakdown the square wave form offers more acceptable parameters which include a broader working range with shorter pulse durations that yield more viable protoplasts. For experiments where the degree of introduction of foreign material is more critical than the number of viable protoplasts the exponential wave form will be a better option. Although an apparatus that delivers a square wave is more expensive, the duration of a pulse can be easily determined based on the beginning and ending of the pulse voltage and the pulse duration can be changed by simply turning a calibration dial. The quick surge accompanied by a exponential decline, due to external and internal conditions found in the exponential wave form pulse, makes measuring the absolute termination of the pulse duration difficult. The duration of the exponential pulse can be adjusted by changing the capacitance.

POLLEN ELECTROTRANSFORMATION

The genetic engineering of crop plants promises to increase plant productivity and the quality of the plant product. One method for gene transfer with great potential, pollen electrotransformation, utilizes pollen as the transformation vehicle. In this system, pollen is allowed to germinate so that the pollen tube begins to extrude from the thick-walled pollen grain. Foreign DNA is taken up through the pollen tube membrane by electroporation and this pollen is used to pollinate emasculated flowers. Plants grown from the resultant seed are screened to identify transformants. Pollen transformation is limited only when the plant pollen is ephemeral and/or proper germination conditions have not been met.

The advantages of pollen transformation are numerous. No protoplast regeneration, lengthy tissue culture techniques, or infection by host limiting vectors, such as *Agrobacterium tumefaciens*, are required. Since plants are obtained directly from seed, it is less likely that vigor and fertility will be compromised as compared to transformed plants obtained through protoplast regeneration. The risk of tissue culture induced variation is limited. Importantly, gametic transformation is obtained instead of chimeric transformation; the latter occurs with protoplast, *Agrobacterium*, or particle gun transformations. Thus, pollen electrotransformation offers a convenient, economical procedure for rapidly producing genetically engineered plants and can be successfully applied to virtually any crop plant without the prerequisite of detailed tissue culture studies.

The success of the electroporation method in pollen depends, in part, on optimizing the same parameters relative to the membrane, the DNA, and the electric field as discussed for protoplast electroporation. The idea to use pollen to effect genetic modification of subsequent progeny has been suggested in the literature for some time and the term pollen transformation was coined in the 1970's (Hess, 1987). Several researchers have reported that DNA, when added to pollen in either a solution or a paste, can be taken up (Hess, 1987) and expressed in progeny (De Wet *et al.*, 1985; Ohta, 1986). Pandey (1978, 1980) provided evidence for pollen transformation by using denucleated pollen as a DNA vector. None of these reports included a mechanism for introducing the DNA into the pollen nor did they provide conclusive molecular proof of gene transfer. Sanford *et al.* (1985) and others pointed out the lack of molecular evidence and could not repeat the results of these early studies. Additionally, Matousek and Tupy (1985) and Roeckel *et al.*, (1988) described the release of DNA nucleases from germinating pollen grains and implied that foreign DNA, present in a mixture of germinating pollen, would be rapidly degraded. These negative reports have hindered efforts to transform pollen as no mechanism was identified to rapidly incorporate foreign

DNA into the pollen grain prior to its degradation by nuclease.

Despite the negative reports on pollen transformation we are using pollen as a transformation vector after introducing DNA into germinating pollen by electroporation. The use of germinating pollen as a method of gene transfer coupled with electroporation was implicated by Mishra *et al.*, (1987), who observed the release of fluorescent dyes from electroporated pollen and concluded that the induced membrane permeability brought about by electroporation may allow the uptake of macromolecules into viable pollen. The pollen tube has a less substantial cell wall than the pollen grain or a plant cell (Cass and Peteya, 1979; Picton and Steer, 1982). The lack of cell wall on the pollen tube presents a situation analogous to a plant protoplast, so that foreign DNA can be directly introduced into the gametic cell by electroporation. The genetic manipulation of plant germplasm by introducing DNA into the pollen is noteworthy since pollen is the natural carrier of DNA to the ovary in flower fertilization. DNA transport by pollen can occur within the same flower, between flowers on the same plant or on different plants, and expression of the introduced DNA in the seed can be easily detected when a stably expressed marker gene is used.

In our work with pollen we have optimized pollen storage, pollen germination and pollen electroporation conditions with insignificant loss of pollen viability for tobacco *(Nicotiana gossei* L. Domin) (Abdul-Baki *et al.*, 1990; Saunders *et al.*, 1989a), sweet corn *(Zea mays,* hybrid golden queen) and alfalfa *(Medicago sativa,* Saranac). The field strengths necessary for uptake of DNA into tobacco pollen (8-9 Kvolts/cm) are much higher than optimal conditions for protoplast electroporation and are attributed to the presence of a cell wall on the pollen grain as well as the higher conductivity of the electroporation medium used for the germinating pollen (Saunders *et al.*, 1989b). The induction of pores through electroporation is directly related to the diameter of the cell used (Bates *et al.*, 1987). Thus, larger cells reach the critical voltage for pore formation at a lower field strength than smaller cells. It should be pointed out that tobacco pollen grains are larger than most animal cells and comparable to plant protoplasts, yet they require a higher field strength and a longer pulse duration that mammalian cells or plant protoplasts for optimum electroporation. Higher rates of DNA incorporation into germinating pollen, as measured by uptake of labelled DNA, were attained when the electroporated germinating tobacco pollen grains were allowed to remain undisturbed in the electroporation medium containing labelled DNA for 45 min after electroporation (Abdul-Baki *et al.*, 1990). This indicated that the time required to reseal induced pores formed by electroporation is an important variable. Uptake of labelled DNA by electroporated pollen was also dependent upon pollen concentration, reaching a maximum of DNA uptake at 4 mg of pollen/300 μl of electroporation medium (Abdul-Baki *et al.*, 1990). The decline in DNA uptake at higher pollen concentrations demonstrates that the DNA is indeed incorporated into the pollen and not adsorbed to the grain surface. In fact, extensive washing of the pollen with DNase for 2.5 h did not remove the DNA from the germinating pollen.

We have investigated the effect of nucleases released by germinating pollen on the integrity of added foreign DNA. Tobacco, corn and alfalfa pollen released nucleases which significantly degraded the plasmid when it was added to germinated pollen suspensions for 0 to 60 min before precipitation of the plasmid from the germination medium. Agarose gel electrophoresis of the DNA's showed an increase in degradation with time, but little difference in degradation when the pollen/DNA suspensions were electroporated using optimal conditions. Addition of EDTA, a nuclease inhibitor, to the germinated pollen before the addition of plasmid DNA, reduced nuclease activity. Exchanging the germination medium with fresh germination medium prior to addition of DNA and electroporation yielded the least degradation. The conclusion from these experiments is that nucleases released from the germinating pollen can be inhibited by EDTA or removed by washing and, thus, we can expect the transforming DNA to be intact during electroporation.

We have used transient expression assays to determine whether or not introduced DNA can be detected and expressed within the pollen grain. The plasmids pBI221 (Figure 2A) and pLAT52-7 (Twell *et al.*, 1989; Figure 2B) were transferred into germinating tobacco pollen using optimal electroporation conditions. Both plasmids contain the *uidA* coding region, which encodes the *E. coli* enzyme ß-glucuronidase (GUS) as described by Jefferson (1985). In pLAT52-7 expression is driven by a pollen-specific promoter from tomato; in pBI221 expression is driven by the CaMV35S promoter. Successful transfer of these plasmids can be monitored by assaying GUS activity in pollen histochemically and fluorometrically (Jefferson *et al.*, 1987) and by Southern blot analysis (Southern, 1975) of DNA extracted from pollen.

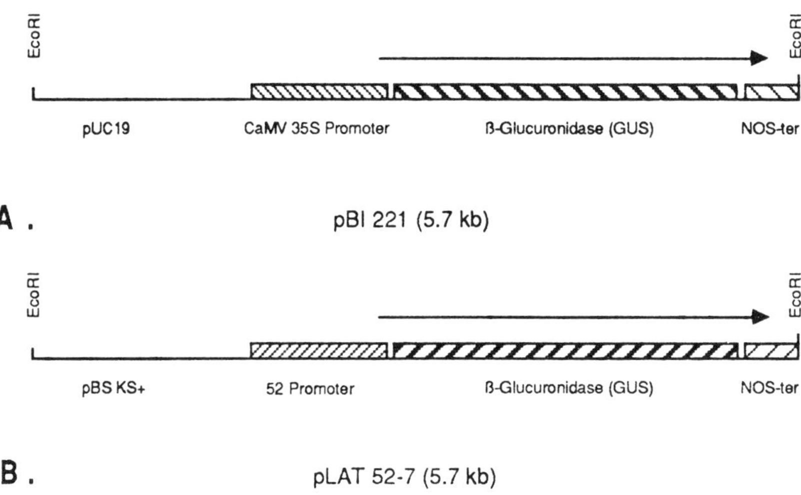

A .

B .

Figure 2A and 2B: The plasmid pBI221 and pLAT 52-7 were electroporated into germinating tobacco pollen. Both plasmids contain the *uidA* coding region for ß-glucuronidase (GUS).

For the histochemical assay 5-bromo-4-chloro-3-indoyl glucuronide was added to the tobacco pollen/DNA suspension 1 h after electroporation. Incubation was continued at 30 C for up to 24 h. Blue color indicative of a positive transformation was visible in pollen/pLAT52-7 suspensions three hours after electroporation. In controls, no color development was seen until 24 hr after the treatment. At 24 hr the pollen/pLAT52-7 suspensions were very dark blue while the pollen/pBI221 suspensions were darker than the controls but not as dark as those containing pLAT52-7 carrying the pollen-specific promoter. The difference in expression in the pollen/pLAT52-7 and pollen/pBI221 suspensions suggest that the pollen specific promoter is much more effective in pollen than the CaMV35S promoter.

Transient GUS activity was also demonstrated using the fluorometric assay (Jefferson *et al.*, 1987) with pollen incubate for 24 h after electroporation in the germination medium. Using plasmid pBI221, increases in GUS activity as high as 60-fold have been detected fluorometrically in electroporated pollen compared to controls (Matthews *et al.,* 1990). This indicated that pBI221 was taken up by the pollen and transiently expressed.

Further evidence for the uptake of plasmid DNA by pollen was provided by Southern hybridization analysis of DNA extracted from pollen 1 h and 24 h after electroporation in the presence of pBI221 (Matthews *et al.,* 1990). The pollen was treated with DNase and extensively washed before analysis. EcoRI digested DNA from pollen treatments and controls were compared by agarose gel electrophoresis. The plasmid DNA appeared as a distinct band of DNA the same size as pBI221 (5.7Kb) in extracts 1 h after electroporation but not in control extracts or extracts 24 h after electroporation. A Southern blot of this gel hybridized with ^{32}P-labeled pBI221 showed no hybridization to control pollen DNA. Some of the hybridizing DNA was of high molecular weight, suggesting that pBI221 had integrated into the pollen genome by 24 h after electroporation. These data conclusively show that pBI221 was taken up by pollen via electroporation, remained intact, provided transient expression of GUS activity and suggested the incorporation of the plasmid into the pollen genome.

The ultimate goal of this research is to produce viable seed with a genetically modified genome. Although DNA can be incorporated into pollen and expressed in pollen, the next question is whether electroporated and genetically modified pollen can fertilize receptive emasculated flowers to produce viable, genetically modified seed. We can obtain viable tobacco (Abdul-Baki *et al.,* 1990), corn and alfalfa seed by pollinating the appropriate emasculated flowers with electroporated, germinating pollen (Table I). These seeds were shown to be more than 90% viable when germinated on moistened filter paper. We have not optimized the pollination conditions, except to determine the number of days following emasculation which lead to the greatest production of viable seed after application of electroporated pollen to emasculated flowers. For tobacco, the largest number of viable seeds were obtained by pollinating 4 days after emasculation (Abdul-Baki *et al.,* 1990); for alfalfa the seed number was equivalent when pollination occurred 0 to 2 days after emasculation.

TABLE I. Viable seed production from flowers pollinated with germinated and electroporated pollen grains.

VIABLE SEED PRODUCED PER FLOWER

	Tobacco	Corn	Alfalfa
POLLEN TREATMENT			
Not-electroporated	332	159	4
Electroporated	214	72	2

We have analyzed tobacco plants from seed derived from flowers fertilized with pollen electroporated with pBI221. Leaves from some of these plants expressed higher amounts of ß-glucuronidase activity than the control. Southern blot analysis confirmed the presence of the plasmid in leaf DNA extracts from a plant with high GUS activity. R1 plants, grown from seed obtained by selfing the DNA and GUS positive plants, will be examined to determine the stability and inheritance pattern of foreign DNA. Currently we are using other plasmids in our pollen electrotransformation studies. These plasmids contain other selectable markers which make it easier and more efficient to screen tobacco, corn and alfalfa seed and identify transformants.

APPLICATIONS OF POLLEN TRANSFORMATION

There are numerous applications for a reliable, efficient pollen transformation system, when developed. Current transformation procedures require regeneration of fertile plants from protoplasts or use of cell culture techniques. Protoplast transformation systems using electroporation or chemical transformation techniques have been generally quite successful (Shillito *et al.,* 1985; Lin *et al.,* 1987; Schocher *et al.,* 1986). Certainly, transient expression of the inserted gene can be detected in the protoplasts in many cases. However, the long tedious process of plant regeneration from protoplasts makes this an arduous method for obtaining transformed plants and is not routine for several economically important crop plants, such as corn, wheat, rice, or soybean. Recently, soybean protoplasts were regenerated into plants (Wei and Xu, 1988) Lin *et al.*, (1987) demonstrated transformation of soybean protoplasts using a polyethylene glycol-electroporation technique. However, regeneration of fertile soybean plants from protoplasts is difficult, time consuming and the method is not yet routine.

The use of *Agrobacterium tumefaciens*, a natural plant pathogen, as a transformation tool has been quite successful (Schell, 1987; Jones, 1985; Draper *et al.*, 1988). Disarmed, non-pathogenic DNA constructs, containing portions of the tumor inducing plasmid, have been successfully used to transform a variety of cell and tissue types of a number of plant species. It does not readily infect some of the major crop plants, but recently has been used to transform soybean (Byrne *et al.*, 1987; Parrott, *et al.* 1989). This method, although promising, is still under development, especially for the major cereal crops and soybean. As with protoplast transformation *Agrobacterium* transformation requires cell or organ culture, and therefore, long periods of time are necessary for plant regeneration.

A new method, which should be generally applicable to the transformation of cells from all plant species, uses particle bombardment. Microspheres are coated with DNA and are shot into target plant tissues using a biolistic gun. The development of the biolistic gun to bombard and transform plant cells has lead to a new round of excitement for transformation of major crop plants. With the use of the biolistic gun (Christou *et al.*,1989; McCabe *et al.*, 1988; Christou, 1990), embryogenic soybean cells can now be transformed and regenerated. Fertile plants can be produced and the length of time required for tissue culture is reduced as compared to transformation of protoplasts. However, chimeric plants are usually produced, from the regeneration of plants from adjacent transformed and non-transformed cells. The state of transformation of rice, corn and wheat is of a similar nature, wherein there is still no routine, reliable method to consistently transform these plants and obtain fertile offspring.

Pollen transformation could also be useful for screening to identify genes encoding important agronomic traits. With current plant genome mapping techniques (Bonierbale *et al.*, 1988; Nam *et al.*, 1989; Helentjaris *et al.*, 1986) molecular markers can be identified which are often within a megabase or two of a locus encoding an important agronomic trait. Thus, important traits can be localized according to their map position. Identification of the actual DNA sequence encoding the trait, however, is extremely difficult because several genes may reside within the delineated region. Pollen transformation could become an important tool in conjunction with current mapping techniques for identifying genes mapped within a circumscribed region of the chromosome. A prerequisite for this procedure would be the efficient transformation of pollen with very large DNA pieces, which is unproven at this time.

Certainly a pollen transformation system would be useful for inserting gene construct encoding disease resistance, insect resistance, herbicide tolerance and resistances to other biotic and abiotic stresses into sexually reproducing plant species. Pollen transformation would eliminate labor-intensive protoplast and tissue culture procedures, allowing larger numbers of DNA constructs to be used in transformation studies. Large numbers of novel DNA constructs could be screened to identify functions of DNA sequences and understand structure-function relationships. Since maintenance of tissue cultures would be eliminated, large numbers of transformants can be obtained and maintained as seed. This would allow transformants to be stored at low cost for a long time before being examined or mailed to other laboratories for examination. The investigator would not be under as stringent labor and cost constraints when doing plant transformations, hence could work with greater efficiency.

Consistent, reliable pollen transformation would provide a very simple and rapid procedure for transformation. Little equipment and training would be needed. If a plant breeder had the desired DNA constructs in his laboratory, he could transform plants at the benchtop or in the greenhouse at low cost and with little labor. This method would also be desirable for Third World countries having a shortage of money and sophisticated cell culture equipment.

REFERENCES

Abdul-Baki AA, Saunders JA, Matthews BF, and Pittarelli GW (1990): DNA uptake during electroporation of germinating pollen grains. *Plant Science* 70:181-190.

Austin S, Baer MA, and Helgeson JP (1985): Transfer of resistance to potato leaf roll virus from *Solanum brevidens* into *Solanum tuberosum* by somatic fusion. *Plant Science* 39:75-82.

Barsby L, Shepard JF, Kemble RJ, and Wong R (1984): Somatic hybridization in the genus *Solanum: S. tuberosum* and *S. brevidens*. *Plant Cell Rep* 3:165-167.

Bates, GW (1985): Electrical fusion for the optimal formation of protoplast heterokaryons in *Nicotiana*. *Planta* 165: 217-224.

Bates GW, Saunders JA and Sowers AE (1987): Electrofusion: Principles and Applications. In: *Cell Fusion*, Sowers AE, ed. New York: Plenum Press.

Binding H, Jain SM, Finger J, Mordhorst G, Nehls R, and Gressel J (1982): Somatic hybridization of an atrazine-resistant biotype of *Solanum nigrum* with *Solanum tuberosum*. *Theor Appl Genet* 63:273-277.

Bonierbale MW, Plaisted RL and Tanksley SD (1988): RFLP maps based on a common set of clones reveal modes of chromosomal evolution in potato and tomato. *Genet* 120:1095-1103.

Butenko RG, and Kuchko AA (1980): Somatic hybridization of *Solanum tuberosum* and *Solanum chacoense* by protoplast fusion. In: Ferenczy L, Farkas GL, eds. Advances in Protoplast Research. Permagon Press, Oxford. pp 293-300.

Butenko RG, Kuchko AA, and Komarnitsky I (1982): Some features of somatic hybrids between *Solanum tuberosum* and *S. chacoense* and its F_1 sexual progeny. p. 643-644. In: A Fujiwara ed. Proc. 5th Intern. Cong. Plant Tissue and Cell Culture, Japan. Assoc. Plant Tissue Culture, Tokyo.

Byrne MC, McDonnell RE, Wright MS and Carnes MG (1987): Strain and cultivar specificity in the *Agrobacterium*-soybean interaction. *Plant Cell, Tissue and Organ Culture* 8:3-15.

Cass DD and Peteya DJ (1979): Growth of barley pollen tubes *in vivo*. I. Ultrastructural aspects of early tube growth in the stigmatic hair. *Can J Botany* 57: 386-396.

Christou P (1990): Morphological description of transgenic soybean chimeras created by the delivery, integration and expression of foreign DNA using electric discharge particle acceleration. *Ann Botany* 66:379-386.

Christou P, Swain WF, Yang NS and McCabe DE (1989): Inheritance and expression of foreign genes in transgenic soybean plants. *Proc Natl Aca Sci USA* 86:7500-7504.

Deimling S, Zitlsperger J, and Wenzel G (1988): Somatic fusion for breeding tetraploid potatoes. *Plant Breeding* 101:181-189.

Dekeyser RA, Claes B, De Rycke RMU, Habets ME, Van Montagu MC, and Caplan AB (1990): Transient gene expression in intact and organized rice tissues. *Plant Cell* 2, 591-602.

De Wet JMJ, Bergquist RR, Harlan JF, Brink DE, Cohen CE, Newell CA and De Wet AE (1985): Exogenous gene transfer in maize *(Zea mays)* using DNA-treated pollen. In: *Experimental manipulation of ovule tissues* Chapman GP, Mantell SH and Daniles RW, eds. London: Longman.

Draper J, Scott R, Armitage P, and Walden R (1988): Plant genetic transformation and gene expression. A laboratory manual. Blackwell Scientific Publications, Oxford.

Fish N, Karp A, and Jones MGK (1988): Production of somatic hybrids by electrofusion in *Solanum. Theor Appl Genet* 76:260-266.

Gibson RW, Jones MGK, and Fish N (1988): Resistance to potato leaf roll virus and potato virus Y in somatic hybrids between dihaploid *Solanum tuberosum* and *S. brevidens. Theor Appl Genet* 76:113-117.

Helentjaris T, Slocum M, Wright S, Schaefer A and Hienhuis J (1986): Construction of genetic linkage maps in maize and tomato using restriction fragment length polymorphisms. *Theor Appl Genet* 72:761-769.

Helgeson JP, Hunt GJ, Haberlach GT, and Austin S (1988): Somatic hybrids between *Solanum brevidens* and *Solanum tuberosum*: expression of a late blight resistance gene and potato leaf roll resistance. *Plant Cell Rep.* 3:212-215.

Hess D (1987): Pollen-based techniques in gentic manipulation. *Int Rev Cyt* 107: 367-395.

Jefferson RA (1985): PhD Dissertation, U Colorodo, Boulder, CO.

Jefferson RA, Kavanagh TA and Bevan MW (1987): GUS fusions: B-glucuronidase as a sensitive and versatile gene fusion marker in higher plants. *EMBO J* 6: 3901-3907.

Jones, MGK (1985): Transformation of cereal crops by direct gene transfer. *Nature* 317:579-580.

Lin W, Odell JT, and Schreiner RM (1987): Soybean protoplast culture and direct gene uptake and expression by cultured soybean protoplasts. *Plant Physiol.* 84:856-861.

Matthews BF, Abdul-Baki AA, and Saunders JA (1990): Expression of a foreign gene in electroporated pollen grains of tobacco. *Sexual Plant Reproduction* 3:147-151.

Matusek J and Tupy J (1985): The release and some properties of nuclease from various pollen species. *J Plant Physiol* 119:169-178.

McCabe DE, Swain WF, Martinell BJ, and Christou P (1988): Stable transformation of soybean *Glycine max)* by particle acceleration. *Biotechnology* 6:923-926

Melchers G, Sacristan MD, and Holder AA (1978): Somatic hybrid plants of potato and tomato regenerated from fused protoplasts. *Carlsberg Research Communications* 43:203-227.

Mishra KP, Joshua DC and Bhatia CR (1987): *In vitro* electroporation of tobacco pollen. *Plant Science* 52:135-139.

Nam HG, Giraudat J, den Boer B, Moonan F, Loos WDB, Hauge BM, and Goodman HM (1989): Restriction fragment length polymorphism linkage map of *Arabidopsis thaliana*. *Plant Cell* 1:699-705.

Ohta Y (1986): High-efficiency genetic transformation of maize by a mixture of pollen and exogenous DNA. *Proc Natl Acad Sci USA* 83: 715-719.

Pandey KK (1978): Gametic gene transfer in *Nicotiana* by means of irradiated pollen. *Genetica* 49:53-69.

Pandey KK (1980): Further evidence for egg transformation in *Nicotiana*. *Heredity* 45:15-29.

Picton JM and Steer MW (1982): A model for the mechanism of tip extension in pollen tubes. *J Theor Biol* 98: 15-20.

Potter H, Weir L and Leder P (1984): Enhancer-dependent expression of human K immunoglobin genes introduced into mouse pre-B lymphocytes by electroporation. *Proc Natl Acad Sci USA* 81: 7161-7165.

Parrott WA, Hoffman LM, Hildebrand DF, Williams EG and Collins GB (1989): Recovery of primary transformants of soybean. *Plant Cell Rep.* 7:615-617.

Perl A, Aviv D, and Galun E (1990): Protoplast-fusion-derived *Solanum* cybrids: application and phylogenetic limitations. *Theor. Appl. Genet.* 79:632-640.

Puite KJ, Roest S, and Pijnacker LP (1986): Somatic hybrid potato plants after electrofusion of diploid *Solanum tuberosum* and *Solanum phureja*. *Plant Cell Reports*. 5:262-265.

Riggs CD and Bates GW (1986): Stable transformation of tobacco by electroporation: Evidence for plasmid concatenation. *Proc Natl Acad Sci USA* 83: 5602-5606.

Roeckel P, Heizmann P, Dubois M and Dumas C (1988): Attempts to transform *Zea mays* via pollen grains. Effect of pollen and stigma nuclease activities. *Sex Plant Repro* 1: 156-163.

Sanford JC, Skubik KA and Reisch BI (1985): Attempted pollen-mediated plant transformation employing genomic donor DNA. *Theor Appl Genet* 69: 571-574.

Saunders JA, Matthews BF and Miller PD (1989a): Plant gene transfer using electrofusion and electroporation. In: *Electroporation and Electrofusion in Cell Biology*, Neumann E, Sowers AE and Jordan C, eds. New York: Plenum Press.

Saunders JA, Smith CR, and Kaper JM (1989b): Effects of electroporation pulse wave on the incorporation of viral RNA into tobacco protoplasts. *BioTechniques* 7(10): 1124-1131.

Schell JS (1987): Transgenic plants as tools to study the molecular organization of plant genes. *Science* 237:1176-1182.

Schocher RJ, Shillito RD, Saul MW, Paszkowski J and Potrykus I (1986): Co-transformation of unlinked foreign genes into plants by direct gene transfer. *Biotechnology* 4:1093-1096.

Shillito RD, Saul MW, Paszkowski J, Muller M and Potrykus I (1985): High efficiency direct gene transfer to plants. *Biotechnology* 3:1099-1103.

Smithies O, Gregg RG, Boggs, SS, Koralewaski MM and Kucherlapati RS (1985): Insertion of DNA sequences into the human chromosomal B-globin locus by homologous recombination. *Nature* 317: 230-234.

Southern E (1975): Detection of specific sequences among DNA fragments separated by gel electrophoresis. *J Mol Biol* 98: 503-517.

Stopper H, Zimmermann U and Wecker E (1985): High yields of DNA transfer into mouse L-cells by electropermeabilization. *Z Naturforsch* 40c: 929-932.

Tempelaar MJ, Jones MGK (1985): Fusion characteristics of plant protoplasts in electric fields. *Planta* 4: 205-216.

Tsong TY and Kinosita K (1985): Use of voltage pulses for the pore opening and drug loading, and subsequent resealing of red blood cells. *Biblio Haematol* 51: 108-114.

Twell D, Klein TM, Fromm ME and McCormick S (1989): Transient expression of chimeric genes delivered into pollen by microprojectile bombardment. *Plant Phys* 91:1270-1274.

Wei, ZM and Xu ZH (1988): Plant regeneration from protoplasts of soybean *(Glycine max)*. *Plant Cell Reports* 7:348-351

Wenzel GO, Meyer C, Przewozny T, Uhrig H, and Scheider O (1982). Incorporation of microspore and protoplast techniques into potato breeding programs. In: ED Earle and Y Demarly (eds.), Variability in Plants Regenerated from Tissue Culture, p. 290-302. Praeger Publishers, New York.

Wenzel GO, Scheider O, Przewozny T, Sopory SK, and Melchers G (1979): Comparison of single cell culture derived *Solanum tuberosum* L. plants and a model for their application in breeding programs. *Theor. Appl. Genet.* 55:49-55.

Zimmerman U, and Scheurich P (1981): High frequency fusion of plant protoplasts by electric fields. *Planta* 151:26-32.

EFFECTS OF ION RESONANCE TUNED MAGNETIC FIELDS ON N-18 MURINE NEUROBLASTOMA CELLS

Stephen D. Smith

Department of Anatomy and Neurobiology

University of Kentucky

Abraham R. Liboff

Department of Physics

Oakland University

Bruce R. McLeod

Department of Electrical Engineering

Montana State University

Elsie J. Barr

Department of Anatomy and Neurobiology

University of Kentucky, Lexington

INTRODUCTION:

Numerous experiments by various laboratories have demonstrated that the effects of ELF magnetic fields on living systems may be dependent upon resonance effects. Bawin and Adey (1976) and Blackman, *et. al.* (1984) observed such responses for calcium efflux from chick brains. Dutta, *et. al.* (1984) saw similar effects in neuroblastoma cells. Liboff (1985) suggested that these effects might be due to cyclotron resonance effects on transmembrane movement of ions. Since then, the theory has been expanded and refined a number of times, and the interested reader is directed to one of the recent theoretical papers by Liboff and McLeod (e.g. 1988) for an analytical discussion. In sum, the theory states that transport will be affected if the combined ac and static magnetic fields satisfy the cyclotron resonance conditions for a particular ion as given by the formula: $2\pi f_c = (q/m)(B)$, where;

f_c = fundamental resonance frequency in Hz

q/m = charge (Coulombs) to mass (Kg) ratio of the ion

B = static magnetic field (Tesla {1 T = 1 X 10^4 Gauss})

The satisfaction of this relationship in a number of experimental systems has proven to be effective in influencing movement of diatoms (Smith, *et. al.*, 1987), transport of calcium ions into lymphocytes (Liboff, *et. al.*, 1987), and development of chick femurs *in vitro* (Smith, *et. al.*, 1991). A capsule review can be found in Liboff, McLeod and Smith (1990).

Given the above findings, especially the observation by Dutta that calcium ion transport in neuroblastoma cells can be affected by resonant fields, we chose to explore the possibility that enhanced transport of cations might influence the proliferation and differentiation of neuroblastoma cells. Hence, we exposed N-18 murine neuroblastoma cells to a variety of tuned fields in these experiments, in the expectation, based on previous experience with diatoms and chick femurs, that calcium ions would be stimulatory and potassium ions would be inhibitory. As may be seen by further perusal of these experiments, we obtained some rather surprising and unexpected results, indicating some possible fundamental differences between the control mechanisms in normal vs. malignant cells.

MATERIALS AND METHODS:

I. Cell Culture Technique; Cells of N-18 Neuroblastoma, obtained from Dr. Betty F. Sisken, were maintained in Dulbecco's Modified Eagle Medium (DMEM - GIBCO or Whittaker Bioproducts) supplemented with 10% fetal bovine serum (FBS - GIBCO, Certified), 1% vol/vol of 200 mM glutamine (SIGMA), 0.25% dextrose, and 1% vol/vol 100X Antibiotic Antimycotic solution (GIBCO) in T-75 flasks (Corning). The flasks were kept at 38^{0} C and 5.5% CO_2 in 100% humidified air in an Hotpack incubator. Prior to experimentation, the cells were adapted to 2% FBS in otherwise identical medium for at least 2 weeks to produce cells which could survive and multiply without large amounts of serum growth promoters and ions.

For the experiments, the adapted cells were diluted to 1.25 X 10^{4}/ml, and 2 ml of cells and medium were introduced to each of the 12 wells of Linbro No. 76-053-05 multiwell plates (Flow Laboratories, McLean, VA) and allowed to attach and replicate for 48 hours prior to testing. The plates were then reintroduced into the incubator in either a control (no experimental field) or experimental (combined parallel ac and static magnetic fields) position, which were separated by appropriate shielding from the cell maintenance area. A minor temperature gradient within the incubator kept the experimentals 1^{0} C cooler than the controls (36 vs 37^{0}). Within 2 hours after the third exposure to the fields, the plates were removed, and 2 ml of 5% glutaraldehyde in 0.2 M pH7.2 Sorensen's buffer were added to each well, bringing the fixative concentration to 2.5%. The cells were then counted on a Nikon Diaphot inverted phase contrast microscope. Each experiment consisted of 36 control and 36 experimental wells, run as three separate sets of 12 wells each on three separate batches of cells.

II. Field Exposures: The experimental plates were placed in the center of the space enclosed by paired parallel 18 cm diameter coils consisting of 400 turns each of No. 26 AWG magnet wire, spaced 9 cm apart (Helmholtz configuration). On 3 subsequent days, the plates were exposed to combined ac and static magnetic fields for 0.5 Hr/day, with 23.5 Hr between exposures. The fields were generated by a Beckman FG-2 function generator. The field parameters were set and measured using a Beckman UC-10 frequency counter, Schonstedt D2220-S5 single-axis fluxgate magnetometer, and a Tektronix 502A oscilloscope, all of which had been recalibrated just prior to the initiation of the experiments.

The ac field was adjusted to 16 Hz and 200 mG peak. The static fields were set at 150, 209, 250, 300, 350, and 408 mG using the dc offset adjustment of the function generator (with the ac field off) to augment the naturally occurring horizontal geomagnetic field component, which was very small (5 mG), since the coil axis was oriented nearly exactly to the East/West (Y) geomagnetic axis. The ambient 60 Hz ac field parallel to the applied fields was 5 mG peak. No attempt was made to control the ac or static fields in the vertical (Z) axis (5 mG peak 60 Hz and 515 mG static) or North-South (X) axis (0.5 mG peak 60 Hz and 110 mG Static), since our prior experiences had demonstrated that cells did not respond to such ambient fields at the levels and exposure times encountered in these experiments. We anticipated that the 209 mG and 408 mG static fields, which were exactly on resonance for calcium and potassium ions (see introduction) would be stimulatory or inhibitory, respectively, given our past experience with diatoms. The other fields were chosen as intermediate points to demonstrate that no effects would be seen off-resonance. As will be seen, the results were far from our expectations.

The magnetic field fluxes were applied parallel to each other and parallel to a horizontal (Y) axis. This was done to avoid the problem of variable induced electrical fields. When fields are applied to dishes in the vertical (Z) axis, the induced field is proportional to the radius of the dish. Thus, cells at the edges of the dish would have experienced larger induced electrical fields than those in the center, a complicating factor. When the fields are applied parallel to the liquid surface and base of the dish, the induced electrical fields at any point are proportional to 1/2 the height of the dish, increasing as the surface and bottom are approached, and are reasonably uniform along any particular plane within the dish, except at the extreme margins.

III. Data Collection: After fixation, each well was positioned so that the microscope objective was approximately beneath the center of the dish. This was done by external inspection to ensure no bias in selecting fields to be measured (other than the approximate central position). The microscope was then put into proper focus, and the attached cells within a 10.3 mm^2 field were counted. The data were expressed as cells/mm^2, as an assessment of cell density/replication. We also counted the number of cells producing processes which were at least 1 cell diameter long, expressing this as the percentage of cells with processes. We used this as an approximation of the state of differentiation of the cells, assuming that cells with extensive processes were more differentiated than those without processes.

Paired experimental and control data from within the same experiment were statistically compared by two-tailed t-tests, adjusted for finite sample size.

RESULTS:

The cell proliferation data are presented in TABLE I and in FIGURE 1. If one looks at these data for cell proliferation (total cells/mm^2), it is obvious that calcium tuning produced only a slight increase in cell proliferation (+17%), which did not reach the 5% level of significance. In fact, the p-level is approximately 0.10. Neither did tuning for potassium produce an inhibition of proliferation, as had been expected. However, there was a dramatic rise in cell replication at 300 mG (+60%, p<.001) which is well illustrated by FIGURE 1.

TABLE I
RESONANCE SCAN
16 Hz AC FIELD, 400 Mg P-P
TOTAL CELLS/MM^2

B-Field	n	Control	Exp't.	t	p	%E/C
150 mG	36	11.45/6.73	11.97/7.03	0.3161	NS	+4
209 mG	36	5.98/3.01	7.00/4.61	1.1106	NS	+17
250 mG	36	8.20/4.72	9.27/4.91	0.9500	NS	+13
300 mG	36	6.32/4.90	10.08/5.61	3.4934	<.001	+60
350 mG	36	10.02/5.08	10.09/5.79	0.0538	NS	+0.7
408 mG	36	15.68/5.57	16.19/6.99	0.0344	NS	+3

number after slash = S.D.

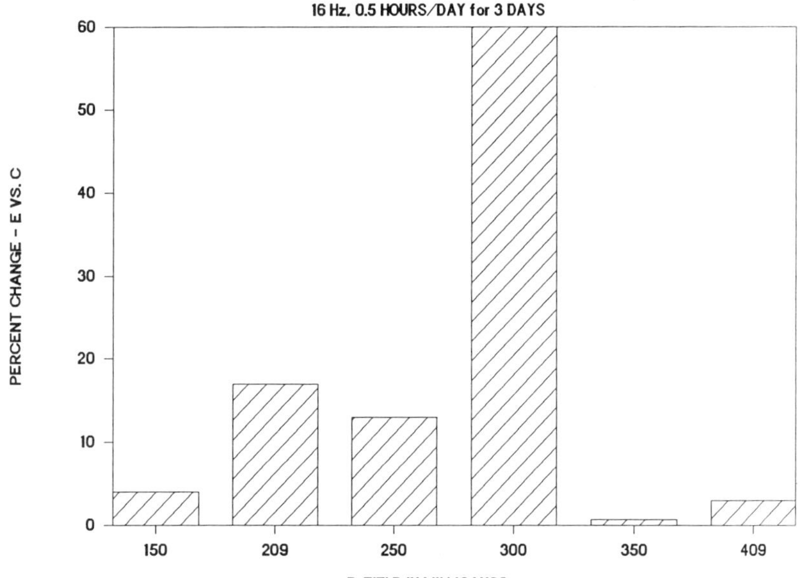

FIGURE 1. Effects of fields on cell proliferation. The bars represent percentage, change, experimental vs. control. The greatest effect occurred at 300 mG, with much lesser effects at 209 and 250 mG. Only 300 mG is statistically significant.

Now, if we turn to the matter of process formation, (TABLE II), the results are a bit different, though still not what we had expected. TABLE II demonstrates that tuning for calcium and potassium ions had no significant effects on cell process formation. However, there were striking and significant decreases when the B-field was set at 250 and

300 mG. FIGURE 2 illustrates the results graphically.

TABLE II

RESONANCE SCAN

16 Hz AC FIELD, 400 mG P–P

PERCENT CELLS WITH PROCESSES

B-Field	n	Control	Exp't.	t	p	%E/C
150 mG	36	0.35/0.64	0.44/1.88	0.2681	NS	+25
209 mG	36	1.40/1.91	1.10/2.03	0.6419	NS	−21
250 mG	36	2.48/2.52	0.91/1.31	3.3201	<.005	−63
300 mG	36	1.68/2.37	0.65/1.28	2.645	<.01	−61
350 mG	36	0.27/0.52	0.20/0.44	0.6147	NS	−26
408 mG	36	0.76/1.19	0.59/0.95	0.6460	NS	−22

number after slash = S.D.

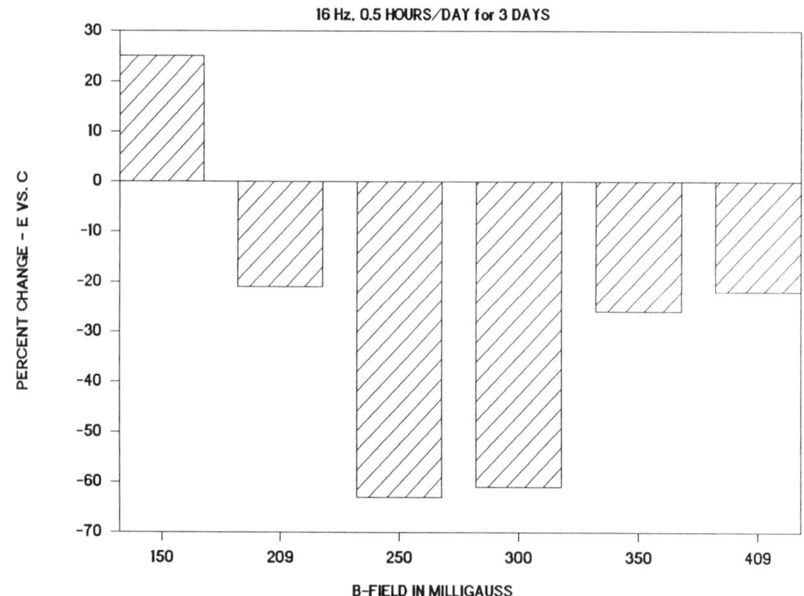

FIGURE 2. Effects of fields on cell process elaboration. The bars represent percentage change, experimental vs. control. Here, the obvious effects are seen at 250 and 300 mG.

DISCUSSION:

These experiments were revealing, if somewhat puzzling. Contrary to our expectation, fields tuned for calcium and potassium ions had only a minor effect (calcium), or none (potassium) on proliferation. The results at 350 mG indicate that the effect of the minor temperature gradient in the incubator was negligible. However, there was a strong response at 300 mG. That point, when combined with a 16 Hz ac field corresponds approximately to the resonance points for 2+ cobaltous ions (307 mG) and 2+ ferrous ions (291 mG). There is certainly no question that cobalt is necessary for cell proliferation, in its association as part of the vitamin B-12 molecule (cyanocobalamin). Unfortunately, we also cannot see how ion resonant tuned fields could affect cobalt transport into the cell, since vitamin B-12 is generally thought to be taken up whole. The same is generally true for ferrous ions. They are certainly critical for cell respiration, and thus for any aerobic biochemical process, but their uptake via channels is not generally assumed. We thus suggest that these data support one of the alternate theories of resonance effects, such as that of Lednev (1991), who has postulated that the resonance interaction occurs at the ion's binding sites. Such a suggestion is plausible here, if the resonance effect could be traced to enzymatic activation or binding of the moiety to the cell membrane transport site. These experiments provide no real evidence in support of such possibilities, but they do suggest that further effort in the area might prove quite fruitful and enlightening.

We were also very surprised at the results for cell process formation. We had expected considerable suppression with calcium (as a concomitant of the usual inverse relationship between expression of differentiation *vs.* cell proliferation), and some increase in elaboration with potassium for the same reason. Evidently, that did not occur. Rather, we did see the suppression of process formation at 300 mG, which we suggest does agree with the assumption that cell proliferation and differentiation are mutually exclusive. No particular theoretical or mechanistic difficulties arise. However, two of the data points are quite interesting from a theoretical point of view.

First, process formation was strongly suppressed at 250 mG. This point corresponds closely to the resonance frequency for sodium ions (240 mG). Reference to the review by Liboff *et. al.* (1990) reveals that generally speaking, the rolloff characteristics of resonance effects are such that measurable changes usually occur if the fields are within 10% of the resonance values. Hence, if there is an effect by sodium ions, which we know to be transported through channels, we might expect to see it at 16 Hz and 240 mG. We must then ask the question, "What effect could sodium ions have on differentiation of these cells?" The answer perhaps lies in the relatively old studies of Barth and Barth (1969), who discovered that isolated epidermal cells, when pretreated with as little as 0.07 mM calcium or 7.5 mM magnesium ions will differentiate into neurons if the external sodium ion concentration is 0.88 mM, but become mucous or ciliated cells at an external sodium ion concentration of 0.44 mM. The same is true for induction of neurons by chordamesoderm in explants of amphibian tissues. Hence, we can postulate here that modification of sodium ion transport has affected the internal/external sodium ion balance in the neuroblastoma cells in such a way as to suppress neuronal differentiation in favor of an epithelial (or epithelioid) pathway. These data are merely suggestive, but support the notion that further exploration of the possibilities would be worthwhile.

Second, the elaboration of processes was enhanced by a static field of 150 mG, representing a reversal of the effect at every other point. The resonance point for Mg2+ ions at 16 Hz is 127 mG. Thus, though the shift at 150 mG is not yet significant, it suggests the possibility that tuning for the exact resonance for magnesium ions, which are transported via channels might well produce enhancement of differentiation for the same reasons quoted above. Barth and Barth found that enhancement of external magnesium ion concentration could induce neuronal differentiation in the presence of adequate external sodium, which is certainly present in DMEM. Hence, modification of the internal/external magnesium ion balance might well strongly influence differentiation, so long as any positive effect of Mg ions on proliferation does not overwhelm its effect on differentiation. The balance point may be rather critical, and seems worth pursuing.

In any event, it appears that the control of cell proliferation in these cells does not follow the same ionic control rules as does control of normal eukaryotic cells in our laboratory and others. If we can elucidate the rules by experiments such as these, we may well be nearer a better understanding of the difference between the fundamental control mechanisms for malignant and normal cells, a useful addition to our basic knowledge and perhaps to clinical oncology as well.

This research was supported in part by funds supplied by IatroMed, Inc.

REFERENCES

Barth LG and Barth LJ (1969): Sodium dependence of embryonic induction. Dev Biol 20:236–262.

Bawin SW and Adey WR (1976): Sensitivity of calcium binding in cerebral tissue to weak environmental electric fields oscillating at low frequency. *Proc Nat Acad Sci USA* 73:1999–2003.

Blackman CF, Benane, SG, Rabiniowitz, JR, House DE and Jones, WT (1985): A role for the magnetic field in the radiation-induced efflux of calcium ions from brain tissue in vitro. *Bioelectromagnetics* 6:327–337.

Dutta SK, Subramoniam A, Ghosh B and Parshad R (1984): Microwave radiation-induced calcium efflux from human neuroblastoma cells in culture. *Bioelectromagnetics* 5:71–78.

Lednev VV (1991): Possible mechanism for the influence of weak magnetic fields on biological systems. *Bioelectromagnetics* 12:71–75.

Liboff AR (1985): Cyclotron resonance in membrane transport. In: *Interactions Between Electromagnetic Fields and Cells,* Chiabrera A, Nicolini C and Schwan HP, eds. New York: Plenum Press.

Liboff AR and McLeod BR (1988): Kinetics of channelized membrane ions in magnetic fields. *Bioelectromagnetics* 9:39–51.

Liboff AR, McLeod BR and Smith SD (1990): Ion cyclotron resonance effects of ELF in biological systems. In:*Extremely Low Frequency Electromagnetic Fields: THE QUESTION OF CANCER,* Wilson BW, Stevens RG and Anderson LE, eds. Columbus: Battelle Press.

Smith SD, McLeod BR, Liboff AR and Cooksey, KE (1987): Calcium cyclotron resonance and diatom mobility. *Bioelectromagnetics* 8:216–227.

Smith SD, McLeod BR and Liboff AR (1991) Effects of resonant magnetic fields on chick femoral development *in vitro. J Bioelect* 10:(in press).

EXACT SOLUTIONS OF A STOCHASTIC MODEL OF ELECTROPORATION.

ISTVAN P. SUGAR

Departments of Biomathematical Sciences and Physiology & Biophysics, The Mount Sinai Medical Center, New York, NY 10029

1. INTRODUCTION

Two basically different approaches have been suggested to describe the experimentally observed properties of electroporation: the electromechanical model and the statistical model of pore expansion. These models have been reviewed by Dimitrov and Jain (1984). This paper considers a simple but exactly solvable statistical model of electroporation for a one-component planar lipid bilayer membrane. Here we concentrate on solutions of the model which are comparable with available experimental data of electroporation such as: size of electropores in thermodynamically stable membranes; lifetime of thermodynamically metastable membranes in the presence of a constant electric field, and conductivity of the planar lipid bilayer membrane during the formation and the shrinking of the electropores.

2. THE MODEL

On the basis of energetic considerations, different pore structures have been proposed: hydrophobic pore and inverted pore model (Abidor et al., 1979), partially hydrophobic pore (Pastushenko and Petrov, 1984), and periodic block model (Sugar and Neumann, 1984). We can describe the present electroporation model without making detailed assumptions about the molecular structure of the pores:

we assume (a.) that the pores are circular, and (b.) that they are independent of each other (the consequences of the pore-pore interaction, pore coalescence, and integral proteins are discussed elsewhere (Sugar et al., 1987)). With these two general assumptions, the process of electroporation can be described by a single stochastic variable a:

$$a = INT(2r\pi/l) \tag{1}$$

where $INT(p)$ is the largest integer of a magnitude not exceeding the magnitude of p combined with the sign of p; r is the pore radius, and l is the characteristic length along the pore circumference. The value of the characteristic length depends on the molecular structure of the pore; for example, in the case of the "periodic block model" (Sugar and Neumann, 1984), l is twice the cross-sectional diameter of a lipid molecule.

Since the pores are considered to be independent from each other the membrane electroporation can be described by the opening and closing of a single membrane pore. From a physical aspect, the pore opening and closing take place as a consequence of the lateral diffusion of lipid molecules to and from the pore wall, respectively (see Fig.1).

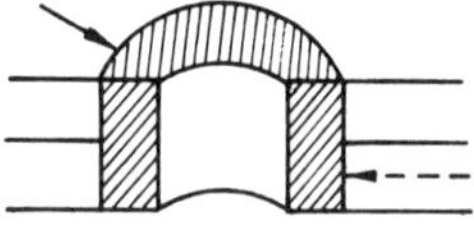

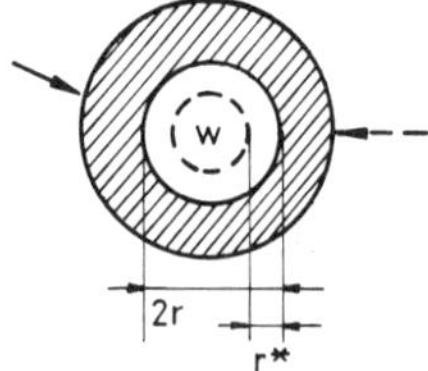

FIGURE 1. Side and top views of the general pore structure. The water filled pore interior (w) is surrounded by the pore wall (shaded area), which in turn surrounded by the bulk bilayer lipid membrane. r is the pore radius; r^* is the thickness of the region of the larger electric field; solid and dashed arrows show possible sites of lipid transfer from the upper and lower monolayer of the bilayer to the pore wall, respectively.

Within a short time interval, Δt, the stochastic variable a may change by unity or may remain unchanged, while the probabilities of higher order changes are negligibly small.

Mathematically, the opening and closing process of a single membrane pore may be treated similarly to the stochastic birth-death processes (Goel and Richter-Dyn, 1974). The <u>forward master equation</u> of the electroporation model is:

$$\frac{P_{a,a_o}(t)}{dt} = w(a, a-1)P_{a-1,a_o}(t) + w(a, a+1)P_{a+1,a_o}(t) - $$
$$[w(a-1, a) + w(a+1, a)]P_{a,a_o}(t) \tag{2}$$

where $P_{a,a_o}(t)$ is the conditional probability of occurrence of the pore in state a at time t given that its state was a_o at time $t = 0$; and $w(a+1, a)$, $w(a-1, a)$ are the <u>transition frequencies</u> for the $a \to (a+1)$ pore opening and $a \to (a-1)$ pore closing steps, respectively. Within the general mathematical framework of the master equation, the physics of electroporation are concentrated in the functions of the transition frequencies, w's.

As a consequence of the *principle of detailed balance* (Gardiner, 1985) the transition frequencies can be found in the following exponential form:

$$w(a+1, a) = \nu \exp -[(\Delta \bar{G}^*(a+1, a) - \Delta \bar{G}(a))/kT] \tag{3}$$

$$w(a, a+1) = \nu \exp -[(\Delta \bar{G}^*(a+1, a) - \Delta \bar{G}(a+1))/kT]$$
$$= \nu \exp -[(\Delta \bar{G}^*(a+1, a) - \Delta \bar{G}(a)) - (\Delta \bar{G}(a+1) - \Delta \bar{G}(a))/kT] \tag{4}$$

where ν is the characteristic frequency of the pore opening/closing processes (number of trials per unit time), T is the absolute temperature, k is the Boltzmann constant, $\Delta \bar{G}(a) = \bar{G}(a) - \bar{G}(0)$ is the Gibbs energy change of the membrane/solution system in the presence of an electric field when a single pore of state a forms in the bilayer, and $\Delta \bar{G}^*(a+1, a)$ is the Gibbs energy change when an activated pore structure forms between state a and $a+1$ in the bilayer.

The activation Gibbs energies are given by

$$\Delta \bar{G}^*(a+1, a) - \Delta \bar{G}(a) = \alpha - kT \ln(\beta a)^2 \tag{5}$$

$$\Delta \bar{G}^*(1, 0) - \Delta \bar{G}(0) = \alpha \tag{6}$$

where α is the activation energy. The second term in Eq.5 contains the activation entropy. For the sake of simplicity the activation energy is independent of the pore state. The transfer of lipid molecules from one monolayer to the pore wall can take place at different sites. The number of possible sites, βa, is proportional to the pore circumference, and also to the stochastic variable a Eq.1, and β is the proportionality constant. The sites of transfer from the two monolayers to the pore wall are assumed to be independent (see Fig.1), and the square of βa gives the thermodynamic weight of the activated pore between a and $a + 1$ states.

The Gibbs energy change of the membrane/solution system when a single pore of radius r forms at zero electric field is

$$\Delta G(r) = \Delta g \Delta A_m + \gamma \Delta A_p / d \tag{7}$$

where the first term refers to the change of the <u>membrane surface tension</u> during pore formation and the second term gives the change in
<u>surface tension of the pore wall</u> during pore formation (Abidor et al., 1979). In the case of thermodynamically metastable black lipid membranes (BLM's) the membrane surface area decreases during pore formation, i.e. $\Delta A_m = -r^2\pi$; however, the surface area of the thermodynamically stable cell (or vesicle) membrane remains unchanged during pore formation, i.e. $\Delta A_m = 0$ (Sugar and Neumann, 1984). The surface area change of the pore wall during formation of the circular pore is: $\Delta A_p = 2r\pi d$, where d is the membrane thickness.

In the presence of a transmembrane electric field, one has to introduce additional energy terms into the Gibbs energy function (Eq.7) describing the change in <u>electric polarization energy</u> of the membrane/solution system $\Delta G_{el}(r)$ when a single pore of radius r forms in the bilayer (Sugar and Neumann, 1984). In the case of small pores, $(r \leq r^*)$:

$$\Delta G_{el}(r) = 0.5\epsilon_o(\epsilon_m - \epsilon_w)\pi r^2 U^2 / d \tag{8}$$

while in the case of large pores, $(r > r^*)$:

$$\Delta G_{el}(r) = 0.5\epsilon_o U^2 \pi [(\epsilon_m - 1)r^2 - (\epsilon_w - 1)(r^2 - (r - r^*)^2)] / d \tag{9}$$

where U is the transmembrane voltage, r^* is defined in the legend of Fig.1, and ϵ_o, ϵ_m and ϵ_w are the vacuum dielectric permittivity, relative dielectric permittivity of the membrane and relative dielectric permittivity of the water, respectively. The detailed derivation and the conditions of Eqs.8,9 are given by Sugar and Neumann (1984).

Introducing the stochastic variable a, defined in Eq.1 into Eqs.7-9, the Gibbs energy function is given by $\Delta\bar{G}(a) = \Delta G(a) + \Delta G_{el}(a)$. Now, Eqs. 3-9 permit the calculation of the transition frequencies as a function of the stochastic variable a and transmembrane voltage U. For this purpose, the numerical values of the oxidized cholesterol bilayer system were used: $\gamma = 1.25 \cdot 10^{-11} N$ (Abidor et al., 1979), $l = 1.8nm$ (Sugar and Neumann, 1984), $T = 313K, \epsilon_m = 2.1, \epsilon_w = 80, r^* = 0.86nm$ (Sugar and Neumann, 1984), $\Delta g = 0.001 N/m$ (Tien, 1974).

By using the above transition frequency functions, exact solutions of the forward master equation Eq.2 can be obtained in particular cases. In Eqs.10-12 solutions are given for the equilibrium pore size distribution and for the "first passage time", respectively. These solutions of the master equations were taken from Goel and Richter-Dyn (1974).

The equilibrium distribution of the pore state is

$$P_{a,a_o}(\infty) =$$
$$\frac{w(a,a-1)w(a-1,a-2)...w(1,0)}{w(0,1)w(1,2)...w(a-1,a)} \bigg/ \sum_{i=1}^{a^*} \frac{w(i,i-1)w(i-1,i-2)...w(1,0)}{w(0,1)w(1,2)...w(i-1,i)} \tag{10}$$

where $0 \leq a_o \leq a^*$.

The "first passage time" $T_{a,a_o}^{(1)}$ is the average time of the $a \leftarrow a_o$ pore-closing process. The "first passage time" is the following function of the transition rates:

$$T_{a,a_o}^{(1)} = \sum_{i=a_o}^{a^*-1} \sum_{n=a+1}^{i} w(n+1,n)^{-1}\Pi_{n+1,i}R_{a,n} -$$
$$R_{a,a_o} \sum_{i=a+1}^{a^*-1} \sum_{n=a+1}^{i} w(n+1,n)^{-1}\Pi_{n+1,i}R_{a,n} \tag{11}$$

where $0 \leq a < a_o < a^*$ and

The "first passage time" of the $a_o \to a$ pore opening process is:

$$T_{a,a_o}^{(1)} = \sum_{i=a_o}^{a-1} \sum_{n=o}^{i} w(n+1,n)^{-1}\Pi_{n+1,i} \tag{12}$$

where $0 \leq a_o \leq a < \infty$. In Eqs.11,12 $\Pi_{i,j}$ and R_{a,a_o} are defined as follows:

$$\Pi_{i,j} = \frac{w(i-1,i)w(i,i+1)...w(j-1,j)}{w(j+1,j)w(j,j-1)...w(i+1,i)}$$

if $i \leq j$ and

$$\Pi_{i,i-1} = 1$$

if $j = i - 1$, and finally

$$R_{a,a_o} = \sum_{i=a_o}^{a^*-1} \Pi_{a+1,i} / \sum_{i=a}^{a^*-1} \Pi_{a+1,i}$$

Eqs.10-12 contain three specific values of the stochastic variable: $0, a_o$ and a^*. The state of the closed pore, $a = 0$ is called *reflecting state* of the system, because the system always returns from the closed pore state to an opened pore state; i.e., there is no negative value of the stochastic variable. a_o is the initial pore state as we defined before. a^* is the state of the largest pore in the thermodynamically metastable membrane. When the pore exceeds this limit state, the membrane becomes unstable and the pore size further increases until the membrane ruptures. Mathematically, this limit state is called the *absorbing state*, because from this state there is no return to the $(0, a^*)$ interval of the stochastic variable. The limit of metastability, a^* decreases with increasing transmembrane voltage (see Fig.2 and Sugar,1989). In the case of thermodynamically stable membranes (cells or vesicles) there is no limit state of the pore; i.e., any $a^* < \infty$ can be chosen.

3. RESULTS AND DISCUSSION

3.1 STATIONARY ELECTROPORES

By means of Eq.10 we were able to determine the stationary solutions $P_{a,a_o}(\infty)$ of the master equations of the electroporation. In light of our knowledge of the pore state distribution, we calculated the average pore state $< a >$ and fluctuation $< \Delta a >= \sqrt{(< a- < a >)^2 >}$. Fig. 2 shows the average pore states and $< a > +3 < \Delta a >$ at different transmembrane voltages.

Fig.2a refers to the case of a thermodynamically stable membrane ($\Delta A_m = 0$ in Eq.7). According to this calculation the closed pore opens up within a narrow interval of the transmembrane voltage centered at 0.4V. At the same time the relative fluctuation of the pore state ($< \Delta a > / < a >$) drops by two orders of magnitude. The average state of the opened pore is $a = 100$. By using Eq.1 the radius of the transmembrane voltage-induced pore is $28nm$. Chang and Reese (1990) have determined the radius of electropores, induced in human erythrocyte membranes, by time-resolved freeze-fracture electron microscopy. The size of the most frequently detected pores, $r = 20-25nm$ is in agreement with our calculated pore radius. However, in contrast to the calculations the pore size distribution is from pore radius 10 to $60nm$ wide. The wide pore size distribution may be the consequence of the inhomogenity of the membrane material of erythrocyte, i.e. of the formation of compositional domains in the membrane. Coalescence of the pores may result in larger pores too. The above factors are not included in our model because it considers one component membrane and non-interacting electropores.

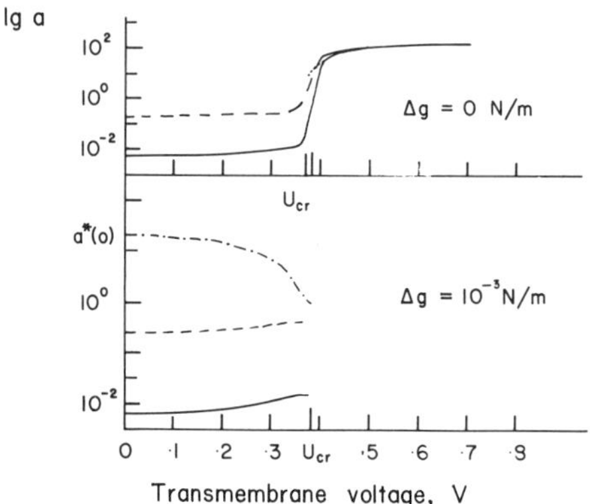

FIGURE 2. Characteristic stationary quantities of electropores as a function of transmembrane voltage. (a) Stable membrane; (b) metastable membrane. Solid line: average pore state; dashed line: fluctuation of the pore state; dot-dashed line: limit state of pores in metastable membrane.

In Fig.2b the voltage dependencies of the average pore state and pore state fluctuation are shown in the case of a thermodynamically metastable membrane. The average pore state is close to zero, while the pore state fluctuation is two orders of magnitude larger. The dot-dashed line shows the limit state of the pore a^* at different transmembrane voltages. At 0.4 V the dashed and the dot-dashed line intercept; i.e., the limit state of the pore can be reached easily by means of the thermal fluctuations of the pore size. At this transmembrane voltage the lifetime of the metastable membrane is close to zero. This prediction of our model is in accordance with the measured short, $10^{-4}s$ lifetime of UO_2^{2+} modified bilayers of azolectin at 0.4 V (Chernomordik et al., 1987).

By comparing the stationary solutions for thermodynamically metastable and stable membranes we can observe similarities and fundamental differences in the nature of the electropores. In both cases, below the critical voltage $U_{cr}(= 0.4V)$ the opening of the pores is always for only a short time. These unstable pores with short lifetimes are the result of thermal fluctuation; however, most of the time the pores are in a closed state. Above U_{cr} in thermodynamically stable membranes the opened pores are stable formations and the role of the fluctuations is negligibly small ($< a >\gg< \Delta a >$). These stable pores are much larger than the fluctuating pores at $U < U_{cr}$. The above comparison implies that experiments on metastable BLM's may teach us about the properties of the electropores in the stable cell or vesicle membranes when $U < U_{cr}$, but the analogy between these systems does not hold above the critical voltage.

Finally, it is important to note that our model predicts similar critical voltages in the case of stable and metastable membranes. Lacking precise experimental data we cannot test this prediction carefully. The critical voltages measured on BLM's are in the range of $0.3 - 0.4V$ (Benz et al., 1979; Chernomordik et al., 1983, 1987), while in the case of small and large unilamellar vesicles $U_{cr} = 0.25 - 1V$ have been obtained (Teissie and Tsong, 1981; Needham and Hochmuth, 1989).

3.2 LIFETIME OF METASTABLE MEMBRANES

In the case of thermodynamically metastable membranes, sooner or later the system becomes unstable. This takes place when one of the fluctuating pores exceeds the limit state of metastability a^*. By means of Eq.12 one can determine the "first passage time", $T^{(1)}_{a,a_o}$ from the closed pore state ($a_o = 0$) to the limit state ($a = a^*$). The "first passage time" is proportional to the lifetime of the membrane (Arakelyan et al.,1979).

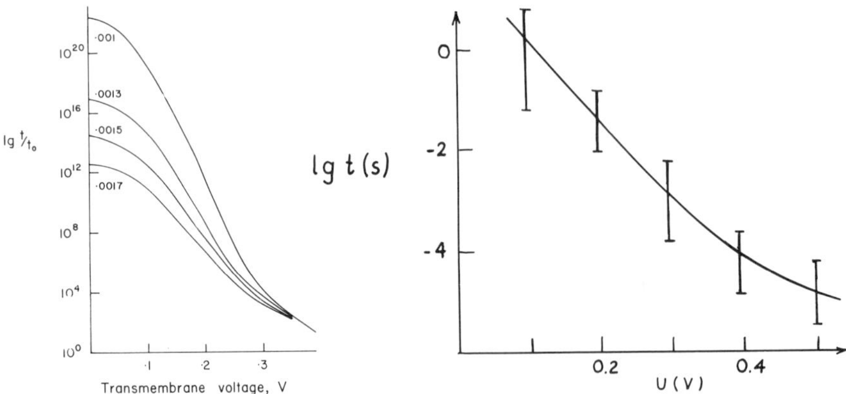

FIGURE 3. Average lifetime of metastable membranes at different transmembrane voltages. (a) Membrane lifetimes calculated at different values of Δg shown at each curve in N/m; $t = T^{(1)}_{a^*,0}$ and $t_o = \beta^2 \nu \exp -(\alpha/kT)$. (b) Lifetimes of oxidized cholesterol BLM's in 1 M KCl, from Chernomordik et al.,(1987).

In Fig.3a the calculated "first passage time" as a function of the transmembrane voltage is shown at different membrane surface energies, Δg. Chernomordik et al. (1987) measured the lifetime of oxidized cholesterol BLM's at different transmembrane voltages as it is shown in Fig.3b. There are obvious similarities between the calculated and experimental results. It would be interesting to check experimentally the existence of the predicted convex section of the lifetime curve at very small transmembrane voltages, $U < 0.1V$.

3.3 PORE OPENING KINETICS

By means of Eq.12 one can determine the "first passage time" $T^{(1)}_{a,a_o}$ as a function of the final pore state, a. The inverse of this function is physically analogous to the time dependence of the average pore state. In this way one can determine pore opening kinetics. In Fig.4a calculated pore opening kinetics are shown at different supercritical voltages and at an initial state $a_o = 0$. The square of the pore state which is proportional to the surface area of the pore is plotted on the ordinate. In every case, the opening of the pore begins with a gradual increase in pore size. The growth rate of the pore size depends on the transmembrane voltage, leading to a subsequent jump in the size of the pore. The moment and the height of the jump depend on the transmembrane voltage. After the jump occurs, the pore size gradually increases until the membrane ruptures.

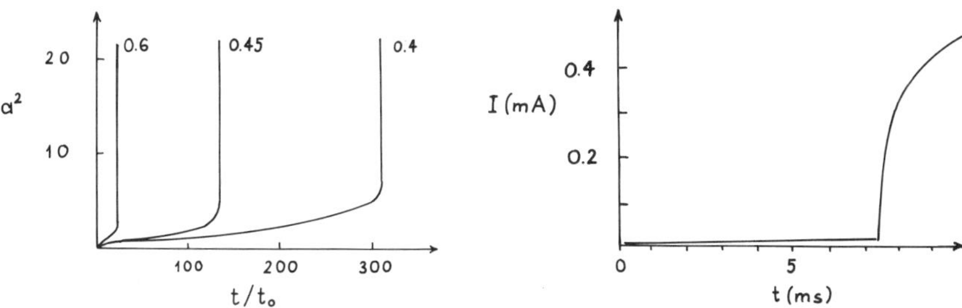

FIGURE 4. Kinetics of pore opening. (a) Calculated pore opening process in an unstable membrane. Transmembrane voltage is shown for each curve in volts; $t = T^{(1)}_{a*,0}$ and $t_o = \beta^2 \nu \exp -(\alpha/kT)$. (b)Current oscillogram of the oxidized cholesterol BLM (voltage pulse amplitudes of U=0.2 V, and pulse durations of 10 ms) from Chernomordik et al.,(1983).

The calculated pore opening kinetics can be compared with the experimental data of Chernomordik et al. (1983). This comparison is based on the assumption that the measured membrane conductance is proportional to the surface area of the pore. This is a good assumption in the case of large pores when the ion-pore wall interaction is negligible (Chernomordik et al., 1987). Fig.4b shows the conductance increase of an oxidized cholesterol BLM from the stage of the initial gradual increase to the stage of an abrupt jump leading to the rupture of the membrane. In accordance with calculated kinetics, after the abrupt jump the conductance increases gradually again. The initial gradual increase of the membrane conductance depends on the transmembrane voltage (see Fig.3 in Ref. Chernomordik et al.,1987)

3.4 PORE SHRINKING KINETICS

By means similar to determining the kinetics of the pore opening by using Eq.11, we can calculate the pore shrinking kinetics. In Fig.5a shrinking kinetics calculated at two different initial pore states are shown. The square of the pore state is plotted in the ordinate. These calculated kinetics can be compared with the conductance change measured during the recovery after the reversible breakdown of UO_2^{+2} modified azolectin BLM's (Chernomordik et al.,1987).

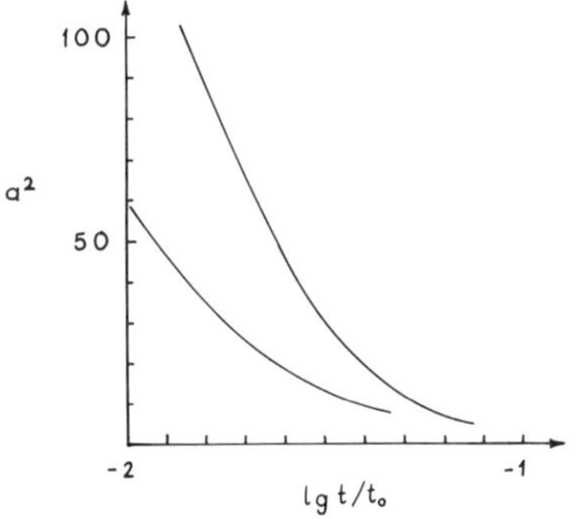

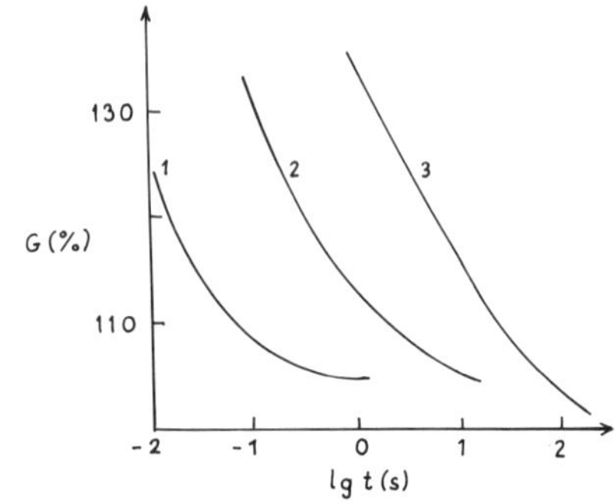

FIGURE 5. Pore shrinking process. (a) Calculated pore shrinking process in a metastable membrane at two different initial states ($a_o < a_*$); $t = T_{a_*,0}^{(1)}$ and $t_o = \beta^2 \nu \exp -(\alpha/kT)$. (b)Measured conductance of UO_2^{2+} modified azolectin BLM's after the following voltage pulses: (1) 0.8 V, 1 ms; (2) 0.68 V, 10 ms; (3) 0.59 V, 100 ms, from Chernomordik et al.,(1987).

In summary the electroporation in both thermodynamically stable and metastable membranes can be described by a simple stochastic model. The exact solutions of the model are comparable with the available conductance and electron microscopy data on electroporated membranes. This, to the best of our knowledge is the first model that provides pore sizes in agreement with recent experimental data by Chang and Reese (1990).

ACKNOWLEDGEMENT

The author would like to thank Dr. R. Schmukler for his encouragement and support, and T. Hill for her excellent technical assistance.

REFERENCES

Abidor IG, Arakelyan VB, Chernomordik LV, Chizmadzhev YA, Pastushenko VF and Tarasevich MR (1979): Electric breakdown of bilayer lipid membranes I. *Bioelectrochem. Bioenerg.* 6: 37-52.

Arakelyan VB, Chizmadzhev YA and Pastushenko VF (1979): Electric breakdown of bilayer lipid membranes V. *Bioelectrochem. Bioenerg.* 6: 81-87.

Benz R, Beckers F and Zimmermann U (1979): Reversible electrical breakdown of lipid bilayer membranes: A charge-pulse relaxation study. *J. Membr. Biol.* 48: 181-204.

Chang DC and Reese TS (1990): Changes in membrane structure induced by electroporation as revealed by rapid-freezing electron microscopy. *Biophys. J.* 58: 1-12.

Chernomordik LV, Sukharev SI, Abidor IG and Chizmadzhev YA (1983): Breakdown of lipid bilayer membranes in an electric field. *Biochim. Biophys. Acta* 736: 203-213.

Chernomordik LV, Sukharev SI, Popov SV, Pastushenko VF, Sokirko AV, Abidor IG and Chizmadzhev YA (1987): The electrical breakdown of cell and lipid membranes: the similarity of phenomenologies. *Biochim. Biophys. Acta* 902: 360-373.

Dimitrov DS and Jain RK (1984): Membrane stability. *Biochim. Biophys. Acta* 779: 437-468.

Gardiner CW (1985): In: *Handbook of Stochastic Methods*, Haken H, ed., New York, Berlin, Heidelberg: Springer-Verlag.

Goel NS and Richter-Dyn N (1974): In: *Stochastic Model in Biology*. New York:

Academic Press

Needham D and Hochmuth RM (1989): Electro-mechanical permeabilization of lipid vesicles. *Biophys. J.* 55: 1001-1009.

Pastushenko VF and Petrov AG (1984): Electro-mechanical mechanism of pore formation in bilayer lipid membranes. In: *Biophysics of Membrane Transport*, School Proceedings, Poland, pp. 70-91.

Sugar IP (1989): Stochastic model of electric field-induced membrane pores. In: *Electroporation and Electrofusion in Cell Biology*, Neumann E, Sowers A, Jordan C, eds., New York: Plenum pp. 97-110.

Sugar IP, Forster W and Neumann E (1987): Model of cell electrofusion: Membrane electroporation, pore coalescence and percolation. *Biophys. Chem.* 26: 321-337.

Sugar IP and Neumann E (1984): Stochastic model for electric field-induced membrane pores: Electroporation. *Biophys. Chem.* 91: 211-225.

Teissie J and Tsong TY (1981): Electric field induced transient pores in phospholipid bilayer vesicles. *Biochemistry* 20:1548-1554.

Tien HT (1974): In: *Bilayer Lipid Membranes*, New York: Dekker.

TIME COURSE OF ELECTROPERMEABILIZATION

TEISSIE, J.

Centre de Biochimie et de Génétique Cellulaires du CNRS

118, route de Narbonne, 31062 TOULOUSE Cédex, FRANCE.

Cell membrane selective permeability is an advantage by preserving the genome and the cellular machinery from the action of exogenous agression but in the same time is a strong limitation for the experimental manipulation of the cytoplasm. Spontaneous introduction of foreign proteins or genes is not possible and artificial methods must be found which of course should alter the membrane permeability but not affect the viability. **Electropulsation**, i.e. submitting cells to strong shortlived electric field pulses, is a new methodology, which is now proved to be highly efficient to give access to the cytoplasm. **Electropermeabilization** and **electrotransformation** are very potent tools for bioengineering.

The proper and clever use of a physical method is first to understand the molecular processes involved at the cell level in order to know the limits of the tool. From that point of view, one must consider that indeed very few things are known on electropermeabilization. A very crude picture is to say that holes are induced in the membrane but up to now these pores have never been detected by optical or electron microscopy when the cell viability is preserved. In the present paper, we try to describe the thermodynamical aspects of electropermeabilization. It should be emphasized that most of the results have been obtained on mammalian cells but the conclusions have been shown to be fully valid on plant protoplasts and with some limitations in the case on walled microorganisms (bacteria, yeasts).

<u>I -The external field alters the membrane potential difference of the pulsed cell.</u>

From a theoretical point of view, Laplace equation predicts that the membrane potential difference (MPD) in a cell can be manipulated by submitting it to an external field (Neumann, 1989) · This effect is due to the dielectric character of the membrane inducing a deformation of the lines of the field applied onto the cells. As described in (Neumann, 1989; Bernhardt and Pauly, 1973; Kinosita and Tsong, 1977), the change in MPD, DV, is related to the external field by the following expression =

$$\Delta V = F \ r \ E \ \cos \theta \qquad (a)$$

where F is a parameter characteristic of the cell

r is the radius of the cell (if it is a sphere)

E is the intensity of the applied field

θ is the angle between the direction of the field and the normal to the cell surface at the point of interest (M).

The MPD change is then dependent on the position at the cell surface. The cell MDP is :

$$V(M) = V_O(M) + \ \Delta V(M) \qquad (b)$$

$V_O(M)$ being the resting MPD of the cell.

One side of the cell is going to be hyperpolarized but the other will be depolarized, the equator being unaffected. Two gradients in potential are present at the level of the membrane : a transversal one described by (b) and a lateral one due to the fact that the cytoplasmic side is isopotential and as shown in (a) ΔV is position dependent.

This theoretical approach has been confirmed experimentally by use of digitalized video microscopy (Gross et al, 1986; Ehrenberg et al, 1987) .

A point is very often neglected when evaluating the field induced MPD change. The parameter F is playing a decisive role in the definition of DV. As described in (Neumann, 1989) , its expression is complex :

$$F = \ 3/(2(1 + \lambda_m(2 + \lambda_i /\lambda_o) /(2\lambda_i d/r))) \qquad (c)$$

(with the restricting assumption that the shape of the cell is a sphere).

λ_m is the specific conductivity of the cell membrane

λ_o is the specific conductivity of the external buffer

λ_i is the specific conductivity of the internal cell volume

d is the thickness of the membrane

In many cases, the membrane is taken as a pure dielectric and the MPD change is calculated to be :

$$\Delta V = 1.5 \ r \ E \ Cos \ \theta \qquad (d)$$

This expression must be taken as a theoretical limit because in many cases a cell is not spherical and the membrane is not a pure dielectric. It was indeed shown that increasing the membrane conductivity λ_m induces a drop in the electric field mediated MPD change (Lojewska et al, 1989).

II- The external field induces a reversible membrane permeabilization.

Studies on BLM showed that a lipid membrane was dramatically affected by an increase in its MPD. The conductance which is very low at low MPD shows a burst if its MPD is brought to values about 200 mV. In most cases, a rupture of the BLM follows.

These observations on model systems lead to a theory where the permeability of a membrane is very low under resting conditions but it is strongly increased when the MPD is brought up to values about 200 mV.

This theory would predict that the application of calibrated electric field pulses would induce an increase in permeability. It is just needed that the field induces a MPD change $\Delta V(M)$ such as to give a MPD $V(M)$ larger than the permeabilizing threshold V_p inducing the permeabilization.This was indeed observed in 1972 (Neumann and Rosenheck, 1972) . A DC pulse larger than 18 kV/cm (exponential decay) induces a release of the content in catecholamines when applied to a suspension of chromaffin granules. This leakage was only transient but lasted nevertheless much longer than the pulse duration. Since then, this process has been applied to many other systems : erythrocytes

(Kinosita and Tsong, 1977), plant protoplasts (Zimmermann, 1982), mammalian cells (Neumann et al, 1982) and bacteria (Dower et al, 1988). In all cases, cell electropulsation was giving access to the cytoplasm in a reversible way keeping the cells viable.

III- Methods

Cell electropermeabilization can be followed by the inflow of exogenous molecules such as dyes or by the leakage of metabolites. The inflow can be monitored by the detection and quantification of the entrapped molecules ; this is easy with dyes (Trypan blue) or fluorescent probes (Ethidium Bromide, calcein) by using cell imaging under a microscope (Rols and Teissié, 1990a). This can be detected by a back effect of the penetration as in the case of Ca^{++}. Electropermeabilization mediated Ca^{++} penetration leads to a cytoplasmic unbalance and the cell lysis. The outflow can be followed by the detection of the leaked molecules or ions. This is very easy in the case of ATP by the Luciferin-Luciferase assay (Teissié and Rols, 1988) or in the case of the ions when working in a low ionic content pulsing buffer by following the increase in conductivity of cell suspension (Kinosita and Tsong, 1979). This last approach is of great convenience for the fast kinetic resolution of the induction of permeabilization. Time resolution down to a fraction of a microsecond can be obtained. Events at the cellular level can be obtained by submicrosecond imaging under a pulsed laser microscope (Hibino et al, 1991; Kinosita et al, 1990).

IV- Electric field parameters control the extend of permeabilization

All approaches bring to the same conclusions which are as follows. In our group, we are using square wave electric field pulses where we can control simultaneously the field strength, the pulse duration, the number of pulses and the delay between the pulses. Similar electrical pulsing conditions are obtained whatever the pulsing buffer ionic content. As shown in Fig.1, at given pulse number, duration and delay, the percentage of permeabilized cells is experimentally a function of the field intensity E. As long as the field strength is less than a threshold E_p, no cell is stained i.e. permeabilized. Then with stronger fields, the percentage is observed to increase ; this increase can be quantified by the first derivative (dP/dE)(P = 50%). Using even stronger fields, all cells are permeabilized. The electropermeabilization of cells is then dependent on the field strength. But at a given field strength, the percentage of permeabilized cells depends on the pulse duration and number. Increasing the pulse duration (or number) shifts the P versus E plot towards lower field intensities by decreasing Ep and increasing dP/dE (P = 50 %). Nevertheless, a limit in E_p, E_O, is present. This can be obtained as shown in Fig.3 by plotting E_p as a function of the reciprocal of the cumulated pulse duration (n x T). This plot is a straight line which crosses the ordinate axis at E_O. E_O is then the threshold in field intensity which is needed to detect the electropermeabilization. dP/dE (P = 50%) reaches such a large value when pulses are accumulated that the P versus E plot turns to be a staircase. This threshold E_O is a thermodynamic characteristic of the cell. It is indicative of a threshold in MPD in the induction of the membrane permeabilization. This threshold in MPD can be obtained from (a) and (b). But one can noticed that it is obtained first in the region facing the electrodes where θ is equal to zero and Cos θ to 1. As a consequence, permeabilization would appear first in this region. This is observed experimentally by following the process by fluorescence microscopy (Tekle et al, 1990).

The extend of permeabilized surface extends with an increase in the field intensity. This is confirmed by the direct observation of the process. It should be noticed that as the permeabilizing MPD is roughly constant, the permeabilizing field strength, which is related to the MPD through the reciprocal of the cell size, would be dependent on the cell specie. It is then possible in a mixture of different cells to specifically permeabilize the larger ones (Sixou and Teissié, 1990). But when working with a cell population where a morphological and

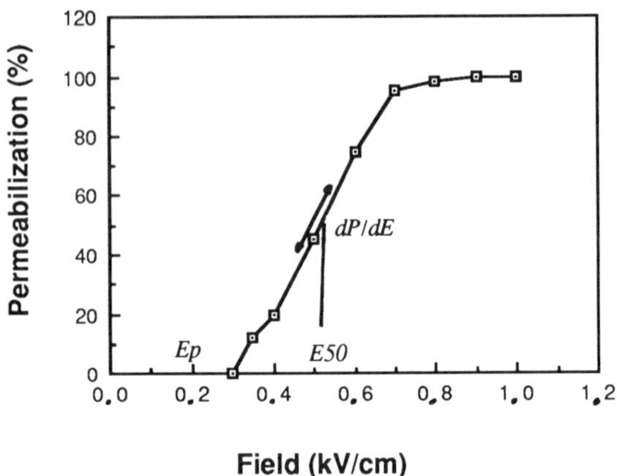

Fig. 1: Permeabilization

CHO cells were pulsed 10 times (pulse duration 0.1 mS, frequency 1 Hz). Permeation was assayed by TB staining

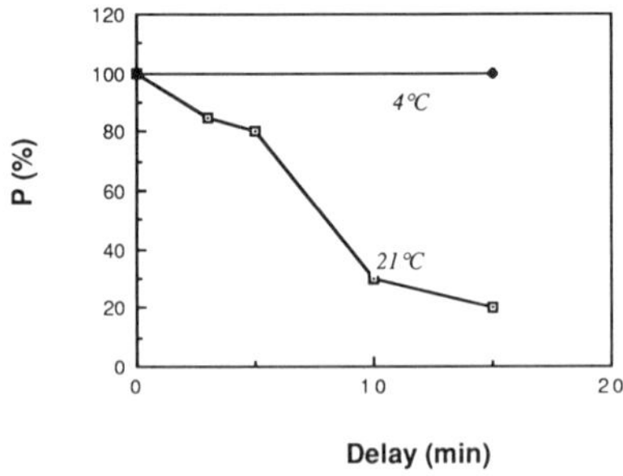

Fig. 2 : Resealing Annihilation of CHO permeabilization was assayed by TB after pulsing (1 kV/cm, 0.1 mS, 10 times)

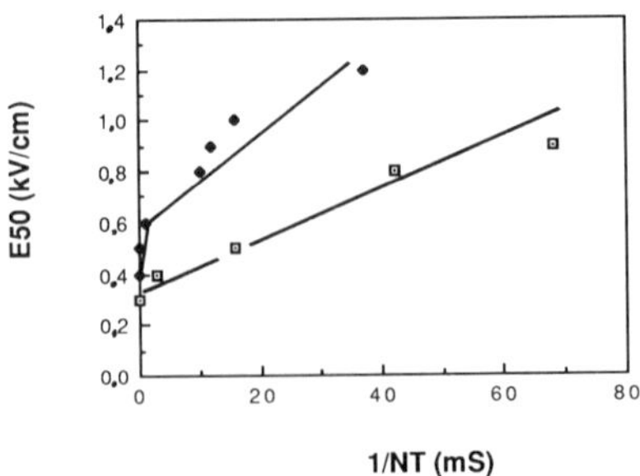

Fig. 3: E_{50} for ATP leakage (□) and for calcein uptake (◆) in the case of tobacco protoplats

may be a physiological distribution is present, we observe that the increase in permeabilization is dependent on the nature of the permeant species (Rols and Teissié, 1989). Ca^{++} is observed to penetrate more easily than Trypan Blue (MW 960). There is indeed an apparent molecular size limitation in the definition of molecules which are able to cross the electropermeabilized membrane. This can be explained by two different but not conflicting and may be complementary ways. 1) The permeability coefficient of a electropermeabilized membrane can be dependent on the nature of the specie S which is crossing it and 2) the limit of detection of the assay used to detect the penetration of the endogeneous molecule can be very dependent on its nature. For example, one copy of a toxin is enough to kill a cell and this explains why electropermeabilization is experimentally proved to allow the penetration of such a molecule (Orlowski et al, 1988).

The extend of permeabilization is dependent on the pulse duration. The kinetic of the induction of electropermeabilization can be monitored by following the conductance change of the cell suspension (Kinosita and Tsong, 1979). By such a way, it is the permeability to small ions such as K^+ which is detected. Permeabilization is induced in less than 200 nS by the external field. This time limit being the one of the electronic detection (Teissié and Tsong, unpublished). This fast step in the process is called **induction** and occurs a soon as the field intensity is larger than E_0. It can be detected only when a very sensitive assay is available such as the conductance change. It is associated to a threshold in the induced MPD. At the cell level, this means that induction is triggered as soon as the cell MPD is brought to a characteristic value. In the case of DPPC LUV, it was computed to be of the order of 200 mV (Teissié and Tsong, 1981).

This induction step is followed by a continuous increase of the conductance of the cell suspension (Kinosita and Tsong, 1979). As far as permeability is concerned, an increase with the pulse duration is observed. This can be explained by the occurrence of an **expansion** step affecting the part of the cell surface where the induction has occured.

For a given field strength E (E larger than E_0), the electropermeabilization would occur only in a cone (assuming the shape of the cell to be a sphere) with an angle θ_p such as :

$$E \cos \theta_p = E_0 \quad (d)$$

The expansion step increases the permeability coefficient and as such the exchange of molecules across the membrane.

The flux Φ of molecules S across an electropermeabilized membrane is then mathematically expressed as (Rols and Teissié,1990a) :

$$\Phi(S) = KP_s \, X(T,N) \, (1 - Eo/E) \, \Delta S \quad (e)$$

where P_s is the permeability coefficient of S across the permeabilized membrane

K is a coefficient

E is the strength of the field

E_0 is the permeabilizing field intensity threshold

ΔS is the concentration gradient of S across the membrane. Of course, when S is present only in the buffer before pulsing, then $\Delta S =$ Sext

X (N,T) is the fraction of the really permeabilized area

X is a function of the pulse duration and number and reflects that as soon as the MPD is brought to the permeabilizing value, there is a time dependent transition of the membrane from a state where it is impermeable to S to a new one where P_s is present. From this result, it is clear that the electric field plays a determinant role in bringing the MPD above the threshold needed for permeabilization and in determining what portion of the cell surface may be permeabilized. But in that portion, the extend of the structural transition of the membrane is only controlled by the cumulated pulse duration not by the field intensity.

As a practical conclusion, two approaches are giving a high level of permeabilization (i.e. a high exchange across the membrane) : 1) a strong electric field with a short duration, affecting a large proportion of the cell surface with a small permeability, 2) an electric field just stronger than the threshold, affecting a small surface but with a long duration (or a large number of successive pulses) giving a local high permeability . Limits are linked to the viability of the pulsed sample.

The expansion step is present as long as the external field is present. It is localized on the cell surface where the electric field modulation of the MPD is larger than the critical value, i.e. inside the cone of half angle θ_p . The perturbed loci are not laterally mobile (Sowers, 1987; Rols and Teissié, 1990a). When the modulation drops below the critical value, a very fast **stabilisation** step is present. This was observed indirectly by observing the inflow of fluorescent dextrans (Dimitrov and Sowers, 1990) where the diffusion rate is high only as long as the field is present. More direct evidences are obtained by measuring the change in conductance of the cell membrane electrically on a suspension (Kinosita and Tsong, 1979), on a lipid bilayer (Chernomordik et al, 1983), on a cell membrane by patch-clamp (Chernomordik et al, 1987) or at the single cell level by video microscopy (Kinosita et al, 1991). A long lived permeability to small molecules (molecular weight up to 2-4000) is then present. But in the case of macromolecules such as plasmids, the transfer can be mediated only if present during the pulse. A key process is present during the expansion step.

The electropermeabilized state of the membrane to small molecules is reversible. The natural impermeability can be recovered progressively but spontaneously. This can be observed by pulsing the cells in a dye free medium and adding the dye at different delays after pulsing. As shown in Fig.2, the percentage of permeabilized cells is observed to decrease back to zero with increasing delays. The viability is checked by pulsing in a dye free buffer and by observing the growth of cells. It should be emphasized that reversibility of permeabilization is not always indicative of cell viability. It is only the ability to grow which is the true criterion. The kinetics of the resealing shows that it is a first order process.

The rate constant is dependent only on the cumulated pulse reaction not on the field strength (Rols and Teissié, 1990a). This process is strongly dependent on the incubation temperature. In the case of CHO cells, resealing is observed in less than 1 min at 37°C but may last several hours if cells are kept at 4°C (Fig.2). This is apparently due to the role of proteins because electropermeabilization is clearly short-lived in pure lipid LUV. One must notice that the temperature is not playing a major role in the definition of E_O and $X(N,T)$, i.e. on the induction and expansion steps.

V- Electropermeabilization is associated to a long lived membrane fusogenicity

Cells in close contact would fuse when electropulsed (Neumann et al, 1989). In 1986, it was shown on a model system (erythrocytes ghosts) and on viable cells that pulsing first and then bringing into contact was inducing cell fusion (Sowers, 1986; Teissié and Rols, 1986). This was recently confirmed on plant protoplasts (Montané et al, 1990). The thermodynamical consequence of this observation is of importance in the understanding of electropermeabilization. Repulsive forces preventing a spontaneous cell fusion are weakened and may be abolished by the electropermeabilization associated membrane structural transitions. Some of these forces are of electrostatic origin due to the surface charges but at a very close contact between cells a much stronger repulsion is present due to the regular organization of interfacial water molecules, which dipoles are electrically oriented by the local fields arising from the membrane (Marra and Israelachvili, 1985). The observation that spontaneous fusion occurs when electropermeabilized cells are brought into contact either by dielectrophoresis or by a low g centrifugation, i.e. under very mild conditions, is the direct indication that these "hydration" forces are not present anymore. Due to their origin, the regular organization of interfacial water molecules, this conclusion means that this network of hydrogen bonded molecules is not present along electropermeabilized membranes (Fig.4). As the structural order of the interfacial water is due to the local electric field arising from the membrane constituents, the final and most important implication is that there is a new organization, presumably a more random and fluctuating one in the membrane. This is supported by ^{31}P NMR studies which showed that during the electric field pulse, the polar heads of phospholipids in MLV were tilted (Stulen, 1981)· Taking advantage of the long lifetime of the electropermeabilized state in mammalian cells, a ^{31}P NMR investigation of their phospholipids was possible (Lopez et al, 1988). It was observed that they displayed a conformation of their polar heads which was different from what is present in the normal organization. An apparent tilt in comparison with the classical position (almost parallel to the plane of the membrane) or by the induction of short wave ripples in the membrane. This second explanation will be associated by an increase in fluctuations in the membrane organization as suggested in (Deuticke and Schwister, 1989). Enhanced fluctuations in the lipid matrix were proposed to increase the permeability coefficient of a membrane if it is considered that the energy barrier for a molecule to cross a membrane is mainly at the glycerol level than in the hydrocarbon chains regions (Nagle and Scott, 1978; Miller, 1991).

VI- Conclusions

The electropermeabilization associated fusogenicity shows that the energy barrier which must be overcome to trigger electropermeabilization is the same as the one needed to fuse cells (Fig.4). The membrane structural transition is associated to the annihilation of the molecular structures which are organizing the interfacial water molecules in a regular array. This conclusion suggests that any modification which will facilitate such an annihilation would make electropermeabilization easier. This is indeed the case as shown by the effects of the ionic content of the pulsing buffer (Rols and Teissié, 1989), of molecules affecting the order of the membrane (Rols et al, 1990) or of the pulsing buffer osmotic pressure (Rols et Teissié, 1990b). But as it was observed that E_0, the field threshold for permeabilization induction, was not affected by these modifications, the expansion step is where the membrane structure modification is occurring.

Another distinctive property of the long-lived electropermeabilized state is its ability to allow the spontaneous insertion of transmembraneous proteins in the membrane. This has been described for glycophorin in the case of erythrocytes (Mouneimne et al, 1989). This is one more experimental evidence of the dramatic thermodynamical alterations of the membrane induced by its electropulsation.

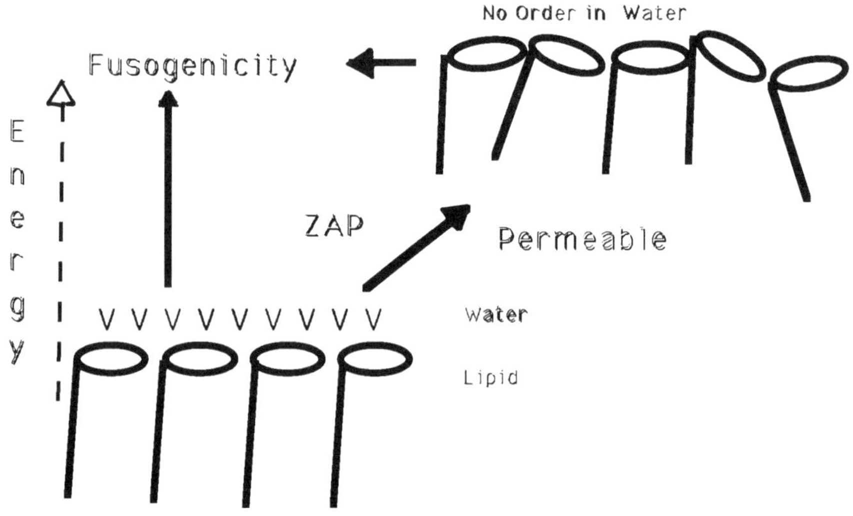

Fig. 4 : Electropermeabilization is mainly associated to a disorganization of the interfacial water network and as such gives a fusogenic character to the membrane

It should be emphasized that very few things are known on the molecular processes affecting the membrane organization along the electropermeabilization.

Models have been proposed using primitive descriptions of the membrane, which was considered as a lipid bilayer. Permeabilization is then due either to the electrocompression of the film (Crowley, 1973; Zimmermann et al, 1974; Dimitrov and Jain, 1984) or to the increase in size of structural defects (Abidor et al, 1979; Weaver et al, 1984; Sugar and Neumann, 1984)· The expansion step can be nicely described by a percolation phenomena (Sugar et al, 1987) . Germs of permeabilized membranes grow in size and can fuse together, such a description being in favor of the description of the electropermeabilization as a phase transition. A membrane is shifted from the regular impermeable organization to a more fluctuating one where molecules can cross it. The role of the MPD is to reach a critical value where this transition would be permitted. The magnitude of the transition would then be controled by the duration of the pulse.

ACKNOWLEDGEMENTS

Thanks are due to my coworkers Dr. Rols, Montané and Sixou for their fruitful comments and to Prof. Neumann for this pionneering work in electropulsation and for many discussions along the German-French Procope project. This work was supported by the CNRS, the MRT, the MEN, the "Région Midi-Pyrénées" and the ANVAR.

REFERENCES

Abidor IG, Arakelyan VB, Chernomordik LV, Chizmadzhev Yu, Pastushenko VF and Tarasevich MR (1979) Electric breakdown of bilayer lipid membranes. I: The main experimental facts and their qualitative discussion, Bioelectrochem. Bioenerg., 6: 37-52

Bernhardt J. and Pauly H (1973) On the generation of potential difference across the membranes of ellipsoidal cells in an alternating electric field, Biophys. J., 10: 89-98

Chernomordik LV, Sukharev SI, Abidor IG and Chizmadzhev Yu (1983) Breakdown of lipid bilayer membranes in an electric field. Biochim. Biophys. Acta 736: 203-213

Chernomordik LV, Sukharev SI, Popov SV, Pastushenko VF, Sokirko AV, Abidor IG and Chizmadzhev Yu (1987) The electric breakdown of cell and lipid membranes; the similarity of phenomenologies. Biochim. Biophys. Acta 902: 360-373

Crowley JM (1973) Electrical breakdown of biomolecular lipid membranes as an electromechanical instability, Biophys. J., 13: 711-724

Deuticke B and Schwister K (1989) Leaks induced by electrical breakdown in the erythrocyte membrane, in: " Electroporation and Electrofusion in Cell Biology", E. Neumann, A.E. Sowers and C. Jordan, eds, Plenum, New York

Dimitrov DS and Jain RK (1984) Membrane stability, Biochim. Biophys. Acta, 779: 437-468

Dimitrov DS and Sowers AE (1990) Membrane electroporation; fast molecular exchange by electroosmosis. Biochim. Biophys. Acta 1022: 381-382

Dower WJ, Miller JF and Ragsdale CW (1988) High efficiency transformation of *E. Coli* by high voltage electroporation, Nucleic Acids Res., 16: 6127-6144

Ehrenberg BD, Farkas DL, Fluhler EN, Lojewska Z and Loew LM (1987) Membrane potential induced by external electric field pulses can be followed by a potentiometric dye, Biophys. J., 51: 833-837

Gross D, Loew LM and Webb WW(1986) Optical imaging of cell membrane potential changes induced by applied electric fields, Biophys. J., 51: 339-348

Hibino M, Shigemori M, Itoh H, Nagayama K and Kinosita K (1991) Membrane conductance of an electroporated cell analyzed by submicrosecond imaging of transmembrane potential. Biophys. J. 59: 209-220

Kinosita K and Tsong TY (1977) Voltage induced pore formation and hemolysis of human erythrocytes, Biochim. Biophys. Acta, 471: 227-242

Kinosita K and Tsong TY (1977) Hemolysis of human erythrocytes by transient electric fields, Proc. Natl. Acad. Sci. USA, 74: 1923-1927

Kinosita K and Tsong TY (1979) Voltage-induced conductance in human erythrocytes membranes, Biochim. Biophys. Acta, 554: 479-494

Kinosita K, Hibino M, Shigemori M, Ashikawa I, Itoh H, Nagayama K and Ikegami K (1990) Submicrosecond imaging under a pulsed laser fluorescence microscope. Electroporation of cell membrane time- and space-resolved; in " Science on Form", Ishizaka S Ed, KTK Publishers, 97-104

Lojewska Z, Farkas D, Ehrenberg B and Loew LM (1989) Analysis of the effect and membrane conductance on the amplitude and kinetics of membrane potentials induced by externaly applied electric fields. Biophys. J. 56: 121-128

Lopez A, Rols MP and Teissié J (1988) ^{31}P NMR Analysis of membrane phospholipid organization in viable, reversibly electropermeabilized chinese hamster ovary cells, Biochemistry, 27: 1222-1228

Marra J and Israelachvili J (1985) Direct measurements of forces between phosphatidylcholine and phosphatidylethanolamine bilayers in aqueous electrolyte solutions, Biochemistry, 24: 4608-4618

Miller DM (1991) Evidence that interfacial transport is rate-limiting during passive cell membrane permeation. Biochim. Biophys. Acta 1065: 75-81

Montané MH, Dupille E, Alibert G and Teissié J (1990), Induction of a long lived fusogenic state in viable plant protoplasts permeabilized by electric fields, Biochim. Biophys. Acta, 1024: 203-207

Mouneimne Y, Tosi PF, Gazitt Y and Nicolau C (1989) Electroinsertion of Xenoglycophorin into red blood cell membranes. Biochem. Biophys. Res Comm. 159: 34-40

Nagle JF and Scott HL (1978) Lateral compressibility of lipid mono- and bilayers. Theory of membrane permeability, Biochim. Biophys. Acta, 513: 236-243

Neumann E (1989) The relaxation hysteresis of membrane electroporation, in: "Electroporation and Electrofusion in Cell Biology", E. Neumann, A.E. Sowers and C. Jordan, eds, Plenum, New York

Neumann E and Rosenheck K (1972) Permeability changes induced by electric impulses in vesicular membranes, J. Membr. Biol., 10: 279-290

Neumann E, Schaefer-Ridder E, Wang Y and Hofschneider PH (1982) Gene transfer into mouse myeloma cells by electroporation in high electric fields, EMBO J., 1: 841-845

Orlowski S, Belehradek J, Paoletti C and Mir L (1988) Transient electropermeabilization of cells in culture. Increase of the cytotoxicity of anticancer drugs, Biochem. Pharmacol. , 37: 4727-4734

Rols MP and Teissié J (1989) Ionic strength modulation of electrically induced permeabilization and associate fusion of mammalian cells, Eur. J. Biochem., 179: 109-115

Rols MP and Teissié J (1990a) Electropermeabilization of mammalian cells: quantitative analysis of the phenomenon, Biophys. J. 58: 1089-1098

Rols MP and Teissié J (1990b) Modulation of electrically Induced permeabilization and fusion of chinese hamster ovary cells by osmotic pressure, Biochemistry, 29: 4561-4567

Rols MP, Dahhou F, Mishra KP and Teissié J (1990) Control of electric field induced cell membrane permeabilization by membrane order, Biochemistry, 29: 2960-2966

Sixou S and Teissié J (1990) Specific electropermeabilization of leucocytes in a blood sample and application to large volumes of cells, Biochim. Biophys. Acta 1028: 154-160

Sowers AE (1986) A long lived fusogenic state is induced in erythrocytes ghosts by electric pulses, J. Cell. Biol., 102: 1358-1362

Sowers AE (1987) The long lived fusogenic state induced in erythrocytes by electric pulses is not laterally mobile. Biophys. J. 52: 1015-1020

Stulen G (1991)Electric field effects on lipid membrane structure, Biochim. Biophys. Acta, 640: 621-627

Sugar IP and Neumann E (1984) Stochastic model for electric field-induced membrane pores- electroporation, Biophys. Chem., 19: 211-225

Sugar IP, Forster W and Neumann E (1987) Model of cell electrofusion-Membrane electroporation, pore coalescence and percolation, Biophys. Chem. 26: 321-335

Teissié J and Rols MP (1986) Fusion of mammalian cells in culture is obtained by creating the contact between the cells after their electropermeabilization, Biochem. Biophys. Res. Comm., 140: 258-266

Teissié J and Rols MP (1988) Electropermeabilization and electrofusion of cells, in: "Dynamics of membrane proteins and cellular energetics", N. Latruffe, Y. Gaudemer, P. Vignais and A. Azzi, eds, Springer, Berlin

Tekle E, Astumian RD and Chock PB (1990) Electropermeabilization of cell membranes: effect of the resting membrane potential. Biochem. Biophys. Res. Comm. 172: 282-287

Weaver JC, Powell KT, Mintzer RA, Ling H and Sloan SR (1984) The electrical capacitance of bilayer membranes: The contribution of transient aqueous pores, Bioelectrochem. Bioenerg., 12: 393-412 Zimmermann U, Pilwat G and Riemann F (1974) Dielectric breakdown of cell membranes, Biophys. J., 14: 881-899

Zimmermann U.(1982) Electric field fusion and related electrical phenomena, Biochim. Biophys. Acta, 694: 227 -277

ELECTRONIC STRUCTURE AND MAGNETIC CIRCULAR DICHROISM STUDIES OF PROTON TRANSFER BY HISTIDINE

Nancy R. Zhang, Sharon R. Cutler,
John A. Kroll, Loyde F. Jones and
Donald D. Shillady

Department of Chemistry,
Virginia Commonwealth University
Richmond, Virginia, USA

INTRODUCTION

Proton transport in many known enzymatic reactions is accompanied by the participation of a histidyl residue in the active site. A useful text on enzymes (Gray, 1971) lists such cases for acetylcholinesterase, alpha-amylase, aspartate aminotransferase, carbonic anhydrase, carboxypeptidase, chymotripsin, creatinine kinase, fructose-diphosphate aldolase, glucose-phosphate isomerase, glyceraldehyde-3-phosphate NAD oxidoreductase, ketosteroid delta-4 delta-5 isomerase, papain, phosphoglucomutase, ribonuclease, subtilopeptidase-A and trypsin. Because proton transfer is critical in such enzymatic mechanisms and because protons are particles of very low mass, one should ask whether magnetic fields can influence and/or interfere with such processes. To the extent that magnetic effects on biosystems have been documented (Allen, Cleary and Hawkridge, 1989), the next question of where to look for a molecular mechanism brings the histidine proton transfer mechanism under scrutiny. The method of choice in this paper is to use magnetic circular dichroism spectroscopy which measures the chirality induced in the electronic structure of a molecule by immersion in a strong, static magnetic field.

It has been known for over 100 years that rather high magnetic fields are required to measure the Faraday Effect, but only with recent perturbation analysis (Buckingham and Stephens, 1966) has a detailed electronic interpretation been possible; in the data reported here a field of 1.5 Tesla was used. These results are limited to environments with rather high magnetic fields, but well within range of alnico magnets occuring commonly in high technology devices.

The main purpose of this study is to estimate whether static magnetic fields in the range of 1 Tesla can have an effect on proton transfer mechanisms in enzymatic reactions involving a histidyl residue in aqueous media over a range of pH near physiological conditions. Since enzymes are intrinsically optically active and have local electric fields which are also chiral, it is a fair comparison _in vitro_ to study aqueous solutions of histidine in a static magnetic field and compare the natural electric field chirality to the chirality induced by a static magnetic field. To the extent that these effects are comparable in magnitude, one can infer that strong magnetic fields can potentially interfere with enzymatic proton transfer mechanisms and then design further experiments to study specific examples. Natural turnover rates for enzymatic reactions (carbonic anhydrase is a good example) may be sufficiently high that even if strong magnetic fields only interfere with the proton transport to a small degree there may be a measurable effect.

METHODS AND PROCEDURES

The experimental circular dichroism (CD) and magnetic circular dichroism (MCD) spectra were measured using a Jasco J-600 CD spectrometer equipped with an electromagnet. The magnetic field was adjusted to 1.5 T with the light beam along the axial field lines for the samples in buffered aqueous solutions placed in 1 cm path cells. The pH of the solutions was adjusted with HCl or NaOH and measured with a pH meter to achieve the stated values on the spectra. L-histidine (S-isomer) and 4-methyl-imidazole were used as purchased from Aldrich Chemical Co.

Two sets of theoretical calculations were carried out on the VCU IBM 3084 mainframe computer. First, <u>ab initio</u> Hartree-Fock-Roothaan-Pople-Nesbet (Pople and Beveridge, 1970) calculations were performed using a good-quality 4-31G basis set for 4-methyl-imidazole and it's protonated cation; the geometry of the molecules was optimized by minimizing the gradient of the energy to less than 0.001 hartree/bohr. These results were obtained using the HONDO5 program (Dupuis, Rys and King, 1976) as slightly modified for the IBM 3084 mainframe (the only substantial change was to redefine the bohr length in the program to 0.529177 angstroms in accordance with the best available data).

Second, the geometry of histidine and its protonated form was adopted from the SYBYL molecular modeling program (Marshall, 1990) without further optimization and the UV spectrum was estimated using the CNDO/S method (Del Bene and Jaffe, 1968). The MCD parameters were then estimated using a FORTRAN program written several years ago to study the MCD of serotonin and substituted indoles (Sprinkel, Shillady and Strickland, 1975). In this case the electronic excited states of the compounds were approximated by 50 singly-excited configurations based on the deorthogonalized orbitals (Shillady, Billingsley and Bloor, 1971) from the CNDO/S semiempirical Fock matrix. The MCD parameters were then calculated in the corresponding Slater-type basis including all two-center matrix elements in both the angular momentum operators for the magnetic transition moments and the dipole-velocity operator for the electric transition moments; details of the computational method are given in an earlier paper (Richardson, Shillady and Bloor, 1971). The methods used have been shown to be qualitatively reliable with respect to wavelength and the sign and relative magnitude of the MCD spectral features for substituted indoles (Sprinkel, Shillady and Strickland, 1975).

RESULTS AND DISCUSSION

The total gradient-optimized energies of the neutral and protonated forms of 4-methyl imidazole are useful to estimate the relative stability of the ring-protonated form of histidine. The first thing to be noted in Table I is that the protonated form is a cation and that the basicity of the second nitrogen is sufficient to bind the second proton. The sum of the energy of neutral 4-methyl imidazole and a hydrogen atom with an electron in the same gaussian basis set indicates that the protonated form is almost as stable as if an extra electron were added to bind the second proton in a covalent N-H bond! <u>This indicates that it would only take an electron of 2.66 ev to reduce the cation to a neutral ring and a free H atom.</u> This explains the ease with which histidyl proton transfer occurs. We infer that this is the primary mechanism of proton transfer by histidyl groups in a number of enzymes such as chymotrypsin (Blow, 1976; Rebek, 1988) and hormones such as thyroliberin (Giralt, Ludevid and Pedroso, 1986) as previously suggested by other workers. Evidently a number of biosystems make use of this low energy pathway as evidenced by the large number of enzymes containing histidine in the active site (Gray, 1971). Note that <u>a 2.66 ev electron controls proton transfer</u>; a magnetic perturbation of the "reducing electron" thus can effect the transfer of a proton which is some 1828 times more massive!

<u>TABLE I Ab Initio Energies with the 4-31G</u>
<u>Gaussian Basis using HONDO5</u>

Molecule	Total Energy (au)
Energy Gradient	
4-methyl imidazole	-263.455064
(Energy Gradient = 0.0005354) (44 e)	
hydrogen atom (1 e)	-0.499278
cation fragment sum (45 e)	-263.954342
protonated 4-methyl imidazole	-263.856832
(Energy Gradient = 0.0000595) (44 e)	
cation "reduction potential" (au)	0.097510
(ev)	2.653
(kcal/mole)	61.18

The next question one can ask of the <u>ab
initio</u> calculations is what orbitals are most
effected by the protonation of the second
nitrogen site in the metastable intermediate.
Table II gives some results for the one-
electron orbital energies which provides some
insight to the electronic effects caused by
ring protonation; especially on the highest-
occupied molecular orbital (HO) and the two
lowest-unoccupied-molecular-orbitals (LU and
LU+1). Within the molecule there is very
little change in the energy differences that
lead to ultraviolet absorption bands even when
protonation shifts the orbital energies down
by roughly the 0.5 au expected due to the
hydrogen atom nucleus that was added.
However, it is very important to note that
<u>relative to neighboring residues in an enzyme
the protonation would make the lowest orbitals
of a histidyl group lower in energy by about
0.5 au or roughly 13.6 ev!</u> Thus most of the
valence orbitals of the neutral imidazole ring
are stabilized by ring protonation, but the
lowest excited orbitals are also lowered in

energy. It is especially noteworthy that the protonated imidazole virtual orbitals are near zero energy in the SCF calculations

TABLE II 4-31G Orbital Energies (Hartrees)

Molecule	E(HO)	E(LU)	E(LU+1)
4-methyl imidazole	-0.3166	0.1879	0.2271
protonated imidazole	-0.5345	-0.0451	0.0212

1 Hartree = 627.43 kcal/mole

because they are then easily subject to low energy perturbations. The 4-31G basis calculations are generally considered to be very good quality representations of the electronic structure of a molecule and these results should be regarded as "semi-quantitative".

The next set of calculations involved a _magnetic perturbation_ summed over as many as 50 electronic excited states and several approximations were made at this point. First, the excited states of histidine and 4-methyl-imidazole were approximated as single-excitations from an SCF calculation. Second, the SCF calculation was itself the CNDO/S parametric scheme (Del Bene and Jaffe, 1968) in which the neglect of two-center matrix elements occurs and the energy of the 2p-pi orbitals is adjusted to be less than the 2s-sigma orbitals by a factor of 0.585 so that the pi-pi* excitations are fitted to those of benzene. Usually this approximation works exceptionally well for aromatic organic molecules and produces results which are accurate to within a few nm in wavelength for ultraviolet absorptions, but experience in this laboratory has shown that it is poor for saturated organic compounds. Accordingly, the histidine molecule represents quite a challenge to this method because of the presence of both the saturated alpha-carbon and the pi-electronic system of the heterocyclic ring. This work used a 0.585 pi-factor for the imidazole ring alone within the spirit of the CNDO/S method when it became clear that no single value of the pi-factor could be used for both the imidazole ring and the carboxyl group which is part of the saturated portion of histidine; results are only presented for the critical imidazole portion of histidine.

An additional complication occurs because L-histidine has three pK values (Windholz, Budavari, BLumeti and Otterbein, 1983), the first at 1.78 is due to the carboxyl group, the second pK at 5.97 is of interest here due to protonation of the second ring nitrogen and the third pK at 8.97 is due to the ionization of the normally protonated ring nitrogen. This study focused on the ring cation intermediate with both ring nitrogen sites protonated which should exist at pH less than 5.97 and the calculations were compared for the completely neutral and completely protonated species respectively.

Table III shows the results of the calculations for the imidazole ring with a methyl group to simulate the connection to the amino acid part of histidine. Although 50 single configurations were included in the magnetic perturbation calculation, only the lowest four excited states are shown as those nearest to the experimental short wavelength limit of 200 nm. Although a number of MCD spectra were measured, the most useful are for 4-methyl imidazole at pH 4.29 (protonated) and at pH 9.95 (neutral, almost anion) and the MCD of L-histidine at pH 7.32 (neutral).

While there are several formulations of the Buckingham-Stephens equations, the program written in this laboratory adopts the "rigid-shift" model (Stephens, Mowery and Schatz, 1971) which assumes the Zeeman effect on the electronic orbital energies is first order in magnetic field, H, and that the magnetic mixing of the energy levels is "rigid" with respect to their relative order and spacing, to first order. This is a very good approximation for the light elements of biochemical interest which have low orbital angular momentum in S or P orbitals and so have only a small interaction with the magnetic field. The equation for the MCD differential absorption of left- and right-circularly polarized light as used here is of the form:

$$k(l)-k(r) = \text{Constants} \times (Axf - Bxg) \times H \quad (1)$$

Here _k_ is the absorption coefficient for left or right circularly polarized light, _g_ is a band shape and _f_ is the derivative of _g_ with respect to wavelength. The quantity _A_ only occurs if there is a degenerate excited state and is absent in this study. The key point to note is that <u>B occurs as a negative quantity in equation (1) and represents the magnetic chirality introduced into each transition by the magnetic field H.</u> The oscillator strengths, f(osc.), in Table III represent the fraction of an electron which "moves" during a transition (0.001 is a weak band, but 0.1 is a strong band) and the quantity _D_ is the electric dipole transition moment; f(osc.) is proportional to DxD. Thus <u>B/D</u> is a measure of magnetic chirality per electron in a given transition; <u>a negative (B/D) is a positive MCD band.</u> The corresponding experimental spectra are shown in a sequence of figures which follow.

FIGURE 1 shows the low pH MCD spectra of 4-methyl-imidazole for the protonated case. FIGURE 2 shows the MCD spectra at a high pH where the positive band at 201 nm had reached its greatest value as the pH was increased; this is higher than the third pK of histidine and may represent the anion, but it is presented here as a limiting case for increasing pH. FIGURE 3 and FIGURE 4 show the CD and MCD spectra respectively for histidine at pH 7.32, slightly above the second pK to stabilize the neutral species.

<u>TABLE III Computed CNDO/S-D MCD Quantities</u>
Neutral 4-methyl-imidazole

<u>nm</u>	<u>f(osc.)</u>	<u>(B/D)x10^5</u>
241	0.05863	+ 2.09
206	0.01343	−33.83
172	0.23969	+ 1.93
155	0.03438	− 4.31

Protonated 4-methyl-imidazole

<u>nm</u>	<u>f(osc.)</u>	<u>(B/D)x10^5</u>
280	0.08021	+ 0.60
202	0.02931	−11.95
160	0.09398	+ 2.51
139	0.11408	− 3.04

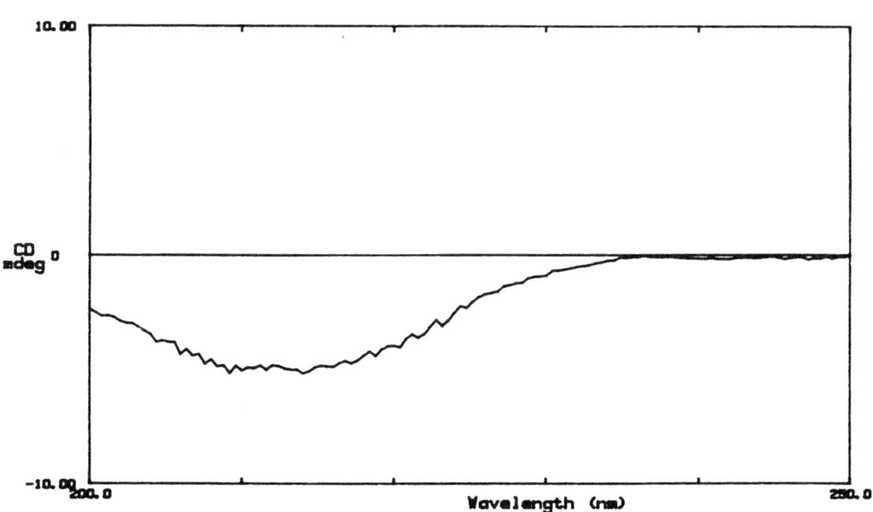

FIGURE 1 Magnetic Circular Dichroism Spectra of 0.0003M 4-methyl imidazole, pH=4.29

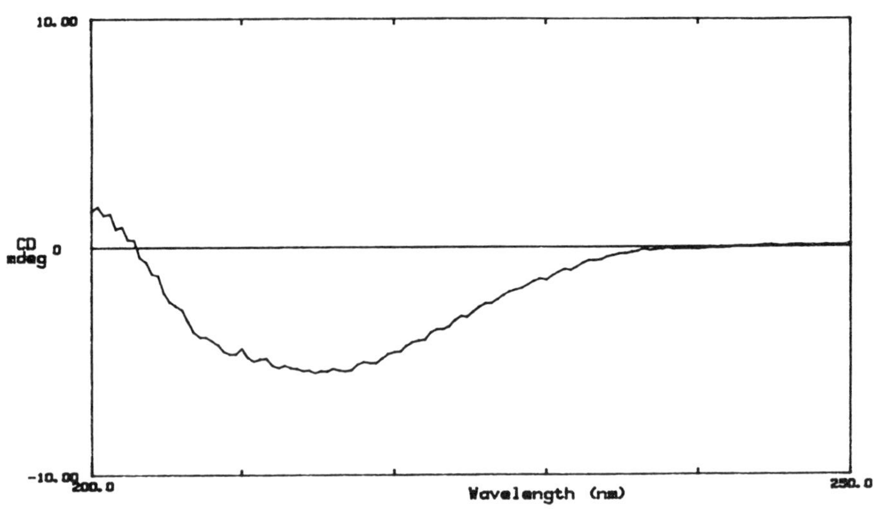

FIGURE 2 Magnetic Circular Dichroism Spectra
of 0.0003M 4-methyl-imidazole, pH=9.95

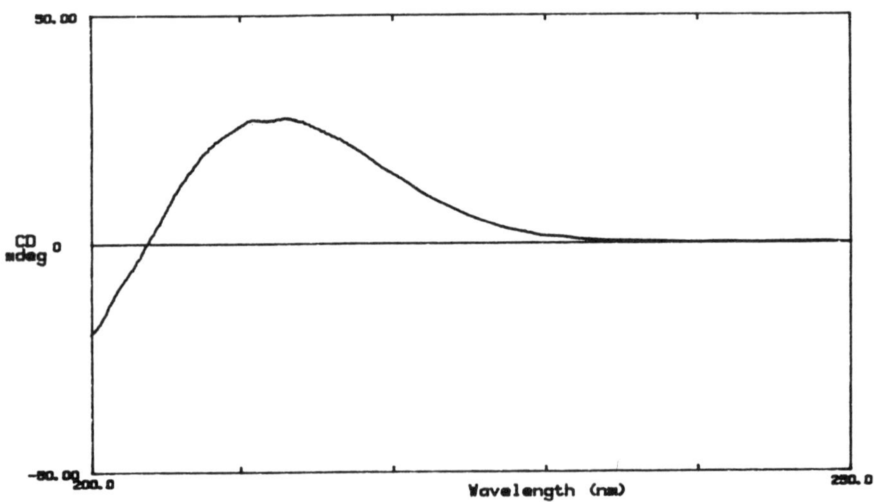

FIGURE 3 Natural Circular Dichroism Spectra
of 0.0003M L-histidine, pH=7.32

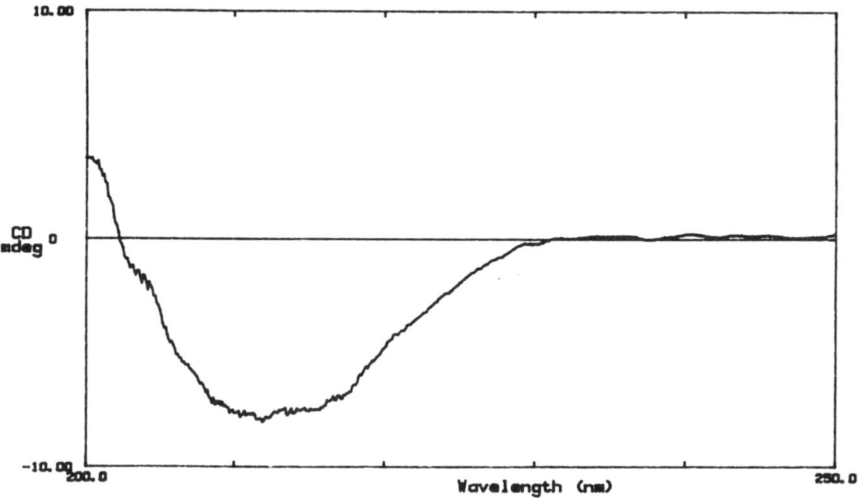

FIGURE 4   Magnetic Circular Dichroism Spectra
of 0.0003M L-Histidine, pH=7.32

Although the computed wavelengths are
several nm to the red of the experimental
bands, complete qualitative agreement is
reached for the main spectral features. The
first band of both low pH and high pH forms is
negative in agreement with a positive (B/D)
value and the second band is positive in
agreement with a negative (B/D) value. The
key feature is the second band at about 200 nm
which is much larger in magnitude for the
neutral species; this is visible in the
histidine spectra as well as in the spectra of
4-methyl imidazole. At 1.5 T the effect of
the magnetic field is about 25% opposite
chirality to the natural CD induced by the
chirality at the alpha-carbon atom as shown in
the last spectra; note vertical scale.

CONCLUSIONS

While it seems obvious to look for biomolecular mechanisms which couple magnetic fields to key enzymes containing iron and other transition metal ions, it is likely that the much larger number of mechanisms utilizing histidyl proton transfer is another "magnetic insult" worth studying. Even if histidyl proton transfer is only slightly effected by magnetic fields, the number of biosystems effected is sure to be at least a "biological stress" of quite general nature due to the wide scope of the effect on many biological mechanisms. At worst, the effect of magnetic fields on histidy proton transfer may lead to an ill-defined "malaise" of an organism which would be difficult to pinpoint specifically because of the many ways the mechanism can effect an organism.

A field of 1.5 T definitely makes a large difference in the chirality of the 200 nm band of the histidine imidazole ring, a transition some 6.2 ev above the electronic ground state. We can infer that the lower energy "reducing electron" of about 2.7 ev would be perturbed to a greater extent by a magnetic field. While the electronic chirality induced by the magnetic field will only distort the molecular geometry to an insignificant amount, <u>the enzymatic-transfer motion of a proton by a histidyl cation in a magnetic field will be distorted by at least (Me/Mp) = 1/1828 and probably much more since the path of the lighter "reducing electron" will also be distorted. This could be significant in systems with high turn-over rates (carbonic anhydrase).</u> Thus the MCD chirality is roughly a direct measure of a magnetically-induced distortion of proton-transfer pathways when the process is controlled by valence electron transfer. Since a 6.2 ev electron transition has been measured here to be strongly effected by a 1.5 Tesla field, we can infer that lower energy valence electron transfers would be effected even more. This preliminary study calls for further experiments to conduct measurements of turnover rates for enzyme systems in magnetic fields where histidyl proton transfer is part of the mechanism.

REFERENCES

Allen M J, SF Cleary and Hawkridge FM, eds. (1989), _Charge and Field Effects in Biosystems-2_, New York, Plenum Press.
Blow D (1976): Structure and Mechanism of Chymotrypsin. _Acct. of Chem. Res._ 9: 145-152.
Buckingham AD and Stephens PJ (1966): Magnetic Optical Activity. _Ann. Rev. Phys. Chem._ 17, 399-432.
Del Bene J and Jaffe HH (1968): Use of the CNDO/2 Method in Spectroscopy. II. Five-Membered Rings. _J Chem. Phys._ 48: 4050-4055.
Dupuis M, Rys J and King HF (1976): Evaluation of Molecular Integrals Over Gaussian Basis Functions. J. Chem. Phys. 65: 111-116.
Marshall G (1990): SYBYL Molecular Modeling Program, Tripos Associates, St. Louis, Mo. USA.
Giralt E, Ludevid MD and Pedroso E (1986): The Relevance of Imidazole Tautomerism for the Hormonal Activity of Histidine-Containing Peptides. _Bioorganic Chemistry_ 14: 405-416.
Gray CJ (1971): _Enzyme-Catalyzed Reactions_, New York, Van Nostrand Reinhold Co.
Pople JA and Beveridge DL (1970): _Approximate Molecular Orbital Theory_, Chapter 2, New York, Mcgraw-Hill Book Co.
Rebek J (1988): On the Structure of Histidine and it's Role in Enzyme Active Sites. _Struct. Chem._ 1: 129-131.

Richardson SF, Shillady DD and Bloor JE
(1971): The Optical Activity of Alkyl-
Substituted Cyclopentanones, INDO Molecular
Orbital Model. J Phys. Chem. 75: 2466-2479.
Shillady DD, Billingsley FP and Bloor JE
(1971): Valence Shell Calculations IV, The
Effect of Deorthogonalization on CNDO/2
Dipole Moments and Charge Distributions.
Theor. Chim. Acta 21: 1-8.
Sprinkel FM, Shillady DD and Strickland RW
(1975): Magnetic Circular Dichroism Studies
of Indole, DL-Tryptophan and Serotonin.
J Amer. Chem. Soc. 97: 6653-6657.
Stephens PJ, Mowery RL and Schatz PN (1971):
Moment Analysis of Magnetic Circular
Dichroism: Diamagnetic Molecular Solutions.
J Chem. Phys. 55: 224-231.
Windholz MS, Budavari S, Blumeti RF and
Otterbein ES (1983): The Merck Index,
Rahway, N.J., Merck and Co. Publishing,
p683.

Photo-Induced Bioelectrochemical Processes

FLAVIN LASER FLASH PHOTOLYSIS STUDIES OF THE ELECTRON TRANSFER MECHANISM IN REDOX PROTEINS

Miguel A. De la Rosa, José A. Navarro, Mercedes Roncel, Antonio Díaz, Manuel Hervás
Instituto de Bioquímica Vegetal y Fotosíntesis, Universidad de Sevilla y CSIC, Apartado 1113, 41080-Sevilla, Spain

Gordon Tollin
Department of Biochemistry, University of Arizona, Tucson, AZ 85721, USA

INTRODUCTION

Most fundamental processes in biology, such as photosynthesis and respiration, involve the transfer of electrons between different substrates and require the active participation of a number of redox proteins (e.g., ferredoxin, plastocyanin, cytochromes) that frequently consist of one or more polypeptide chains binding one or more prosthetic groups (e.g., iron-sulfur clusters, copper ions, hemes). Such redox proteins are usually arranged in a well-ordered system - the biological membranes - and act as effective carriers of electrons, which generally enter at one redox level and leave at another. There also exist multi-center redox proteins with tightly associated subunits (e.g., flavocytochromes, multiheme and molybdoheme enzymes) in which electrons similarly enter at one redox level and leave at another (Marcus and Sutin, 1985; Matthew, 1985; Tollin et al, 1986a; 1986b).

The electron transfer mechanism in such redox systems is not well understood, even though important efforts have been made over the last few years. Recent experimental evidence suggests that the most fundamental features of the electron transfer mechanism in proteins are the same as those of the much simpler non-biological systems. The redox potential difference between the prosthetic groups appears to be one of the most decisive factors determining the rate of biochemical electron transfer reactions. Exposure and steric accessibility of the reaction centers at the protein surface and electrostatic interactions involving charged groups at or near the reaction sites are two more factors of critical importance (Manstein et al, 1988; Marcus and Sutin, 1985; Matthew, 1985; Tollin et al, 1986a).

ELECTRON TRANSFER REACTIONS IN PROTEINS AND ENZYMES

During the last several years, we have been studying the mechanism of electron transfer reactions in a number of redox proteins and enzymes involved in photosynthesis, either in the light-induced electron flow or in the dark reactions. In particular, the heme protein cytochrome $c552$ (a member of class I c-type cytochromes whose generic name is cytochrome $c553$) and the copper protein plastocyanin have been the objects of special investigation. As shown in Fig. 1, these two metalloproteins are functionally equivalent, and play the same physiological role as redox carriers between cytochrome f and the photooxidized chlorophyll a (P700) in the oxidizing side of photosystem I (PSI). Actually, in cyanobacteria and some eukaryotic algae, plastocyanin is replaced by cytochrome $c552$ as electron donor to PSI when the organisms are grown in low copper media (Sandmann and Böger, 1980; Wood, 1977).

Plastocyanin and cytochrome $c552$ have been isolated from the green alga *Monoraphidium braunii*. Although they have different redox centers, both proteins exhibit an acid isoelectric point and have similar redox potentials (E'o,

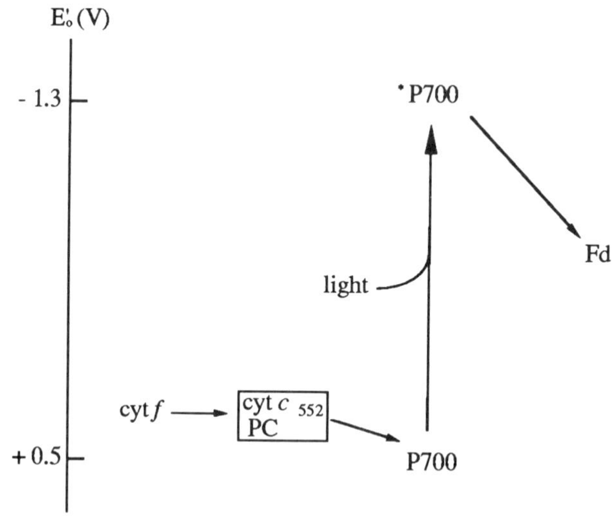

FIGURE 1. Photosynthetic electron flow driven by the photoexcited chlorophyll a (P700) in photosystem I. In some green algae and cyanobacteria, cytochrome $c552$ and plastocyanin can interchangeably act as redox carriers between cytochrome f and P700. Whichever of the two metalloproteins is synthesized depends on the relative availabilities of iron and copper in the culture medium.

approximately +350 mV) and molecular weights (*ca.* 8 kDa), which is in good agreement with their identical physiological function (unpublished data). In principle, one would expect them to have closely analogous electron transfer properties. Experiments carried out both under steady-state conditions (Hervás et al, 1991b) and by laser flash absorption spectroscopy (Hervás et al, 1991a) have actually shown that cytochrome $c552$ and plastocyanin can reduce the photo-oxidized PSI with similar efficiency, the rate constants for the respective electron transfer reactions showing a similar dependence on ionic strength and pH.

Another system of great interest regarding the mechanism of photosynthetic electron transfer reactions is that involving the reduction of nitrate to nitrite. In photosynthetic organisms, the reducing power - that is, energized electrons at the level of ferredoxin or pyridine nucleotides - generated in the light reactions is used to reduce the oxidized bioelements carbon, nitrogen and sulphur. In the case of nitrate assimilated by higher plants and eukaryotic algae, it is the enzyme nitrate reductase which catalyzes the reduction of nitrate to nitrite with electrons donated by reduced pyridine nucleotides, either NADPH or NADH. The so-called NAD(P)H-nitrate reductase is an extraordinarily complex enzyme, which contains FAD, cytochrome $b557$ and molybdenum as prosthetic groups. Oxidation-reduction midpoint potentials have been determined for the theree redox centers of *Chlorella* and spinach nitrate reductases using various techniques including visible, CD, and EPR potentiometric titrations and microcoulometry (Solomonson and Barber, 1990). The values obtained have confirmed previous kinetic studies which established the sequence of electron transfer between the redox centers. The thermodynamic scheme presented in Fig. 2 shows how the electrons enter from reduced pyridine nucleotides into the flavin cofactors, are transferred intramolecularly to the hemes, and then to the molybdenum ions to be finally removed by nitrate in a bimolecular reaction. The global process is strongly exergonic ($\Delta G^{\circ \prime} = -1.48$ eV; $\Delta E'_o = 0.74$ V), the electrons falling down in a cascade inside the enzyme from NAD(P)H to nitrate by hopping from one prosthetic group to the other. The three enzyme components actually form a short electron transport chain within the multisubunit enzyme (De la Rosa et al, 1989; 1991).

It is important to realize that cytochrome $b557$, as well as cytochrome $c552$ and plastocyanin, plays a functional role as a redox carrier between two other redox components in an electron transport chain. However, an important difference is that cytochrome $c552$ and plastocyanin are free mobile proteins, which interact with the membrane-bound cytochrome b_6f complex to be reduced, and then move to another place to interact with the PSI complex, equally embedded in the thylakoid membrane, to be oxidized. Cytochrome $b557$ - or, more properly, the heme $b557$ - is rather a prosthetic group, that is, a fixed element inside the enzymatic complex nitrate reductase, with no possibility to move independently of the other redox centers.

An important question which arises in this context concerning the reaction mechanism of such redox proteins is the following: are the sites for oxidation and reduction the same or different? In proteins having freedom of movement, the same active site could be used, in principle, for electron entry and removal. In

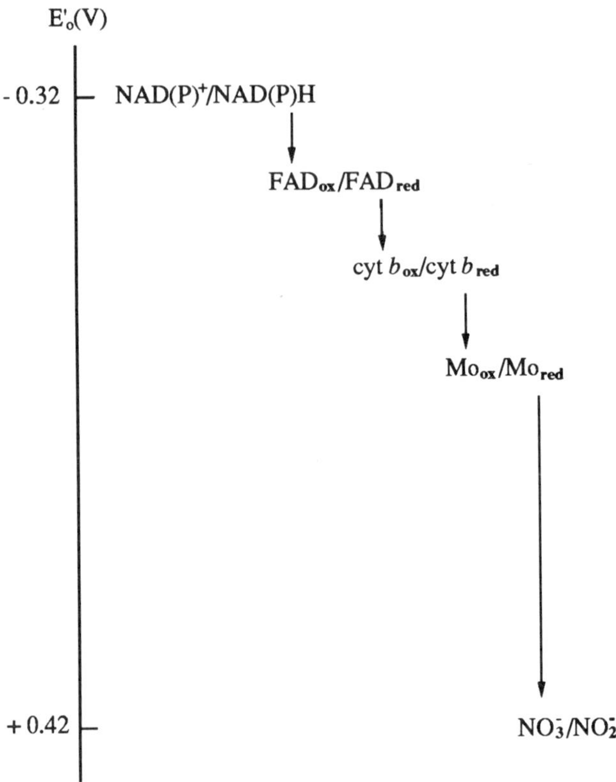

$$NAD(P)H + H^+ + NO_3^- \xrightarrow{2\,e^-} NAD(P)^+ + NO_2^- + H_2O$$

$$\Delta E'_o = 0.74\ V\ ;\quad \Delta G^{o\prime} = -1.48\ eV$$

FIGURE 2. Photosynthetic reduction of nitrate to nitrite with electrons donated by reduced pyridine nucleotides, either NADPH or NADH. This is a strongly exergonic reaction catalyzed by the complex NAD(P)H-nitrate reductase, a polymeric enzyme itself containing FAD, cytochrome $b557$ and molybdenum as redox centers which form a short electron transport chain within the protein. The redox potential values of the three prosthetic groups are based on those reported by Solomonson and Barber (1990).

non-mobile proteins, however, the oxidation and reduction reactions would be required to occur at different sites, unless some rotational movement of the protein subunits (or the redox center) takes place so that the electrons can enter and leave at the same site.

In order to compare the sites for oxidation and reduction in redox proteins, a quite useful light-induced procedure based on the peculiar structural and photochemical properties of flavins (see below) has been developed. Actually, much of our present knowledge on the structure-function relationships in redox proteins has been obtained from experiments on the interactions and electron-transfer reactions between flavins and proteins (Meyer et al, 1983; Navarro et al, 1991a; 1991b; Tollin et al, 1986a; 1986b; Tollin and Hazzard, 1991).

STRUCTURE AND PHOTOCHEMICAL PROPERTIES OF FLAVINS

Flavins are yellow pigments which are based on the heterocycle 7,8-dimethylisoalloxazine, and differ from one another in the side chain bound to N(10) of the heterocycle ring (Fig. 3). Some flavins are very well known in

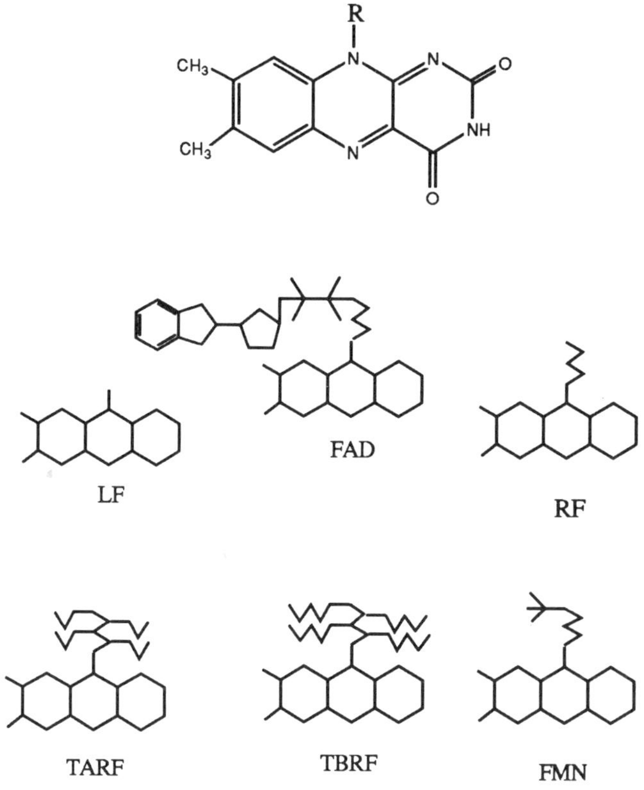

FIGURE 3. Molecular structure of flavins. *Upper*, the 7,8-dimethylisoalloxazine ring from which derive the different flavins, which differ from one another depending on their side chain R. *Lower*, schematic drawing of a series of flavins: LF, lumiflavin; RF, riboflavin; FMN, flavin mononucleotide; TARF, riboflavin-2',3',4',5'-tetraacetate; TBRF, riboflavin-2',3',4',5'-tetrabutyrate; FAD, flavin adenine dinucleotide. LF, RF, FMN and FAD are commercially available, whereas TARF and TBRF were a generous gift of Dr. P.F. Heelis.

biochemistry, such as riboflavin - its side chain is a ribityl radical - which is the familiar vitamin B_2 that constitutes the basis for the synthesis of the redox coenzymes FMN and FAD, important prosthetic groups for a wide number of enzymes commonly known as flavoenzymes, or flavoproteins. In addition to such natural flavins, there exist many artificial flavins, chemically synthesized in the laboratory, with different side chains, both polar and non-polar. Some of them are schematically presented in Fig. 3 to emphasize their different side chains. As a consequence, we can utilize a set of flavin molecules with the same functional group but different side chains, which can thus supply important information regarding the steric accessibility and the electrostatic character of the active site in proteins when studying flavin-protein interactions and electron transfer reactions.

The midpoint redox potential value of flavins in aqueous solutions at neutral pH is *ca.* -0.22 V; that is, reduced flavins behave as rather strong reducing agents (Draper and Ingraham, 1968). As shown in Fig. 4, the flavosemiquinone radical (Fl⁻) can easily donate in the dark the electron lodged in its antibonding pi molecular orbital so as to become oxidized in its ground state (Fl). However, the bright yellow colour of flavins is due to their intense absorbance of blue light

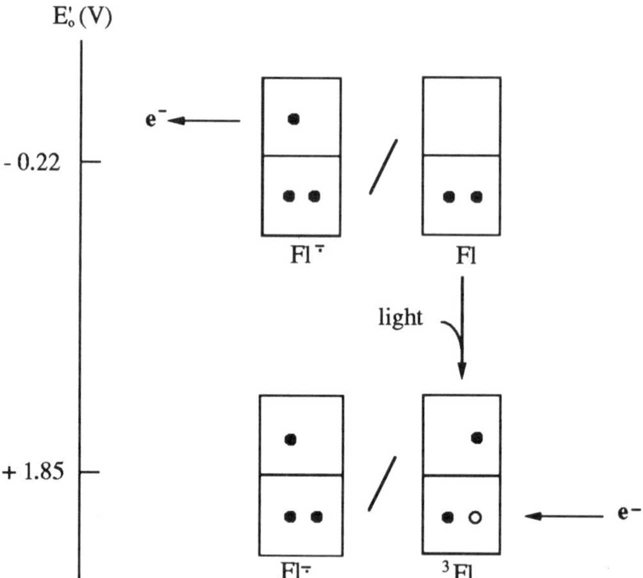

FIGURE 4. Oxidation of flavin semiquinones at low potential in the dark, and reduction of photoexcited triplet flavins at high potential in the light. Photo-excitation of flavins (Fl) involves the promotion of one electron from a filled low-energy molecular orbital to an empty high-energy molecular orbital, the resulting excited triplet state (³Fl) having an additional electronic energy of 2.07 eV over the ground state. As a consequence, the midpoint potential value of the pertinent redox pair is shifted from -0.22 V in the dark to +1.85 V in the light.

(molar absorption coefficient at 450 nm, 1.2 x 10^4 M^{-1} cm^{-1}). Light energization of flavins actually involves the promotion of one electron from a full low-energy bonding pi orbital to an empty high-energy antibonding pi orbital in such a way that their electron affinity is significantly increased. In other words, the midpoint redox potential value of the pertinent redox couple is shifted to +1.85 V because of energization of the oxidized form of the redox pair; the photoexcited triplet state (^{3}Fl) actually has an additional electronic energy of 2.07 eV over the ground state (De la Rosa et al, 1989; Heelis, 1982). Fig. 4 shows how one electron can be easily accomodated in the hole left in the bonding orbital by the photoexcited electron. This is the reason why photoexcited flavins in their triplet state can behave as strong oxidizing agents, accepting electrons from a wide range of molecules such as amino acids, hydroxycarboxylic acids, thiols, aldehydes, and unsaturated hydrocarbons which, on the other hand, are unable to reduce flavins in the dark (Heelis, 1982).

Utilizing these two peculiar characteristics of flavins - that is, their identical functional group but different side chain, and their redox behaviour as reductant or oxidant in the dark and in the light, respectively - we have extensively studied the reaction mechanism of electron transfer - both oxidation and reduction - between flavins and redox proteins.

FLAVIN-PHOTOSENSITIZED OXIDATION AND REDUCTION OF REDOX PROTEINS

As noted above, upon light excitation under anaerobic conditions, flavins become electronically energized to their metastable triplet state (lifetime, 10-100 μs), and can remove electrons from an appropriate reduced protein present in the reaction medium according to the following reactions:

$$Fl + h\nu \longrightarrow {}^3Fl$$

$$^3Fl + Prot_{red} \longrightarrow Fl^- + Prot_{ox}$$

However, if flavin photoexcitation takes place in the presence of high concentrations of a molecule such as EDTA - which is unable to reduce flavins in the dark, but is an effective donor of electrons to the flavin triplet state - photoexcited flavins remove electrons from such an unexpected reducing agent

$$^3Fl + EDTA \longrightarrow Fl^- + \text{oxidation products of EDTA}$$

and the resulting flavin semiquinone radical can in turn reduce an appropriate oxidized protein which is present in the solution

$$Fl^- + Prot_{ox} \longrightarrow Fl + Prot_{red}$$

One can thus follow the reaction kinetics of oxidation or reduction of redox proteins by exciting free flavins either in the absence or in the presence of EDTA, respectively (De la Rosa et al, 1989; 1991). This approach has been applied successfully to the study of the redox photoregulation of the catalytic activity of the enzyme nitrate reductase, which can be inactivated by reduction of its molybdenum centers and reactivated upon their reoxidation (De la Rosa et al 1989; Solomonson and Barber, 1990). Thus, nitrate reductase from the green alga *Monoraphidium braunii* has been demonstrated to be inactivated by irradiation in the presence of flavins and EDTA, whereas it becomes reactivated when illuminated in the presence of flavins alone (De la Rosa et al, 1989; Navarro et al, 1991b). The steady-state kinetics of flavin-photosensitized nitrate reductase inactivation and reactivation were followed using the various flavin species presented in Fig. 3. The reaction rate was found to be mostly dependent on the size of the flavin side chain, thus demonstrating the importance of steric factors in the flavin-protein interactions (Navarro et al, 1991b).

A more profound understanding of the reaction mechanism of flavin-sensitized nitrate reductase redox photoregulation would require high protein concentrations - much higher than those required to determine its catalytic activity - so as to detect by spectrophotometric techniques (visible, EPR) the redox state of its prosthetic groups. However, nitrate reductase is rather scarce inside the cells, and it will thus be necessary to employ the powerful tools of molecular biology to obtain significant amounts of pure nitrate reductase, a difficult task to be undertaken in the near future.

At the present time, and for the sake of simplicity, the study of flavin-sensitized redox reactions was focused on soluble proteins containing a single redox center, such as the photosynthetic c-type cytochromes and copper proteins. Fig. 5 shows the results of a steady-state irradiation experiment which clearly demonstrates the ability of flavins to photosensitize the oxidation of the reduced cytochrome $c552$ and plastocyanin from *Monoraphidium braunii*. In both cases, the absorbance changes observed throughout the experiment indicated that the reduced protein was being oxidized, as effectively demonstrated by the absorption spectrum recorded after irradiation, once the absorbance signal became stabilized, in comparison with the absorption spectrum recorded just before irradiation (Navarro et al, 1991a; Roncel et al, 1990). That this is a triplet state sensitized process was confirmed by experiments using laser flash photolysis, a technique which allows us to follow the kinetics of formation and subsequent disappearance of the transient species formed in the course of the photochemical reactions.

FLAVIN LASER FLASH PHOTOLYSIS STUDIES OF CYTOCHROME $c552$ AND PLASTOCYANIN

By exciting free flavins with a short duration laser pulse (a few nanoseconds), either in the absence or presence of an electron donor like EDTA, it is possible to populate with high yield the triplet and semiquinone flavin states, respectively. Time-resolved spectrophotometry then allows us to follow the kinetics of triplet and semiquinone disappearance upon interaction with the proteins in oxidative and reductive reactions, thus providing valuable information on the active redox site(s) in the proteins.

This procedure was employed to determine the kinetics and mechanism of the flavin-sensitized photooxidation of the cytochrome $c552$ and plastocyanin

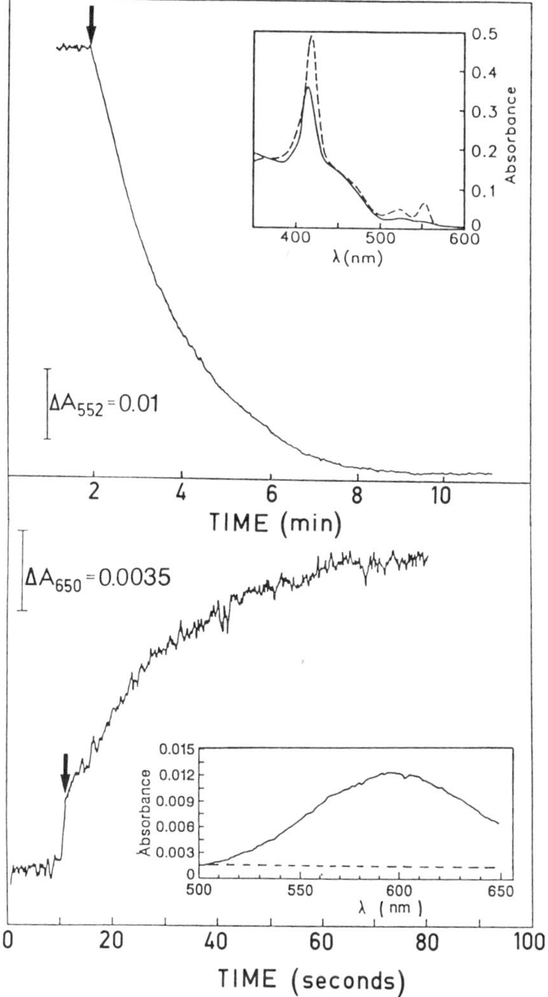

FIGURE 5. Steady-state kinetics of the flavin-sensitized photooxidation of native reduced cytochrome $c552$ (*upper*) and plastocyanin (*lower*) isolated from the green alga *Monoraphidium braunii*. The insets show the electronic absorption spectra of the two redox proteins recorded just before (dashed line) or after (continuous line) irradiation, which correspond to their respective reduced and oxidized states.

considered above (Navarro et al, 1991a; Roncel et al, 1990). Fig. 6 shows the kinetic transients obtained at 640 nm, which are associated with the flavin triplet state, both in the absence and in the presence of the two algal proteins. The initial rise in absorbance at 640 nm is due to the formation of the flavin triplet state, and the subsequent decay corresponds to the triplet disappearance. As expected, the triplet decay was accelerated in the presence of reduced proteins, thus indicating that the algal proteins effectively quench the flavin triplet state.

The parallel absorbance decrease at 552 nm in the case of cytochrome $c552$, as well as a significant absorbance increase at 600 nm in the case of plastocyanin (as the spectral properties of flavin triplet and reduced plastocyanin are quite similar in the red region of the visible spectrum, it is very difficult to monitor independently both species, although the triplet state dominates at 640 nm and the plastocyanin at 600 nm), unequivocally demonstrated that the oxidized protein was being formed in the course of the reaction (Navarro et al, 1991a; Roncel et al, 1990).

Parallel studies were also carried out using ferricyanide-oxidized proteins and flavins in the presence of high concentrations of EDTA. Under these conditions, the flavin triplet state is efficiently quenched by EDTA to generate the flavosemi-quinone radicals, which then transfer one electron to the oxidized cytochrome $c552$ or plastocyanin (see above). Fig. 6 likewise shows kinetic transients

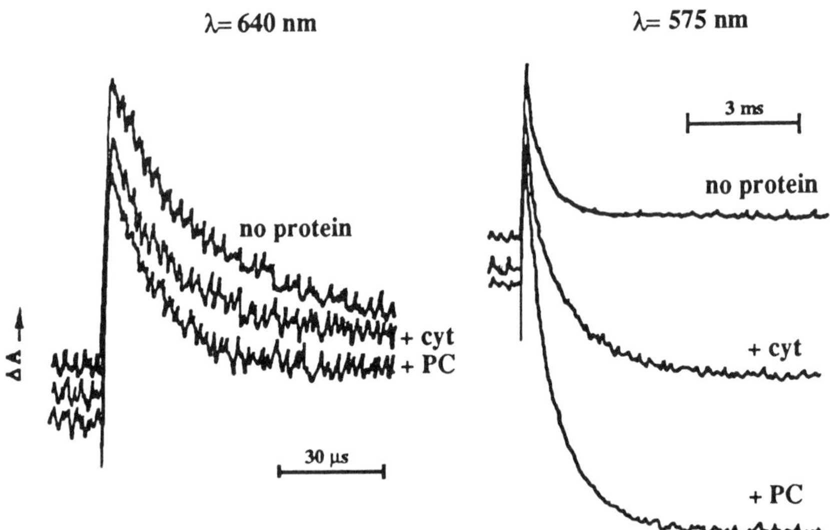

FIGURE 6. Flavin-photosensitized redox reactions of algal cytochrome $c552$ and plastocyanin as studied by laser flash photolysis. *Left*, kinetic traces showing flavin triplet state formation and decay at 640 nm, both in the absence and in the presence of the native reduced proteins. *Right*, kinetic traces showing EDTA-dependent flavin semiquinone formation and subsequent decay at 575 nm, both in the absence and in the presence of the corresponding ferricyanide-oxidized proteins. See text for details.

obtained at 575 nm in solutions containing flavin and EDTA, both in the absence and in the presence of oxidized algal protein. After an initial rapid rise in absorbance due to flavin semiquinone formation, the absorbance decreases as a consequence of either semiquinone disproportionation (in the absence of protein) or electron transfer from the semiquinone to the protein. Again as expected, the semiquinone disappearance kinetics were accelerated in the presence of the oxidized proteins and the absorbance decreased to below the preflash baseline, thus indicating that protein reduction was occurring (Navarro et al, 1991a; Roncel et al, 1990).

The kinetics of algal cytochrome $c552$ and plastocyanin oxidation by flavin triplet states and reduction by flavosemiquinones were followed using different types of flavins (lumiflavin, riboflavin and FMN) so as to determine the steric and electrostatic effects of the flavin side chain on flavin-protein interactions. Table I summarizes the second-order rate constants of the respective redox reactions, including those corresponding to spinach plastocyanin which was used for comparative purposes. Based on the values given in Table I, it is apparent that the two algal proteins are quantitatively very similar in their oxidative and reductive reactions with all three flavin triplets and semiquinones, respectively.

Protein/Flavin	k_2 or k_∞ (M^{-1} s^{-1})	
	Oxidation	Reduction
algal cytochrome $c552$		
lumiflavin	2.6×10^9	11.1×10^7
riboflavin	1.9×10^9	7.0×10^7
FMN	1.2×10^9	7.5×10^7
algal plastocyanin		
lumiflavin	2.2×10^9	6.2×10^7
riboflavin	1.7×10^9	5.0×10^7
FMN	1.5×10^9	5.9×10^7
spinach plastocyanin		
lumiflavin	1.6×10^9	2.4×10^7*
riboflavin	1.1×10^9	2.0×10^7*
FMN	0.7×10^9	1.6×10^7*

TABLE I. Rate constants for oxidation and reduction of redox proteins by flavin triplet states and semiquinones. k_2 is the measured second-order rate constant, and k_∞ is the second-order rate constant extrapolated to infinite ionic strength. All the rate constant values reported in the Table are from Navarro et al (1991a), except those marked by an asterisk which are from Tollin et al (1986b).

This is consistent with their physiological properties, which lends support to the idea that the use of flavin triplet and semiquinone species as probes provides information which is biologically relevant. The spinach plastocyanin was shown to be more sensitive to FMN substitution than is the algal protein.

It is important to realize that the rate constant for reduction are considerably smaller - one or two orders of magnitude - than those for oxidation, which suggests that the triplet state is a more effective reactant than is the semiquinone. This is consistent with the excited state nature of the triplet state, i.e. its high free-energy content which would provide a large thermodynamic driving force for the redox reaction (Marcus and Sutin, 1985).

The study of the ionic strength effect on the FMN-photosensitized oxidation and reduction of the algal cytochrome $c552$ and of the two plastocyanins suggested that the same (or closely adjacent) sites on the three proteins are involved in the two redox reactions, that is, that the site for the electron entry and removal could be the same (Navarro et al, 1991a; Roncel et al, 1990).

In conclusion, the present experiments demonstrate that flavin laser flash photolysis offers a useful probe for comparison of the oxidation and reduction processes of a wide variety of redox proteins, thus supplying important information on the active sites at the protein surface. It will be informative to extend these studies to multi-center redox proteins such as nitrate reductase.

ACKNOWLEDGEMENTS

The authors wish to thank Prof. M. Losada for helpulful advice and criticism. The research from our laboratories reviewed here was supported by grants from The Spanish Ministry of Education and Science (DGICYT, PB87-401), the Andalusian Government, the Division of Chemical Sciences, Office of Basic Energy Sciences, Office of Energy Research, U.S. Department of Energy (DE-FG02-86ER13631), NIH (DK 15057), and NATO (CRG 900065).

REFERENCES

De la Rosa MA, Navarro JA, Roncel M, Hervás M and Tollin G (1991): Flavin-photosensitized oxidation and reduction of redox proteins. In: *Research Trends - Photochemistry and Photobiology*, Menon J, ed. Trivandrum: Council of Scientific Research Integration, in press.

De la Rosa MA, Roncel M and Navarro JA (1989): Flavin-mediated photoregulation of nitrate reductase; a key point of control in inorganic nitrogen photosynthetic metabolism. *Bioelectrochem Bioenerg* 22: 355-364.

Draper RD and Ingraham LL (1968): A potentiometric study of the flavin semiquinone equilibrium. *Arch Biochem Biophys* 125: 802-808.

Heelis PF (1982): The photophysical and photochemical properties of flavins (isoalloxazines). *Chem Soc Rev* 11: 15-39.

Hervás M, De la Rosa MA and Tollin G (1991a): A comparative laser flash absorption spectroscopy study of the kinetics of algal plastocyanin and cytochrome c-552 photooxidation by photosystem I particles from spinach. *Eur J Biochem*, submitted.

Hervás M, Díaz A, Roncel M, Navarro JA and De la Rosa MA (1991b): Cytochrome c552 and plastocyanin from green algae; a comparative study of their respective functional interaction with the PSI complex. *Workshop on Transition Metal Clusters in Biology*, Lübeck.

Manstein DJ, Massey V and Ghisla S (1988): Stereochemistry and accessibility of prosthetic groups in flavoproteins. *Biochemistry* 7: 2300-2305.

Marcus RA and Sutin N (1985): Electron transfers in chemistry and biology. *Biochim Biophys Acta* 811: 265-322.

Matthew JB (1985): Electrostatic effects in proteins. *Annu Rev Biophys Chem* 14: 387-417.

Meyer TE, Przysiecki CT, Watkins JA, Battacharyya A, Simondsen RP, Cusanovich MA and Tollin G (1983): Correlation between rate constant for reduction and redox potential as a basis for systematic investigation of reaction mechanisms of electron transfer proteins. *Proc Natl Acad Sci USA* 80: 6740-6744.

Navarro JA, De la Rosa MA and Tollin G (1991a): Transient kinetics of flavin-photosensitized oxidation of reduced proteins: comparison of c-type cytochromes and plastocyanins. *Eur J Biochem*, in press.

Navarro JA, Roncel M and De la Rosa MA (1991b): On the reaction mechanism of flavin-sensitized photoregulation of *Monoraphidium braunii* nitrate reductase. *J Photochem Photobiol, B: Biol*, in press.

Roncel M, Hervás M, Navarro JA, De la Rosa MA and Tollin G (1990): Flavin-photosensitized oxidation of reduced c-type cytochromes. *Eur J Biochem* 191: 531-536.

Sandmann G and Böger P (1980): Copper-induced exchange of plastocyanin and cytochrome c553 in cultures of *Anabaena variabilis* and *Plectonema boryanum*. *Plant Sci Lett* 17: 417-424.

Solomonson LP and Barber MJ (1990): Assimilatory nitrate reductase; functional properties and regulation. *Annu Rev Plant Physiol Plant Mol Biol* 41: 225-253.

Tollin G and Hazzard JT (1991): Intra- and intermolecular electron transfer processes in redox proteins. *Arch Biochem Biophys* 287: 1-7.

Tollin G, Meyer TE and Cusanovich MA (1986a): Elucidation of the factors which determine reaction-rate constants and biological specificity for electron transfer proteins. *Biochim Biophys Acta* 853: 29-41.

Tollin G, Meyer TE, Cheddar G, Getzoff ED and Cusanovich MA (1986b): Transient kinetics of reduction of blue copper proteins by free flavin and flavodoxin semiquinones. *Biochemistry* 25: 3363-3370.

Wood PM (1977): The roles of c-type cytochromes in algal photosynthesis; extraction from algae of a cytochrome similar to higher plants cytochrome f. *Eur J Biochem* 72: 605-612.

CHARACTERIZATION OF CHARGE SEPARATION IN MEMBRANE SPANNING PROTEIN REACTION CENTERS OF BACTERIAL PHOTOSYNTHESIS[a]

Theodore J. DiMagno[t,*], Chi-Kin Chan[t], Deborah K. Hanson[§],
Marianne Schiffer[§], Graham R. Fleming[t] and James R. Norris[*,t],

[*]Chemistry Division, Argonne National Laboratory, Argonne, Illinois,
60439

[§]Biological and Medical Research Division, Argonne National Laboratory,
Argonne, Illinois, 60439

[t]Department of Chemistry, The University of Chicago, Chicago, Illinois,
60637

INTRODUCTION

The detailed description and understanding of membrane bound
proteins has greatly increased in the past decade. In large part this
advancement resulted from the investigation of the structure-function
relationship for the photosynthetic electron transfer process occurring in
bacterial reaction center proteins. The most important development was
the growth of reaction center protein crystals (Michel, 1982) and the
following structure determination for this reaction center protein complex
(Deisenhofer *et al.*, 1984; Chang *et al.*, 1986; Allen *et al.*, 1987; Chang *et
al.*, 1991; El-Kabbani *et al.*, 1991). The second important breakthrough
was the expression of the genes encoding for the reaction center (Youvan *et
al.*, 1984) and the subsequent development of site-specific mutagenesis
(Zoller and Smith, 1982; Bylina and Youvan, 1987). These genetic
techniques, in conjunction with the x-ray structure, have allowed the
alteration of a specific amino acid in the reaction center protein. Thus a
method for directly testing proposed mechanisms of function involving the
protein environment has been developed. For example, the mechanisms for

[a]This work was supported by the U.S. Department of Energy, Office
of Basic Energy Sciences (JRN), and Office of Health and Environmental
Research (M.S., D.K.H.), Division of Chemical Sciences under contract W-31-
109-Eng-38. M.S. was also supported by Public Health Service Grant
GM36598.

controlling electron transfer, redox potential, optical properties, the role of symmetry, etc. can now be investigated rather routinely. Consequently, the reaction center has become a "model" system for understanding membrane bound proteins as well as electron transfer reactions.

The reaction center protein isolated from the *Rhodospirillaceae* subgroup (*Rhodopsuedomonas viridis*, *Rhodobacter sphaeroides*, and *Rhodobacter capsulatus*) consists of three polypeptide protein subunits, four bacteriochlorophylls, two bacteriopheophytins, two quinones, and a non-heme iron (Straley *et al.*, 1973). Reaction center preparations from *Rp. viridis* also have an additional bound cytochrome subunit which contains four c-type heme groups. The three conserved polypeptide subunits in all these species are labeled H (for heavy), M (for medium), and L (for light) from their apparent molecular weights from SDS polyacrylamide gel electrophoresis (Okamura *et al.*, 1974). The secondary and tertiary structure of the L and M subunits are related by approximate C_2 symmetry. Moreover, the prosthetic groups of the reaction center protein arranged along two branches of the L and M subunits also are related by the C_2 rotation axis which passes through the non-heme iron (Figure 1). Two of the bacteriochlorophylls strongly interact with each other and form the primary electron donor special pair, designated P (Norris *et al.*, 1971). Adjacent to the special pair are the two accessory (bridging) bacteriochlorophylls (B_L and B_M), one along each branch related by the C_2 axis. Next to the monomeric bacteriochlorophylls are the two bacteriopheophytins H_L and H_M (H_L next to B_L and H_M next to B_M), followed by the two quinones Q_A and Q_B (Q_A next to H_L and Q_B next to H_M). The non-heme iron is between the two quinones and lies on the C_2 axis which extends to the midpoints of the two Mg atoms of the special pair.

In about approximately 3 ps at room temperature, the electron is transferred from the singlet excited state of the special pair donor ($^*P$) to the bacteriopheophytin associated with the L subunit (H_L) (Holten *et al.*, 1980). The electron is then transferred to Q_A in approximately 200 ps (Rockley *et al.*, 1975; Kaufmann *et al.*, 1975). There are three major unresolved issues concerning the initial electron transfer in the purple photosynthetic bacteria. One is the unidirectionality of the electron transfer. Even though there are two roughly equivalent branches of chromophores related by a pseudo C_2 symmetry axis, the electron transfer occurs only down the one branch that is most closely associated with the L protein subunit. A second is the role of the accessory bacteriochlorophyll (B_L) in the initial electron transfer mechanism. Does the B_L work through superexchange or is it just a short lived intermediate? The difference between the two mechanisms is that $P^+B_L^-H_L$ is a true chemical intermediate in the sequential mechanism but is not in the superexchange mechanism. The third and last aspect is the temperature dependence of the electron transfer chemistry. This paper addresses only with the issue of

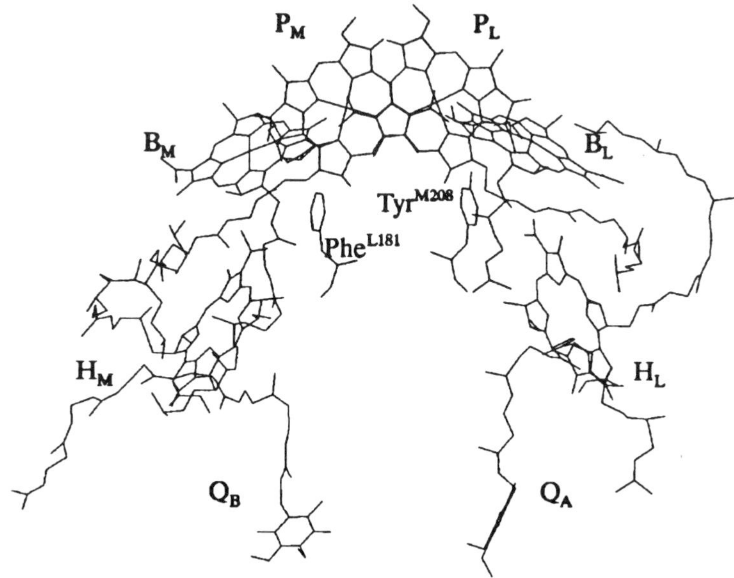

FIGURE 1. Stick drawing of the reaction center showing the two potential chemical pathways and the associated C_2 symmetry.

unidirectionality of the initial 3 ps electron transfer step and neglects subsequent electron transfers.

DISCUSSION

Because the pigments involved are identical and nearly symmetrically arranged along the two possible electron transfer pathways, any major disruption of this symmetry needs to be examined as a contributor to the possible cause for the unidirectional electron transfer. Numerous deviations from exact C_2 symmetry between the L and M branches are caused by different amino acids in symmetry related positions of the protein. One explanation for the unidirectionality of electron transfer is that the relative energy levels for electron transfer intermediates along the two pathways depends on the specific amino acid groups near the pathway, thus affecting the electron transfer rate down each of the two branches. One protein asymmetry which has attracted attention is the tyrosine residue at position M208 and its symmetry related phenylalanine residue at position L181 (Tiede *et al.*, 1988). Molecular dynamics calculations of Parson *et al.* (Parson *et al.*, 1990) indicated that a major reason for the unidirectionality might be caused by the energetics of the two possible intermediate states $P^+B_L^-H_L$ and $P^+B_M^-H_M$. They found that $P^+B_L^-H_L$ was approximately isoenergetic with $^*P$ and that $P^+B_M^-H_M$ was higher in energy by approximately 5 Kcal/mole (1750 cm^{-1}). This large energy difference could effectively make the superexchange and sequential mechanisms both have very small rate constants along the M branch compared to the L branch, and therefore cause the unidirectionality. They also suggested that the main reason the L-branch intermediate energy was much lower than the M-branch intermediate energy was due to the stabilization caused by the tyrosine M208 residue, and therefore this residue may be the main reason for the unidirectional electron transfer in the reaction center.

Through the use of site-specific mutagenesis, this hypothesis can be checked experimentally by changing the tyrosine residue, and in fact has been done by several different groups (Finkele *et al.*, 1990; Nagarajan *et al.*, 1990; Chan *et al.*, 1991). Two of the groups have focused on symmetrizing the two branches at these residues by the substitution Tyr^{M210} $\rightarrow$ Phe in *Rb. sphaeroides* or by making the L branch unfavorable with the substitutions $Tyr^{M210} \rightarrow$ Leu (Finkele *et al.*, 1990) and $Tyr^{M210} \rightarrow$ Iso (Nagarajan *et al.*, 1990). They found that the electron transfer rate was slower for these three mutants and suggested this was a confirmation of their electrostatic calculations (Parson *et al.*, 1990). The third group (Chan *et al.*, 1991) has also studied the mutant $Tyr^{M208} \rightarrow$ Phe in *Rb. capsulatus* (the numbering for *Rb. capsulatus* is the same as *Rp. viridis* for the M subunit, but *Rb. sphaeroides* has two additional amino acids and the numbering scheme is 2 higher at this location), but in addition the two other possible mutants with either a tyrosine or phenylalanine at the positions M208 and L181. With these two additional mutants ($Phe^{L181} \rightarrow$ Tyr and Tyr^{M208}-$Phe^{L181} \rightarrow Phe^{M208}$-$Tyr^{L181}$), the question regarding the importance of the tyrosine for the unidirectionality can be better addressed because the double mutant which reverses the asymmetry at these residues can give a decisive answer depending on which branch the electron transfer occurs. If the break in symmetry between the Tyr^{M208} and Phe^{L181} is solely responsible for the unidirectionality, then the mutant $Tyr^{M208} \rightarrow$ Phe should cause the electron transfer down both branches and the double mutant Tyr^{M208}-$Phe^{L181} \rightarrow Phe^{M208}$-$Tyr^{L181}$ should cause the electron transfer to occur solely down the normally inactive M branch.

For these two mutants, the measured stimulated emission decay time constants were 9.2 ps and 3.5 ps respectively at room temperature (Table 1). To determine along which branch or branches the electron transfer was occurring, the Q_X bacteriopheophytin ground state absorption was monitored (the two bacteriopheophytins are resolved and the active bacteriopheophytin B_L has been determined to have a ground state absorption at 545 nm by linear dichroism of oriented *Rp. viridis* crystals) (Knapp *et al.*, 1985). For electron transfer to occur down the L branch, the ground state bleaching at 545 nm would be observed. Electron transfer down the normally inactive M branch would show ground state bleaching of the bacteriopheophytin at 529 nm. For both of these mutants, only bleaching at 545 nm was observed and none at 529 nm. Even though the symmetry had been reversed in the protein at residues M208 and L181 to favor the inactive M branch, the electron transfer still occurred solely down the L branch. These results clearly indicate that the Tyr^{M208} residue cannot be solely responsible for the unidirectional electron transfer in the reaction center as previously predicted. However, these results also suggest that only a few key amino acid residues could cause the unidirectionality and that the symmetry related residues Tyr^{M208} and Phe^{L181} may be some of these important residues since changes at these sites can affect the electron transfer rate by an order of magnitude.

$$\frac{1}{\tau_{obs}} = \frac{1}{\tau_{L181}} + \frac{1}{\tau_{M208}}$$

TABLE 1. Stimulated Emission Decay Time Constants for *Rb. capsulatus* Mutants

	TEMP (K)	Tau (ps)
Wild-Type	293	2.8 ± 0.4
$\mathbf{Phe^{L181} \rightarrow Tyr}$	296	2.1 ± 0.3
$\mathbf{Tyr^{M208} \rightarrow Phe}$	296	9.2 ± 1.5
$\mathbf{Phe^{L181}\text{-}Tyr^{M208} \rightarrow Tyr^{L181}\text{-}Phe^{M208}}$	298	3.5 ± 0.3
$\mathbf{Tyr^{M208} \rightarrow Thr}$	296	15 ± 3
$\mathbf{Phe^{L181}\text{-}Tyr^{M208} \rightarrow Thr^{L181}\text{-}Thr^{M208}}$	295	29 ± 5

Since the residue Phe^{L181} is further than 10 Å away from both B_L and H_L, changes in this residue should have little effect on either of the two possible electron acceptors on the normally active L branch. But this residue is within van der Waals contact with each half of the special pair, P_L and P_M (Table 2), and could affect the electron transfer directly by affecting the primary donor. Since the lowest energy absorption maximum for the special pair remains approximately the same for this series of mutants, it appears that the charged state $P^+B_L^-H_L$ and $P^+B_LH_L^-$ are most affected by these mutations at L181. An amino acid at the symmetry related M208 position is in van der Walls contact with all the chromophores associated with the electron transfer (P_L, P_M, B_L, and H_L) and can likewise affect the intermediate states $P^+B_L^-H_L$ and $P^+B_LH_L^-$. From the proximity of these two amino acids and the observation that the a mutant with a tyrosine at L181 facilitates right way electron transfer, a parallel resistor model was used to rationalize the changing rates for the different mutants at L181 and M208. In this model, amino acids at both L181 and M208 can act independently to affect the electron transfer rate given by the equation: The agreement between the predicted and experiment rate constants for the wild-type and double mutant (Phe^{L181}-$Tyr^{M208} \rightarrow Tyr^{L181}$-$Phe^{M208}$) based on τ_{Phe} and τ_{Tyr} obtained from the mutants $Tyr^{M208} \rightarrow Phe$ and $Phe^{L181} \rightarrow Tyr$ respectively is quite reasonable considering the approximations and simplicity of this model. This model predicts a time constant of 3.4 ps for the wild-type and double mutant (the same since each organism has one tyrosine and one phenylalanine at the two locations) and experimentally the time constant was found to be 3.5 ps for the double mutant and 2.8 ps for wild-type.

TABLE 2. Distances Between Chromophores and Amino Acid Residues M208 and L181

CHROMOPHORE	RESIDUE	DISTANCE
P_L	L181	4.3 Å
P_M	L181	4.4 Å
B_L	L181	12.5 Å
B_M	L181	4.9 Å
H_L	L181	13.6 Å
H_M	L181	3.6 Å
P_L	M208	4.1 Å
P_M	M208	3.5 Å
B_L	M208	4.2 Å
B_M	M208	12.0 Å
H_L	M208	3.3 Å
H_M	M208	13.5 Å

In addition we have measured the initial electron transfer rate in a $\text{Tyr}^{M208} \to$ Thr mutant and found $\tau_{obs} = 15 \pm 3$ ps at room temperature (Chan *et al.*, 1991). According to our parallel resistor model above, the electron transfer time constant of 15 ps is largely the result of the amino acid Phe^{L181}. This model predicts that the mutant with $\text{Phe}^{L181}\text{-Tyr}^{M208} \to \text{Thr}^{L181}\text{-Thr}^{M208}$ will exhibit a τ_{obs} of less than 39 ps. Since these two symmetry related amino acid sites (i.e., M208 and L181) have too little influence on the overall electron transfer process to control completely the unidirectionality (as demonstrated by these experiments), then other additional factors or residues will contribute and the 39 ps prediction is the upper limit for the electron transfer rate for the double Thr mutant. Recent preliminary experimental results have found that $\tau_{obs} = 29 \pm 5$ ps for this double Thr mutant. This result supports the general view that both the M208 and L181 residues do influence the initial kinetics of electron transfer, but not sufficiently to be the sole cause of the unidirectionality.

CONCLUSIONS

The parallel resistor model can be placed on more conventional theoretical grounds by employing standard electron transfer theory. In a more elaborate treatment we view the effect of the amino acid groups at L181 and M208 as affecting the energy of $\text{P}^+\text{B}_L^-\text{H}_L$. In addition we include the fact that M208 is closer to B_L than B_M and thus the effect of the L181 group is less than the effect of the M208 group on influencing the reaction rate down the active side. All things considered suggest that several additional symmetry breaking groups of the protein are involved in the unidirectionality (Scherz, 1989). Four or five aromatic amino acids are in the nearest protein shell along the prosthetic groups of the active electron transfer network but are missing on the inactive side in more than several photosynthetic organisms. This suggests that new mutants constructed by genetically manipulating these few aromatic amino acids can greatly change the asymmetry of the two branches. Only when a genetically altered reaction center protein is made that has electron transfer occurring down the M branch will the issue of unidirectionality be completely resolved. These experiments are currently in progress. Since the reaction center is a membrane bound protein a better understanding of its structure and function will contribute to the data base for all trans-membrane proteins.

REFERENCES

Allen, J.P.; Feher, G.; Yeates, T.O.; Komiya, H.; Rees, D.C. *Proc. Natl. Acad. Sci. USA* 1987, **84**, 5730-5734.
Bylina, E.J.; Youvan, D.C. *Z. Naturforsch. C* 1987, **42**, 769-774.]

Chan, C.-K.; Chen, L.X.-Q.; DiMagno, T.J.; Hanson, D.K.; Nance, S.L.; Schiffer, M.; Norris, J.R.; Fleming, G.R. *Chem. Phys. Lett.* 1991, **176**, 366-372.

Chang, C.-H.; Tiede, D.; Tang, J.; Smith, U.; Norris, J.R.; Schiffer, M. *FEBS Lett.* 1986, **205**, 82-86.

Chang, C.-H.; El-Kabbani, O.; Tiede, D.; Norris, J.R.; Schiffer, M. *Biochemistry* 1991, **30**, 5352-5360.

Deisenhofer, J.; Epp, O.; Miki, K.; Huber, R.; Michel, H. *J. Mol. Biol.* 1984, **180**, 385-398.

El-Kabbani, O.; Chang, C.-H.; Tiede, D.; Norris, J.R.; Schiffer, M. *Biochemistry* 1991, **30**, 5361-5369.

Finkele, U.; Lauterwasser, C.; Zinth, W.; Gray, K.A.; Oesterhelt, D. *Biochemistry* 1990, **29**, 8517-8521.

Holten, D.; Hoganson, C.; Windsor, M.W.; Schenck, C.C.; Parson, W.W.; Migus, A.; Fork, R.L.; Shank, C.V. *Biochim. Biophys. Acta* 1980, **592**, 461-477.

Kaufmann, K.J.; Dutton, P.L.; Netzel, T.L.; Leigh, J.S.; Rentzepis, P.M. *Science* 1975, **188**, 1301-1304.

Knapp, E.W.; Fischer, S.F.; Zinth, W.; Sander, M.; Kaiser, W.; Deisenhofer, J.; Michel, H. *Proc. Natl. Acad. Sci. USA* 1985, **82**, 8463-8467.

Nagarajan, V.; Parson, W.W.; Gaul, D.; Schenck, C. *Proc. Natl. Acad. Sci. USA* 1990, **87**, 7888.

Norris, J.R.; Uphaus, R.A.; Crespi, H.L.; Katz, J.J. *Proc. Natl. Acad. Sci. USA* 1971, **68**, 625-628.

Okamura, M.Y.; Steiner, L.A.; Feher, G. *Biochemistry* 1974, **13**, 1394.

Parson, W.W.; Chu, Z.-T.; Warshel, A. *Biochem. Biophys. Acta* 1990, **1017**, 251.

Rockley, M.; Windsor, M.W.; Cogdell, R.J.; Parson, W.W. *Proc. Natl. Acad. Sci. USA* 1975, **72**, 2251-2255.

Scherz, A. in *Photochemical Energy Conversion*, 1989, Norris, J.R. and Meisel, D., eds. New York, Elsevier Science Publishing Co.

Straley, S.C.; Parson, W.W.; Mauzerall, D.C.; Clayton, R.C. *Biochim. Biophys. Acta* 1973, **305**, 597-609.

Tiede, D.M.; Budil, D.E.; Tang, J.; El-Kabbani, O.; Norris, J.R.; Chang, C.-H.; Schiffer, M. *NATO ASI Series A, The Photosynthetic Bacterial Reaction Center*, Plenum Press: New York, 1988, 13.

Youvan, D.C.; Bylina, E.J.; Alberti,M.; Begusch,H.; Hearst, J.E. *Cell* 1984, **37**, 949-957.

Zoller, M.J., and Smith, M. *Nucleic Acids Res.* 1982, **10**, 468.

THE INTERACTION OF THE PHOTORECEPTOR CELLS WITH THE CONSTANT
ELECTRIC FIELD

Eugenia Chirieri-Kovacs,Alexandru Dinu,Tudor Savopol

Departament of Biophysics,"Carol Davila" Institute of Medicine and
Pharmacy, Medical Faculty , Bucharest

INTRODUCTION

The rod outer segments (ROS) are specialized structures for
transduction of the light signal into the electric hyperpolarization of
the photoreceptor cell. Their morphology is subordinated to this aim:
the ROS plasma membrane envelops several thousands of superposed double
membraneous disks loaded with rhodopsin (at least 95% of the total disk
membrane protein is rhodopsin). Rhodopsin molecules are known to be
dipoles, randomly distributed in the plane of the disk membrane
(Wolken,1963).
In the dark, a constant current (tens of μA) is flowing into the
ROS, carried mainly by Na^+ ions (Hagins,1970;Owen,1987).
There are morphological as well as spectrophotometric studies
revealing the structural and functional asymmetry of the ROS along its
long axis.It is known that new disks are continuously synthetized at
the outer segment's base while the old ones are pushed to the rod tip
where they degenerate and are phagocytized by the pigment epithelium
(Young,1976).
It was also shown that at the two ends of the outer segment the
regeneration ability of rhodopsin is different (Makino,1987).
The pattern of the dark current may be also regarded as the
cause/consequence of the morphological and functional asymmetry of the
ROS.
Any of the mentioned asymmetries may generate a charge asymmetry
which would determine the behavior of the ROS as an electric dipole.
Our observations have proved that this is the case. In static
electric fields the rods rotate, getting parallel to the electric
field.The tip of the rod seems to be negatively charged since it always
rotates towards the positive electrode. Some of the rods rotate faster
in the dark (in dim red light) than under illumination; the average
speed of rotation is by 40% higher for dark adapted rods (FIG.2).

METHODS

Frogs (Rana ridibunda) were dark-adapted for 12-24 hours and the
retinas were dissected free from eye cup and pigment epithelium in
aerated Ringer solution under dim red light. ROS were harvested by
gentle shaking in a mixture of 0.5 ml Ringer solution (100 mM NaCl,5 mM
$NaHCO_3$, 3.5 mM KCl, 0.18 mM $CaCl_2$, 2 mM $MgCl_2$,20 mM glucose, 20 mM
Tris, pH 7.7) and 0.5 ml saccharose 45 %. The ROS were not subjected
to subsequent purification and washing procedures in order to make the
processes studied relevant to the functional intact rod photoreceptors.

A drop of the ROS suspension has been placed on the microscope
glass between two platinized platinum electrodes and the cover slip
applied on it. The preparation has been observed when applying constant
fields of 1 min. duration; the sign of the field has been alternated
every 1 min. in view of limiting the electrolysis effects.

Those ROS have been chosen for measurements which initially layed
perpendicularly to the field direction; the time of rotation has been
measured for different field intensities. All operations have been
performed in dim red light, the microscope beam being passed through a
red cut-off filter (λ =580 nm).

RESULTS

It has been observed that the rods which could move freely (not
adhering to the microscope glass) slowly rotated in the electric field
so that in several minutes all cells have been oriented along the field
direction (FIG.1).

The rotation was always oriented with the distal end (tip) of the
ROS to the positive pole; thus, when observing the microscope field,
some cells have been apparently rotating to opposite directions,
depending on the initial orientation of their tips. Once oriented along
the field direction, the rods have been sinchronously translating when
the sign of the field was changed; the speed of translation increased
with the field intensity.

In several cases a reversible change of rotation speed was
observed when alternating dark and illumination. The same rod switched
from a fast rotation to a slower one when the red filter was taken off
from the way of the microscope beam.

Usually, after half an hour of observation in the electric field,
the rod membrane elicited fine cuts through which the suspension fluid
penetrated between the disks; the ROS stopped rotating. In most cases a
small local damage of the membrane quickly propagated along the rod as
if the plasma membrane "melted", denuding the disks stack.

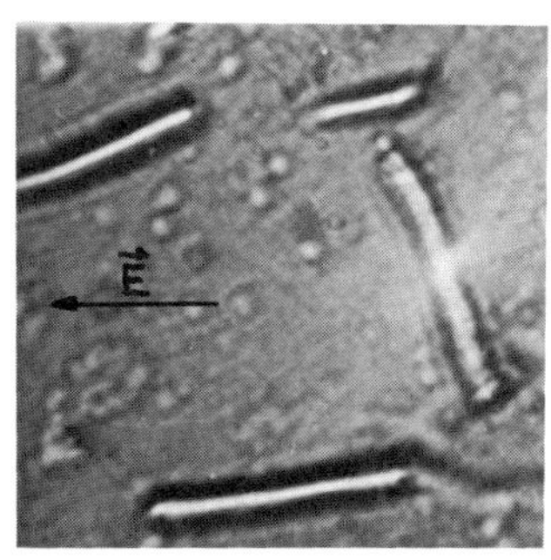

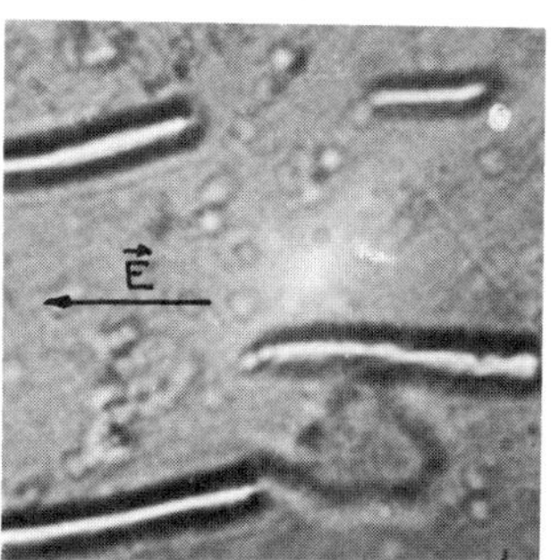

a   b

FIG.1 The orientation of ROS in the electric field.
a) ROS in suspension before applying the field;
b) 2 minutes after applying a field of 8 V/cm.

A simple model (which disregards the complex phenomena taking place at charged interfaces) is proposed to describe the rod rotation in the electric field.

The mechanical rotation of the rod in the electric filed is supposed to result from contributions ofthe electric and frictional forces.

$$M_i = M_e - M_f$$

where M_i is the resulting inertial momentum $M_i = \dfrac{\pi \varrho\, d^2 l^3}{48}\, \ddot{\theta}$

$M_e = p\dfrac{E_0}{K} \sin\theta$ is the moment of the electric force, p is the

dipolar electric moment of the rod, K is the relative dielectric constant of the medium.

The further computations have been made for small angles which allowed the approximation $\sin\theta \cong \theta$

$M_f = \xi\dot{\theta}$ is the "frictional momentum".

where $\xi = \dfrac{16\,\pi\,\eta\,(l/2)^3}{-3+6\,\ln(2l/d)}$ is the rotational friction coefficient,

according to Hong et al.(1971).

Thus, the equation describing the movement of the rod becomes:

$$\frac{\pi \varrho\, d^2 l^3}{48}\,\ddot{\theta} + \xi\dot{\theta} - p\frac{E_0}{K}\theta = 0$$

TABLE I. The times and velocities of rotations by 10 are
given for three rods.As the rotational velocity was observed
to be constant (0),the time of rotation by 10 was deduced
from the time of rotation by 45 .

	Rod 1 (o)		Rod 2(Δ)		Rod 3 ($\square$)	
ξ_r	3.2x10		2.8x10		1.9x10	
$l(\mu)$	40		38		30	
$d(\mu)$	6		6		8	
E(V/m	t(s)	$\dot\Theta(s^{-1})$	t(s)	$\dot\Theta(s^{-1})$	t(s)	$\dot\Theta(s^{-1})$
400	2.00 ± 0.08	0.0873	2.6 ± 0.6	0.0671	1.6 ± 0.2	0.1091
500	1.7 ± 0.1	0.1027	1.9 ± 0.4	0.0919	-	-
600	1.2 ± 0.0	0.1454	1.8 ± 0.2	0.0970	1.30 ± 0.04	0.1343
700	1.1 ± 0.0	0.1587	1.5 ± 0.2	0.1164	-	-
800	1.0 ± 0.0	0.1745	1.2 ± 0.2	0.1454	1.00 ± 0.09	0.1745
900	0.90 ± 0.07	0.1939	1.10 ± 0.05	0.1587	-	-
1000	0.80 ± 0.02	0.2182	0.90 ± 0.06	0.1939	0.76 ± 0.06	0.2296
1100	0.70 ± 0.05	0.2493	0.87 ± 0.10	0.2006	-	-
$a(s^{-1})$	-0.00096		-0.014		0.021	
$b(mVs^{-1})$	0.00022		0.000197		0.0002	
r	0.99		0.99		0.99	

E -intensity of the applied electric field;
-rotational friction coefficient; l -length of the rod;
d -diameter of the rod; a -intercept; b -slope of the graph;
r -correlation coefficient; t,$\dot\Theta$ -time and velocity of
rotation.

As the coefficient of $\ddot{\theta}$ is very small (1.65×10^{-22} kg.m.^{2}) and the rod rotation is observed to be uniform ($\ddot{\theta} \cong 0$), the first term of the equation may be neglected, obtaining:

$$\dot{\theta} = \frac{p\,\theta}{\xi\,K}\,E_o$$

This shows a linear dependence of θ versus E, with a slope

$$b = \frac{p\,\theta}{\xi\,K}$$

(FIG.2). From the experimental values of "b" coefficient (TABLE I) one can calculate the dipolar moments of the rods (K = 75).

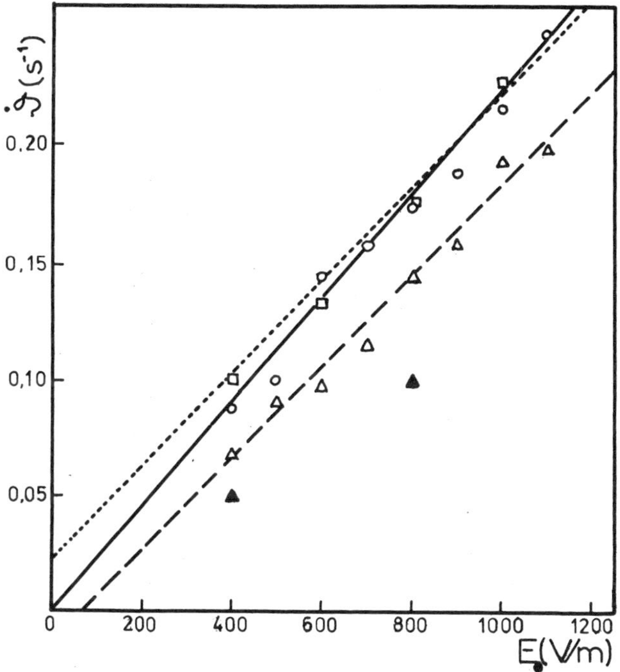

FIG.2 The plot of rotation velocity θ vs. electric field intensity E for three different dark-adapted ($-\!\circ\!-$, $-\!\triangle\!-$, $\cdots\!\square\!\cdots$) and illuminated ($\blacktriangle$) rods.

It can be observed (TABLE II) that the dipolar moments varies with the rod length. If one calculates the dipol charge per disk, a

strikingly reproducible value results for all three rods:
(279 + 21) e/disk (e - electronic charge), assuming tha there are
approximately 40 disks/μ.

TABLE II

.	Rod 1	Rod 2	Rod 3
p (C.m)	3.03×10^{-18}	2.37×10^{-18}	1.64×10^{-18}
q(e)	473438	389813	341688
q/disk (e)	296	256	284

where p - dipolar moment of the rod
q - dipolar charge of the rod
q/disk - dipolar charge of the disk.

DISCUSSION

The slow rod orientation in the magnetic field has been observed
in the early 70's by Chalazonitis and coworkers (1970); this was
explained by the difference between the axial and radial magnetic
susceptibilities of the rod (Hong et al.,1971).
Later on, W. Vaughan (1987) has estimated the maximum induced
dipole moments which bovine disks may develop in a suspension as being
1.054×10^{-21} C.m. Our data lead to an average dipole moment of 1.11×10^{-24}
C.m/disk.
More additional experiments are necessary for indicating the
nature of the rod electric dipole (structural or functional) as well as
its dependence on the rod illumination.
The exploration of these types of effects as well as of the
action of metabolic inhibitors on the rod dipole moment is our charge
for the immediate future.
There might be also a practical interest in preparation of films
of oriented ROS for optical studies as well as for modelation of visual
information processing as suggested in the work of Nakanishi and
Yamaguchi (1990).

ACKNOWLEDGEMENTS

This work was supported by the grant 91CH from the Ministery of
Science and Education. We are grateful to Prof.W.D.Stein for his
stimulating advices as well as to our colleagues Elena Balu and Jean
Vinersan for their helpful comments and help in the manuscriot
preparation.

REFERENCES

Chalazonitis N, Chagneux R and Arvanitaki A (1970): Rotation des
  segments externes des photorecepteurs dans le champ magnetique
  constant. C R Acad Sci Hebd Seances Ser D Sci Nat 271:130.
Hagins WA, Pen RD and Yoshikami S (1970): Dark current and
  photocurrent in retinal rods. Biophys J 10:380
Hong FT, Mauzeroll D and Mauro A (1971): Magnetic anizotropy and
  the orientation of retinal rods in a homogenous magnetic
  field. Proc Natl Acad Sci USA 68:1283
Makino CL, Howard LN and Williams TP (1987): Intracellular
  topography of rhodopsin bleaching. Science 238:1716.
Nakanishi H and Yamaguchi H (1990): Molecular based devices
  modeled on the visual information processing in the retina.
  In Molecular Electronics, Hong FT ed. New York and London,
  Plenum Press: 1752.
Owen WG (1987): Ionic conductances in rod photoreceptors. Ann Rev
  Physiol 49:743.
Vaughan W (1987): Dynamic response on spherical biopolymers to
  electrical field - the Kerr effect of disk membrane vesicles.
  J Mol Liquids 36:1.
Wolkan JJ (1963): Structure and molecular organization of the
  retinal photoreceptors. J Opt Soc Am 163:No.1.
Young RD (1976): Visual cells and the concept of renewal. Invest
  Ophtalmol 15:700.

RESONANCE RAMAN SPECTROSCOPY WITH NEAR ULTRAVIOLET EXCITATION OF PEROXIDASE INTERMEDIATES IN HIGH OXIDATION STATES

V. Palaniappan, Ann M. Sullivan, Melissa M. Fitzgerald, John R. Shifflett and James Terner
Dept. of Chemistry
Virginia Commonwealth University
Richmond, Virginia 23284-2006, U.S.A.

INTRODUCTION

The oxidative function of peroxidases has long been of interest to chemists and biochemists because of the unstable and highly oxidized states that are reversibly assumed by the heme active site during the catalytic cycle. Features of the various peroxidase mechanisms have been postulated to occur in the catalytic cycles of other important heme enzymes such as cytochrome oxidase and cytochrome P-450. Peroxidases are widespread in plants but are also found in numerous animal tissues. An extensively studied peroxidase which is isolated from horseradish root is the main topic of this paper. The catalytic sequence of horseradish peroxidase involves two oxidative intermediates known as compounds I and II (Dunford, 1982), (Dawson, 1988). Compound I, two oxidation equivalents above the resting enzyme, contains a ferryl heme with an additional electron removed from the porphyrin ring, forming a porphyrin π-cation radical (Dolphin and Felton, 1974), (McMurry and Groves, 1986). Compound II, which contains an oxo-ferryl heme, results from the reduction of compound I which restores an electron to the porphyrin ring (Hewson and Hager, 1979).

Structural aspects of compound II and other ferryl hemes have been under active investigation by resonance Raman spectroscopy and other techniques in recent years (Terner et al., 1989). Though the porphyrin π-cation radical formulation has been known for many years, such compounds, and especially those contained within proteins, have been difficult to study by physical methods (Palmer, 1983). Even recently, resonance Raman studies of horseradish peroxidase compound I (Teroaka et al., 1982), (Oertling and Babcock, 1985 & 1988), (Paeng and Kincaid,

1988) have been inconsistent and at variance with resonance Raman spectra of model metalloporphyrin π-radical cations (Oertling and Babcock, 1988), (Kim et al., 1986), (Salehi et al., 1986), (Oertling et al. 1987a,b), (Czernuscewicz et al., 1989). These studies employed experimental procedures that normally provide strong resonance Raman enhancement of signals from ferric and ferryl hemes via excitation of the intense Soret absorption near 400 nm. During our own experiments aimed at resolving some of these issues, we have noticed that resonance Raman signals from compound I are quite weak under Soret excitation and that signals from ferric and ferryl species can dominate, particularly when small amounts of these are formed by the tendency of compound I to be converted to other species by the laser excitation. We found that we could obtain enhancement of resonance Raman signals from compound I, and lessen the enhancement of scattering from interfering ferryl and ferric photoproducts by tuning the laser excitation away from the Soret absorption into the near ultraviolet (Palaniappan and Terner, 1989).

METHODS

Horseradish peroxidase was purified by DEAE- and CM-Sepharose (Pharmacia) ion exchange chromatography according to established procedures (Shannon et al., 1966). Compound I was formed by mixing equal volumes of a 2 molar excess of H_2O_2 (Fisher) with buffered horseradish peroxidase solution (800 μM, in 0.01 M sodium phosphate, pH 6.8) in a Ballou 4-jet mixer fed by two 100 mL syringes driven by a Harvard Bioscience model 975 syringe pump. The exit port of the mixer was a 26 gauge syringe needle that formed the activated sample into a jet stream which was excited transversely by low power (3 mW, 100 μm beam diameter) continuous wave laser excitation (FIGURE 1). The flow rate was 0.2 ml/sec with a deadtime of 30 msec. Compound II was formed in a similar manner either by mixing preformed compound I (400 μM) with a 2-fold excess of buffered ascorbic acid (Sigma) solution, or by premixing a 1 molar equivalent of ascorbic acid with the resting enzyme prior to mixing with oxidant.

The excitation wavelengths used for compound I were the near ultraviolet laser lines of the Kr (3375, 3507 and 3564 Å) and Ar (3511 and 3638 Å) ion lasers. Soret excitation wavelengths at 4067 and 4131 Å (Kr ion laser) were also used, however as mentioned above, these lines generated only weak resonance Raman signals from compound I, which had a

tendency to be overwhelmed by much stronger scattering from ferryl and
ferric photoproducts.

Scattered light was collected with a 55 mm f/1.4 Rolleinar MC lens
and imaged with a quartz singlet (180 mm f/4.5, Melles Griot) onto the
slit of a 0.5 m spectrograph (Spex model 1870). The single grating
spectrograph allowed a maximization of signal intensities, though stray
light needed to be eliminated with baffling or interference filters.
The detection system was an optical multichannel analyzer utilizing an
EG&G Princeton Applied Research Corp. model 1254 SIT vidicon.

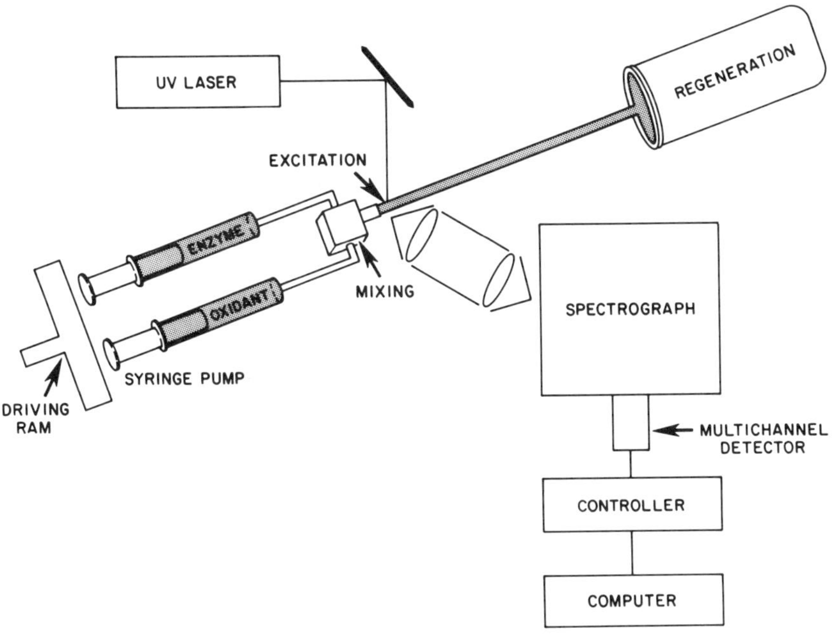

FIGURE 1. Diagram of the experimental apparatus used for the continuous flow mixing experiment.

RESULTS

The electronic absorptions of porphyrins are typified by three
prominent absorption bands (Gouterman, 1978). The most intense is the
Soret absorption (or B-band), centered near 400 nm. A two-banded
visible absorption near 550 nm is known as the α-β or Q-band. A third
absorption known as the N-band, occurs in the near ultraviolet
(Gouterman, 1978), (Makinen and Churg, 1983), (Loew, 1983), (Edwards and

Zerner, 1985). Resonance Raman spectroscopists have made predominant use of excitation within the Soret and α-β bands for several reasons. Strong resonance enhancement of porphyrin modes is realized from the Soret and α-β bands, along with a high degree of mode specificity. A_{1g} modes are preferentially enhanced with Soret excitation while A_{2g}, B_{1g} and B_{2g} modes predominate under α-β excitation (Li et al., 1990).

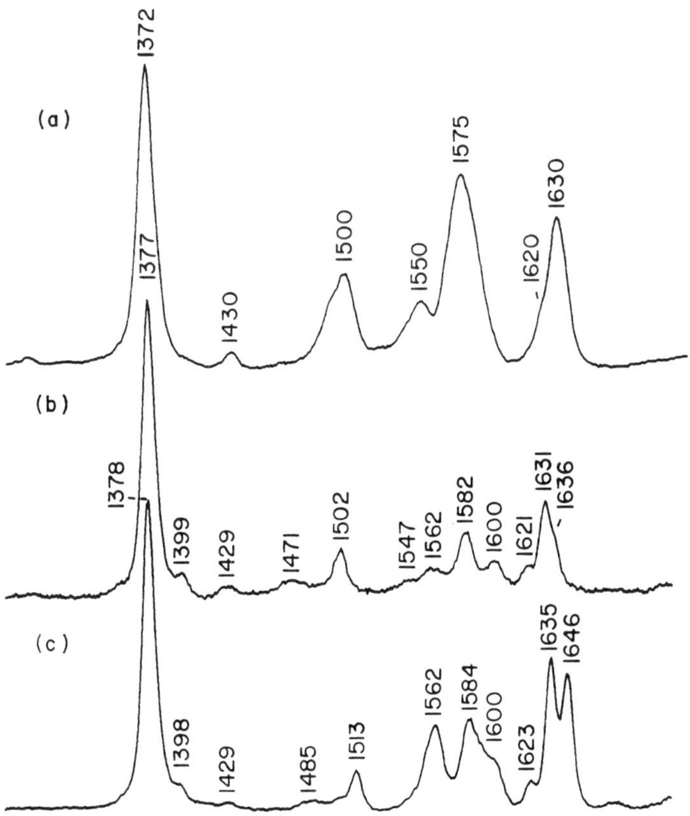

FIGURE 2. Resonance Raman spectra of a) resting horseradish peroxidase, b) NO-ferrous horseradish peroxidase, and c) NO-ferric horseradish peroxidase. All spectra were obtained in a spinning quartz cell with 4067 Å excitation (5 mW).

Resonance enhancement of hemes upon excitation in the ultraviolet, in the vicinity of the N-band has been used only infrequently, since signals are significantly weaker than those obtained by exciting into the Soret band, which lies close by. Nonetheless, we have found near-ultraviolet excitation to be useful since it provides an

enhancement pattern that is complementary to that normally found with
Soret and α-β excitation, and as well as preferential enhancement of
scattering from the porphyrin π-radical cation of compound I.

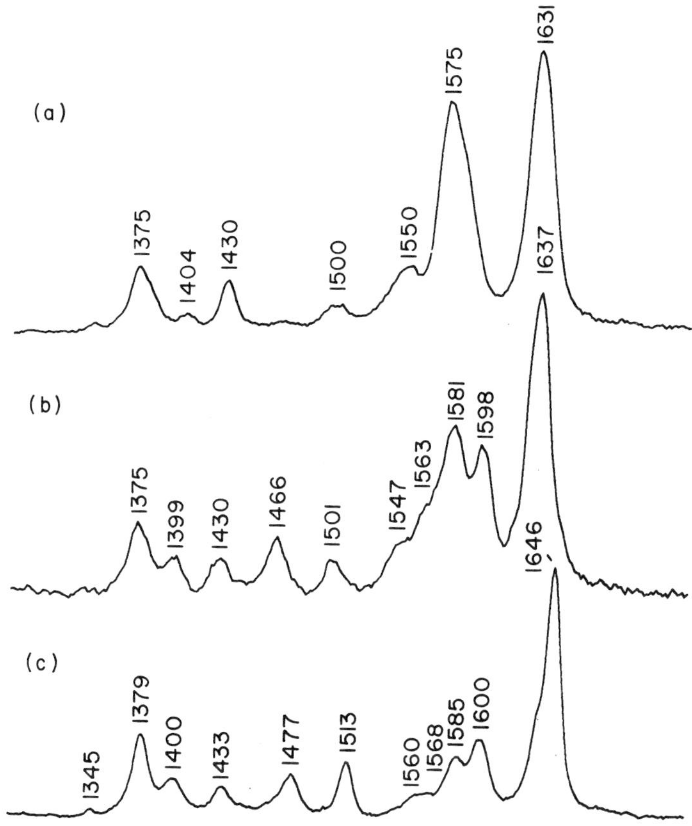

FIGURE 3.   Resonance Raman spectra of a) resting horseradish
peroxidase, b) NO-ferrous horseradish peroxidase, and c) NO-ferric
horseradish peroxidase. All spectra were obtained in a spinning
quartz cell in the same manner as for FIGURE 2, except with 3564 Å
excitation (5 mW).

A comparison of Soret and near-UV excitation is shown in FIGURES 2
and 3 for resonance Raman spectra of resting horseradish peroxidase and
six-coordinate low-spin NO complexes. The differing resonance Raman
data on the NO complexes result from the effects of a change of oxidation state from Fe(II) to Fe(III) on the porphyrin core size (Spiro,
1983). From these traces it can be seen that a modified intensity
pattern relative to that observed under Soret excitation (FIGURE 2) can
be observed with N-band excitation (FIGURE 3). The heme oxidation state

marker, ν_4 (A_{1g}), normally the most intense of the porphyrin modes under Soret excitation (Li et al., 1990), (Spiro, 1983) is frequently very weak under ultraviolet excitation. B_{1g} and B_{2g} modes appear to achieve enhancement through vibronic coupling of the N-band with the intense

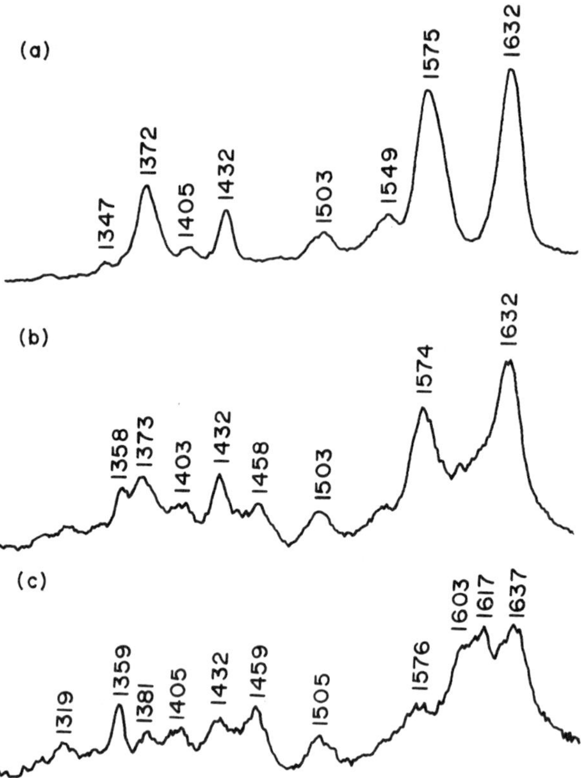

FIGURE 4.   Resonance Raman spectra obtained using the continuous flow mixing apparatus shown in FIGURE 1, using 10 mW 3564 Å excitation.   a) Resting horseradish peroxidase (800 μM, pH 6.8) before mixing with oxidant.   b) After mixing with H_2O_2 c) Same as "b" except laser power lowered to 1 mW.

Soret in a manner that is similar to the mechanism for enhancement of these modes under α-β excitation. There also appears to be a preferential enhancement of modes involving expansion and contraction motions of the outer periphery of the porphyrin ring. We have noticed that under near ultraviolet excitation, ν_2 is very strong, but ν_3, ν_4 and ν_7 can be very weak, even though all of these are A_{1g} modes.   Heme resonance

Raman spectra with near-ultraviolet (N-band) excitation appear superficially similar to spectra obtained with α-β excitation because of the dominance of ν_{10} and weakness of ν_4 and ν_3, however there are some marked differences. The dominance of ν_2 in the UV-excited spectra is not observed under α-β excitation. Rather, the α-β excited spectra tend to be dominated by an A_{2g} mode, ν_{19}, especially when excitation occurs between the α and β (Q_0 and Q_1) substructures (Li et al., 1990). Though we have frequently observed A_{2g} modes under near-ultraviolet excitation, these have much less intensity than the strong ν_2 and ν_{10} bands.

The porphyrin π-cation radical has a strong tendency to be photoreduced by laser excitation. We found it important to minimize the photoalteration parameter (Mathies, 1979) by keeping the sample concentration high, the laser power low, and rapidly flowing the sample. An illustrative set of resonance Raman data is shown in FIGURE 4. FIGURE 4a shows the resting enzyme before the addition of oxidant. In FIGURE 4b, oxidant has been added, and features of the porphyrin π-radical cation are evident (e.g. 1358 cm^{-1} and the valley between 1575 and 1632 cm^{-1}), however these are superimposed on the resonance Raman spectrum of the resting enzyme which dominates the spectrum. In FIGURE 4c, we adjusted conditions (lowered the incident laser power) so that the resonance Raman spectrum of the porphyrin π-radical cation of compound I appears clearly.

A UV-excited resonance Raman spectrum of horseradish peroxidase compound I is compared with the unbound ferrous and ferric enzyme, and the ferryl form (compound II) in FIGURE 5. Resonance Raman band assignments for compound I are made from the frequency pattern and band polarizations. The following bands are polarized: ν_{37} (E_u) at 1614 cm^{-1}, ν_2 (A_{1g}) at 1606 cm^{-1}, ν_3 (A_{1g}) at 1504 cm^{-1} and ν_4 (A_{1g}) at 1359 cm^{-1}. The following bands are depolarized: ν_{10} (B_{1g}) at 1636 cm^{-1}, ν_{11} (B_{1g}) at 1570 cm^{-1} and ν_{28} (B_{2g}) at 1458 cm^{-1}. The frequencies of the other derivatives were assigned in a similar manner (TABLE I).

The compound I frequencies are observed to be systematically shifted from the compound II frequencies. Both compounds I and II contain a six-coordinate low-spin (S=1) Fe(IV), though compound I has an electron removed from the porphyrin ring. The most apparent features of the compound I resonance Raman spectra relative to compound II and the six-coordinate low-spin hemes are the substantial downshifts of ν_4 (20 cm^{-1}) and ν_{28} (14 cm^{-1}), and upshifts of ν_2 (19 cm^{-1}), ν_{37} (12 cm^{-1}) and ν_{11} (10 cm^{-1}). Smaller but significant downshifts are seen for ν_{10} (3

cm^{-1}), and ν_3 (5 cm^{-1}). The resonance Raman frequency shifts follow
normal mode compositions which correlate well with the characteristic
shifts previously described for octaethylporphyrin π-cation radical
model compounds (Oertling et al, 1987b). The ν_4 frequency (primarily
C_a-N) exhibits a 20 cm^{-1} downshift. The ν_2, ν_{11} and ν_{37} vibrational

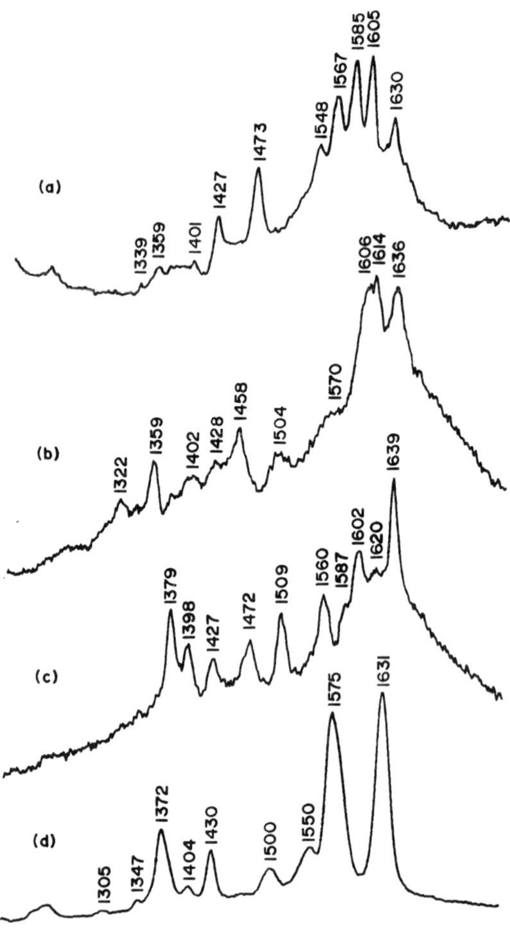

FIGURE 5. Resonance Raman spectra of horseradish peroxidase (3 mW,
3564 Å excitation, pH 6.8, a) ferrous enzyme, b) compound I, c)
compound II, d) resting enzyme.

frequencies whose mode compositions are primarily porphyrin C_b-C_b,
exhibit 10 to 20 cm^{-1} upshifts. Downshifts are observed for the ν_3, ν_{10}
and ν_{28} frequencies, whose mode compositions are primarily C_a-C_m. The

downshifts for ν_3 and ν_{10} are small, 3-6 cm^{-1}, however the downshift for ν_{28} is 14 cm^{-1}.

The 1359 cm^{-1} ν_4 frequency of compound I (FIGURE 5b) is 20 cm^{-1} downshifted from the compound II (FIGURE 5c) frequency of 1379 cm^{-1}. It is coincident with the 1358 cm^{-1} ν_4 frequency of the reduced (ferrous) enzyme (FIGURE 5a). However, the other compound I frequencies of FIGURE 5b do not match with the data for the ferrous enzyme (FIGURE 5a), but exhibit a unique spectral pattern that is systematically shifted from the compound II frequencies as described above.

<u>DISCUSSION</u>

Resonance Raman scattering from porphyrin π-radical cations under Soret and α-β excitation is known to be weak relative to porphyrins containing an unoxidized ring (Czernuscewicz, 1989). Early attempts to obtain resonance Raman spectra of compound I (Woodruff and Spiro, 1974), (Felton et al., 1976), resulted in reports of elevated values of the ν_4 (oxidation state marker) frequency, which were difficult to interpret since removal of an electron from model metallo-tetraphenylporphyrins had been characterized by a lowering of resonance Raman frequencies (Yamaguchi et al., 1982). An apparent low ν_4 frequency of 1359 cm^{-1} for compound I, relative to a ν_4 frequency of 1380 cm^{-1} for compound II, was previously reported by Kitagawa's laboratory in a low temperature resonance Raman study (Teroaka et al., 1982). Unfortunately, these data were demonstrated (Van Wart and Zimmer, 1985) to be due to the ferrous enzyme, generated by photoreduction of compound I. In the data shown in this paper (and Palaniappan and Terner, 1989) we do in fact observe a 20 cm^{-1} downshift of ν_4 for compound I relative to compound II. However, this can be shown not to arise from the ferrous enzyme, since we also observe a large upshift of ν_2, plus other significant frequency shifts which have been shown to be typical of porphyrin π-radical cation model compounds (Salehi et al., 1986), (Oertling et al. 1987a,b), (Czernuscewicz et al., 1989).

The symmetry state of the porphyrin π-radical cation of horseradish peroxidase compound I is usually given the $^2A_{2u}$ designation, versus the $^2A_{1u}$ designation for catalase compound I. This has been discussed extensively in the literature [for example, (Hanson et al., 1981) and Sontum and Case, 1985)]. It is interesting that recent resonance Raman model studies have noted that resonance Raman frequency shifts of model octaethylporphyrin π-radical cations appear to be insensitive to $^2A_{1u}$ or $^2A_{2u}$ radical designation (Oertling et al., 1989), and that

TABLE I. A listing of resonance Raman frequencies (in cm^{-1}) and their assignments (following Czernuscewicz et al., 1989 and Li et al., 1990) for the horseradish peroxidase intermediates and derivatives shown in this paper. (CPD I, compound I: CPD II, compound II; dp, depolarized; p, polarized; ap, anomalously polarized).

mode	polari-ation	resting (ferric)	ferrous	NO-ferrous	NO-ferric	CPD I	CPD II
ν_{10} (B_{1g})	dp	1631	1605	1637	1646	1636	1639
$\nu_{C=C}$ (2)	p	1630	1630	1631	1634		1630
$\nu_{C=C}$ (1)	p	1620	1622	1621	1623		1620
ν_{37} (E_u)			1585	1598	1600	1614	1602
ν_{19} (A_{2g})	ap						
ν_2 (A_{1g})	p	1575	1567	1581	1585	1606	1587
ν_{11} (B_{1g})	dp	1550	1548	1563	1568	1570	1560
ν_{38} (E_u)				1547	1560		1552
ν_3 (A_{1g})	p	1500	1473	1501	1513	1504	1509
ν_{28} (B_{2g})	dp			1466	1477	1458	1472
Vinyl δ_s(=CH$_2$) (1)		1430	1427	1430	1433	1428	1427
ν_{29} (B_{2g})	dp	1404	1401	1399	1400	1402	1398
ν_4 (A_{1g})	p	1372	1359	1375	1379	1359	1379
Vinyl δ_s(=CH$_2$) (2)		1347	1339	1343	1345		
ν_{21} (A_{2g})	ap	1305				1322	

octaethylporphyrin π-radical cations appear to be predominantly $^2A_{1u}$ rather than $^2A_{2u}$ (Czernuscewicz et al., 1989). The octaethylporphyrin radicals were originally assigned on the basis of characteristic optical absorption spectra, as either $^2A_{1u}$ or $^2A_{2u}$, with a reversible switch of symmetry states being attainable in certain cases by an appropriate choice of axial ligands (Dolphin et al., 1973).

The characteristics of the resonance Raman frequency shifts we have observed for horseradish peroxidase compound I are most similar to those reported for the $^2A_{1u}$ type porphyrin π-cation radicals (octaethylporphyrins) which are characterized by large upshifts in the C_b-C_b modes, ν_2 and ν_{11}. This is in contrast to the large downshifts of ν_2 observed for the $^2A_{2u}$ type (meso-tetraphenylporphyrins). The protoporphyrin IX heme of horseradish peroxidase has substituents on the C_b carbons in common with the octaethylporphyrins, in contrast to the meso-positions of the tetraphenylporphyrins, which has been proposed to influence the relative ordering of the a_{1u} and a_{2u} orbitals (Czernuscewicz et al., 1989). Nevertheless, additional effects may be involved such as electron donation from the axial ligands which could influence the symmetry state (Czernuscewicz et al., 1989).

The highly oxidized intermediates of horseradish peroxidase, compounds I and II, had at one time been considered not to contain ferryl hemes since it was felt that the high valent ferryl iron (Fe^{IV}) was too unstable to exist for extended periods of time in aqueous media or within a protein (Peisach et al., 1968). However current physical and chemical methods have allowed considerable advances towards the understanding of these chemical structures. Porphyrin π-cation radicals are known to occur in photosynthesis (Fajer et al., 1970) and other biological processes (Hanson et al. 1982). A refinement of the procedures given in this account should lead to a description of the interaction of these species with other charged groups in the various protein active sites.

Acknowledgements This work was supported by NIH Grant GM34443.

REFERENCES

Czernuscewicz, R.S., Macor, K.A., Li, X.-Y., Kincaid, J.R., and Spiro, T.G. (1989): Resonance Raman Spectroscopy Reveals a_{1u} vs a_{2u} Character and Pseudo-Jahn-Teller Distortion in Radical Cations of Ni(II), Cu(II), and ClFe(III)Octaethyl- and Tetraphenylporphyrins.

J. Amer. Chem. Soc. 111: 3860-3869.

Dawson, J.H. (1988): Probing Structure-Function Relations in Heme-Containing Oxygenases and Peroxidases. Science 240: 433-439.

Dolphin, D., Muljiani, Z., Rousseau, K., Borg, D.C., Fajer, J. and Felton, R.H. (1973): The Chemistry of Porphyrin π-Cations. Ann. NY Acad. Sci. 206: 177-200.

Dolphin, D. and Felton, R.H. (1974): The Biological Significance of Porphyrin π-Cation Radicals. Acc. Chem. Res. 7: 26-32.

Dunford, H.B. (1982): Peroxidases. Adv. Inorg. Biochem. 4: 41-68.

Edwards, W.D. and Zerner, M.C. (1985): A theoretical study of the electronic structure and spectra of metalloporphine cations. Can. J. Chem. 63: 1763-1772.

Fajer, J., Borg, D.C., Forman, A., Dolphin, D. and Felton, R.H. (1970): π-Cation Radicals and Dications of Metalloporphyrins. J. Amer. Chem. Soc. 92: 3451-3459

Felton, R.H., Romans, A.Y., Yu, N.-T. and Schonbaum, G.R. (1976): Laser Raman Spectra of Oxidized Hydroperoxidases. Biochim. Biophys. Acta 434: 82-89.

Gouterman, M. (1978): Optical Spectra and Electronic Structure of Porphyrins and Related Rings. In: The Porphyrins, Volume IIIA Dolphin, D., ed., Academic Press, New York, pp. 1-156.

Hanson, L.K., Chang, C.K., Davis, M.S., and Fajer, J. (1981): Electron Pathways in Catalase and Peroxidase Enzymic Catalysis. Metal and Macrocycle Oxidations of Iron Porphyrins and Chlorins. J. Amer. Chem. Soc. 103: 663-670.

Hanson, L.K., Chang, C.K., Davis, M.S., and Fajer, J. (1982): Role of π-Cation Radicals in the Enzymatic Cycles of Peroxidases, Catalases, and Nitrite and Sulfite Reductases. In: Electron Transport and Oxygen Utilization, Ho, C., ed., Elsevier North Holland, Amsterdam pp. 245-252.

Hewson, W.D. and Hager L.P. (1979): Peroxidases, Catalases, and Chloroperoxidase. In: The Porphyrins, Volume VII, Dolphin, D., ed. Academic Press, New York, pp. 295-332.

Kim, D., Miller, L.A., Rakhit, G. and Spiro, T.G. (1986): Resonance Raman Spectra of Metallooctaethylporphyrin Cation Radicals with a_{1u} and a_{2u} Orbital Character. J. Phys. Chem. 90: 3320-3325.

Loew, G.H. (1983): Theoretical Investigations of Iron Porphyrin. In: Iron Porphyrins, Part One, Lever, A.B.P. and Gray, H.B. eds. Addison-Wesley, Reading, MA pp. 1-87.

Li, X.-Y., Czernuszewicz, R.S., Kincaid, J.R., Stein, P. and Spiro, T.G. (1990): Consistent Porphyrin Force Field. 2. Nickel Octaethylporphyrin Skeletal and Substituent Mode Assignments from ^{15}N, Meso-d_4, and Methylene-d_{16} Raman and Infrared Isotope Shifts. J. Phys. Chem. 94: 47-61.

Makinen, M.W. and Churg, A.W. (1983): Structural and Analytical Aspects of the Electronic Spectra of Hemeproteins. In: Iron Porphyrins, Part One, Lever, A.B.P. and Gray, H.B. eds., Addison-Wesley, Reading, MA pp. 141-235.

Mathies, R. (1979): Biological Applications of Resonance Raman Spectroscopy in the Visible and Ultraviolet: Visual Pigments, Purple Membrane and Nucleic Acids. In: Chemical and Biological Applications of Lasers. Moore, C.B., ed., Academic Press, New York, pp. 55-99.

McMurry, J. and Groves, J.T. (1986): Metalloporphyrin Models for Cytochrome P-450 In: Cytochrome P-450, Structure Mechanism and Biochemistry, Ortiz de Montellano, P.R. ed., Plenum Press, New York, pp. 1-28.

Oertling, W.A. and Babcock, G.T. (1985): Resonance Raman Scattering from Horseradish Peroxidase Compound I. J. Amer. Chem. Soc. 107: 6406-6407.

Oertling, W.A., Salehi, A., Chang, C.K. and Babcock, G.T. (1987a): Resonance Raman Spectroscopic Detection of Demetallation of Metalloporphyrin π-Cation Radicals. J. Phys. Chem. 91: 3114-3116.

Oertling, W.A., Salehi, A., Chung, Y.C., Leroi, G.E., Chang, C.K. and Babcock, G.T. (1987b): Vibrational, Electronic, and Structural Properties of Cobalt, Copper, and Zinc Octaethylporphyrin π-Cation Radicals. J. Phys. Chem. 91: 5887-5898.

Oertling, W.A. and Babcock, G.T. (1988): Time-Resolved and Static Resonance Raman Spectroscopy of Horseradish Peroxidase Intermediates. Biochemistry 27: 3331-3338.

Oertling, W.A, Salehi, A., Chang, C.K. and Babcock, G.T. (1989): Resonance Raman Vibrational Analysis of Cu(II), Fe(III), and Co(III) Porphyrin π-Cation Radicals and their Meso-Deuteriated Analogues. J. Phys. Chem. 93: 1311-1319.

Ogura, T. and Kitagawa, T. (1987): Device for Simultaneous Measurements of Transient Raman and Absorption Spectra of Enzymic Reactions: Application to Compound I of Horseradish Peroxidase. J. Amer. Chem, Soc. 109: 2177-2179.

Paeng, K.-J. and Kincaid, J.R. (1988): The Resonance Raman Spectrum of

Horseradish Peroxidase Compound I. <u>J. Amer. Chem. Soc.</u> 110: 7913-7915.

Palaniappan, V. and Terner, J. (1989): Resonance Raman Spectroscopy of Horseradish Peroxidase Derivatives and Intermediates with Excitation in the Near Ultraviolet. <u>J. Biol. Chem.</u> 264: 16046-16053.

Palmer, G. (1983): Electron Paramagnetic Resonance of Hemeproteins. In: <u>Iron Porphyrins, Part Two</u>, Lever, A.B.P. and Gray, H.B. eds. Addison Wesley, Reading, MA pp. 43-88.

Peisach, J., Blumberg, W.E., Wittenberg, B.A., and Wittenberg, J.B. (1968): The Electronic Structure of Protoheme Proteins, III. Configuration of the Heme and Its Ligands. <u>J. Biol. Chem.</u> 243: 1871-1880.

Salehi, A., Oertling, W.A., Babcock, G.T. and Chang, C.K. (1986): One-Electron Oxidation of the Porphyrin Ring of Cobaltous Octaethylporphyrin [Co(II)OEP]. Absorption and Resonance Raman Spectral Characteristics of the $Co(II)OEP^{+} \cdot ClO_4^{-}$ π-Cation Radical. <u>J. Amer. Chem. Soc.</u> 108: 5630-5631.

Shannon, L.M., Kay, E. and Lew, J.Y. (1966): Peroxidase Isozymes from Horseradish Roots. <u>J. Biol. Chem.</u> 241: 2166-2172.

Sontum, S.F. and Case, D.A. (1985): Electronic Structures of Active Site Models for Compounds I and II of Peroxidase. <u>J. Amer. Chem. Soc.</u> 107: 4013-4015.

Spiro, T.G. (1983): The Resonance Raman Spectroscopy of Metalloporphyins and Heme Proteins. In: <u>Iron Porphyrins, Part Two</u>, Lever, A.B.P., and Gray, H.B., eds., Addison- Wesley, Reading, MA pp. 89-160.

Terner, J., Sitter, A.J. and Shifflett, J.R. (1989): Resonance Raman Characterization of the Oxidation of the Horseradish Peroxidase Active Site. <u>Charge and Field Effects in Biosystems-2</u>, Allen, M.J., Cleary, S.F. and Hawkridge, F.M., eds., Plenum Press, New York, pp. 31-42.

Teroaka, J., Ogura, T. and Kitagawa, T. (1982): Resonance Raman Spectra of the Reaction Intermediates of Horseradish Peroxidase Catalysis. <u>J. Amer. Chem. Soc.</u> 104: 7354-7356.

Van Wart, H.E. and Zimmer, J. (1985): Resonance Raman Studies of the Photoreduction of Horseradish Peroxidase Compounds I and II. <u>J. Amer. Chem. Soc.</u> 107: 3379-3381.

Woodruff, W.H. and Spiro, T.G. (1974): Resonance Raman Spectroscopy of Reaction Intermediates by a Rapid Mixing Continuous Flow Technique. <u>Applied Spectroscopy</u> 28: 576-578.

Yamaguchi, H., Nakano, M. and Itoh, K. (1982): Resonance Raman
Scattering Study on the π-Cation Radicals of Magnesium, Zinc, and
Copper Tetraphenylporphines. <u>Chem. Lett.</u> 1982: 1397-1400.

PHOTORESPIRATION OF THE MONOLAYERS OF HYDRATED CHLOROPHYLL-A OLIGOMER

Alexander G. Volkov*, Maya I. Gugeshashvili*

Gaetan Munger and Roger M. Leblanc
Centre de recherche en photobiophysique, Universite du Quebec a Trois-Rivieres, 3351 boul.des Forges, C.P.500, Trois-Rivieres, Quebec, G9A 5H7, Canada
*Permanent Addres: A.N.Frumkin Institute of Electrochemistry, Academy of Sciences of the USSR, 31 Leninsky Prospekt, 117071 Moscow, USSR.
Present Address:Chemical Department, Chemistry Building, College of Natural Science, Michigan State University, East Lansing, MI 48824 USA

INTRODUCTION

The photoelectrochemical properties of wet chlorophyll-a in monolayers, multilayers and thin films at oil/water, gas/water and solid/water interfaces have been studied extensively in order to obtain a thorough understanding of its role in the primary events in plant photosynthesis, i.e., light harvesting, energy transfer and charge separation in the photosynthetic unit (Tang and Albrecht,1975; Volkov, 1984).

Photorespiration, the reduction of O_2 under illumination, plays a very important role in natural photosynthesis although the molecular mechanism of this phenomenon is still vague. Photorespiration is a key problem of modern biotechnology, since this process decreases agricultural productivity in a plant photosynthesis due to the dissipation of solar energy. The state of chlorophyll molecules in photosynthetic apparatus, believed to be in various aggregated forms, is still a matter of speculation and investigation. In view of the importance of the in vitro membrane models of photosynthesis, we have investigated light energy conversion in wet chlorophyll-a monolayers at SnO_2 optically transparent electrode (OTE). During this investigation, we have observed the reaction of hydrated oligomer chlorophyll-a photorespiration.

MATERIALS AND METHODS

All the experiments were performed at 20^oC under dim green light. The construction and characteristics of a photoelectrochemical cell, a three-electrode poentiostat, and a monochromator-microcomputer interface have been described in detail elsewhere (Blanchet et al.,1989). The cell was also used for the collection of spectroscopic absorption data. The illuminated area of SnO_2-OTE was 1.4 cm^2. A saturated calomel electrode (SCE) served as the reference, and a small platinum wire as the counter electrode. A gold edge connector was used to establish the linkage with the tin oxide working electrode. The illumination system is made up of a xenon arc lamp coupled with a step driven, f/3.4 monochromator furnished with a digital controller. The electronic shutter is secured to the exit of the monochromator; the monochromator and the shutter are computer controlled. The results are normalized for a photons flux of $1.4x10^{15}cm^2s^{-1}$. Absorption spectra were recorded on a Hewlett-Packard Model 8452A spectrophotometer.

SnO$_2$-OTE plate (Delta Technologies Ltd., Stillwater,MN) of dimension 2.5x6.0 cm^2 with one side coated with an 10^{-5} cm thick SnO_2 layer was used as the solid substrate for monolayer deposition. The SnO_2 layer had a resistivity of 10^{-3} ohm.cm and an optical transmittance of 90% in the visible region above 400 nm. From the slope of the Mott-Schottky plots, the donor density (antimony) of the electrode was estimated to be $7x10^{20}$ cm^{-3}. Prior to the monolayer deposition, the SnO_2-OTE was washed with acetone and hot ethanol, rinsed 15 times with distilled water and dried with nitrogen. The SnO_2-OTE was washed 5 minutes in the boiling solution of detergent, rinsed 15 times with distilled water and dried with nitrogen. Following this treatment, the electrode was immersed in cleaning solution for 3 days. The electrode was then rinsed several times with doubly distilled water and sonicated for 5 minutes. The electrode was dried with argon prior to monolayer deposition. Chlorophyll-a was obtained from Fluka Chemical Corp. For preparation of the wet n-hexane the system consisting of equal volumes of n-hexane and water was equilibrated for 48 hours. After equilibration, the aqueous phase was removed from the vessel. Chlorophyll-a was disolved in wet n-hexane and sonicated for 100 min in a cooled bath. After sonication the formation of hydrated chlorophyll-a oligomer was checked by absorption spectroscopy. A total volume of 1 ml of hydrated oligomer of chlorophyll-a solution at a concentration of 2×10^{-5} M in wet hexane was carefully deposited dropwise from a microsyringe over the water surface in Langmuir trough equipped with Langmuir film balance. The film was compressed until the surface pressure reached 20 mN m^{-1}. The subphase used for monolayer deposition was phosphate buffer (10^{-3} M, pH 8.0). The SnO$_2$ plate was dipped vertically into subphase and then withdrawn under constant surface pressure at a speed 10 mm min^{-1} by means of vibration-free hydraulic system. The chlorophyll-a monolayer at the glass side of the plate was removed after deposition using cotton swabs soaked with acetone.

RESULTS AND DISCUSSION

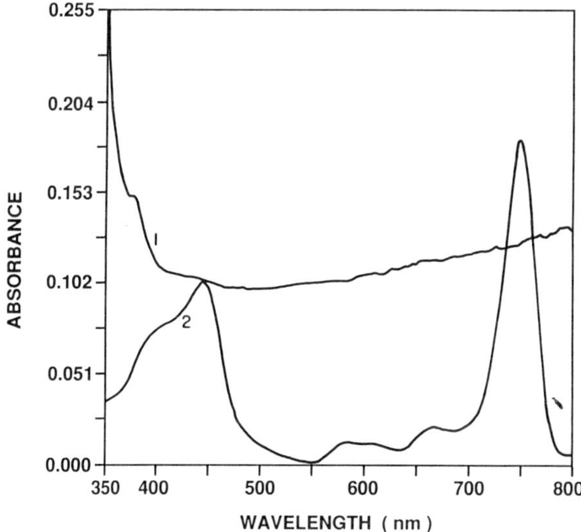

FIGURE 1. Electronic absorption spectra of SnO$_2$-OTE (1) and hydrated oligomer of chlorophyll-a solution in wet n-octane (2).

Optically transparent SnO_2 electrode doped with antimony is a quasimetallic conductor. The SnO_2 electrode is transparent in the visible region of the optical spectrum (Fig.1) and has a high reflectivity for infrared radiation. SnO_2-OTE is used very often as a working electrode in photoelectrochemistry due to its sufficient large potential window and its optical transparency in the visible region. For comparison, the absorption spectrum of hydrated oligomer of chlorophyll-a in n-octane solution is also shown on Fig.1.

Fig.2 shows the electronic absorption spectra of dry and wet chlorophyll-a in n-hexane solutions. The absorption spectrum of dry chlorophyll-a posesses two absorption bands at 428 nm, and 660 nm similar to the typical absorption spectra of chlorophyll-a in non-polar solvents. If chlorophyll-a is dissolved in wet (saturated by water) hexane, the hydrated oligomer of chlorophyll is formed and its absorption spectrum is shifted to lower energy with the absorption bands at 448 nm and 742-5 nm. It is worth to note that intensity ratios of the blue to red absorption bands are 1.3 for dry chlorophyll-a and 0.6 for wet chlorophyll-a. These differences of the considered absorption spectra coupled with the strong decrease in energy of the lowest electronic transition of wet chlorophyll-a ($\Delta \nu = 1700$ cm^{-1}) indicate, that in wet hexane chlorophyll-a exists in well organized aggregate forms.

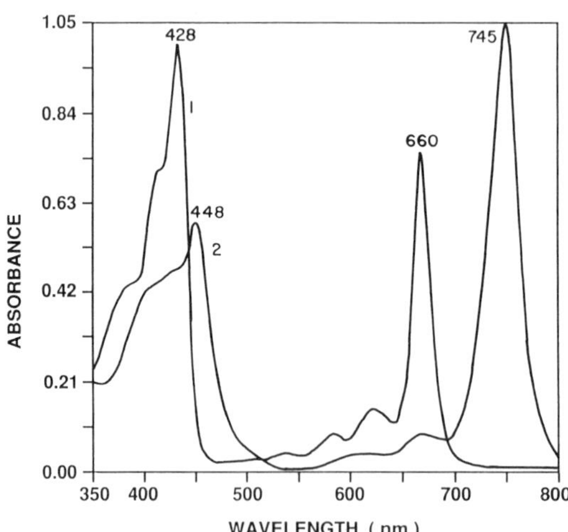

FIGURE 2. Electronic absorption spectra of dry chlorophyll-a (1) and wet chlorophyll-a (2) in n-hexane solution.

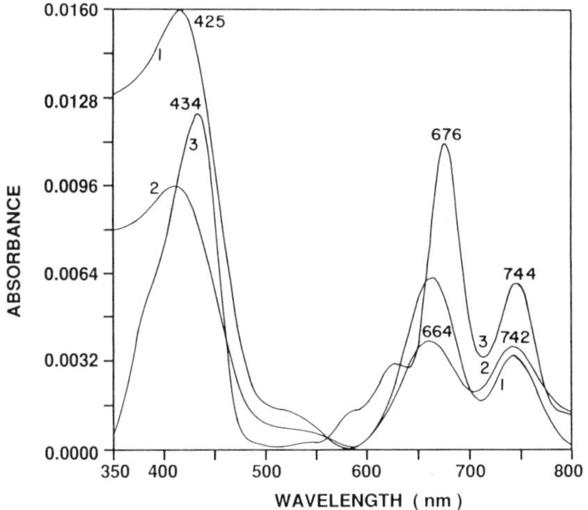

FIGURE 3. Electronic absorption spectra of wet chlorophyll-a monolayers on SnO_2-OTE before contact with electrolyte solution (1), after photoelectrochemical experiments (2) and on the glass side of OTE (3).

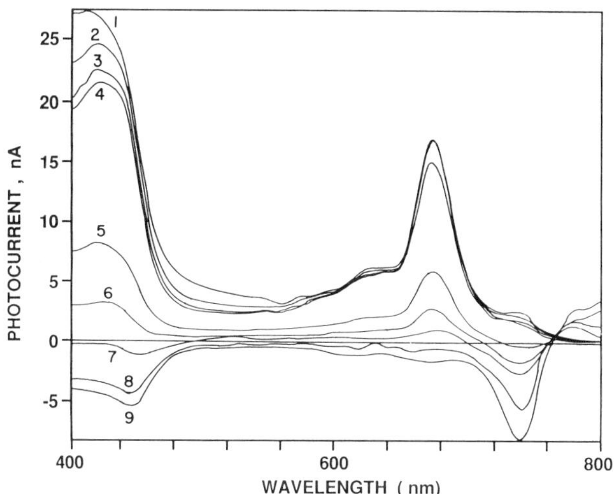

FIGURE 4. Action spectra of wet chlorophyll-a monolayer on SnO_2-OTE. Medium: 0.1 M KCl, 0.05 M QH_2 and 0.025 M NaH_2PO_4 at pH 6.9. Electrode potential: (1) + 200 mV, (2) +150 mV, (3) +100 mV, (4) +50 mV, (5) -50 mV, (6) -100 mV, (7) -150 mV, (8) -200 mV, (9) -250 mV (the aqueous solution was initially saturated with argon).

The electronic absorption spectrum of wet chlorophyll-a monolayer on SnO_2-OTE (Fig.3) posseses a Soret band at 425 nm and low energy band at 664 and 742 nm. The glass side of OTE gives absorption maxima at 434, 676 and 744 nm. Under deposition of one monolayer at SnO_2-OTE, both forms of chlorophyll-a (dry and wet) are represented. After immersion of the SnO_2-OTE, covered with a monolayer of wet chlorophyll-a into an aqueous solution at pH 6.9 (consisting 0.1 M KCl and 0.025 M NaH_2PO_4), the positions of maxima in absorption spectrum are not changed. Howevere, as shown in Figure 3, the absorbance corresponding to the dry form of chlorophyll-a decreases (see curve 2).

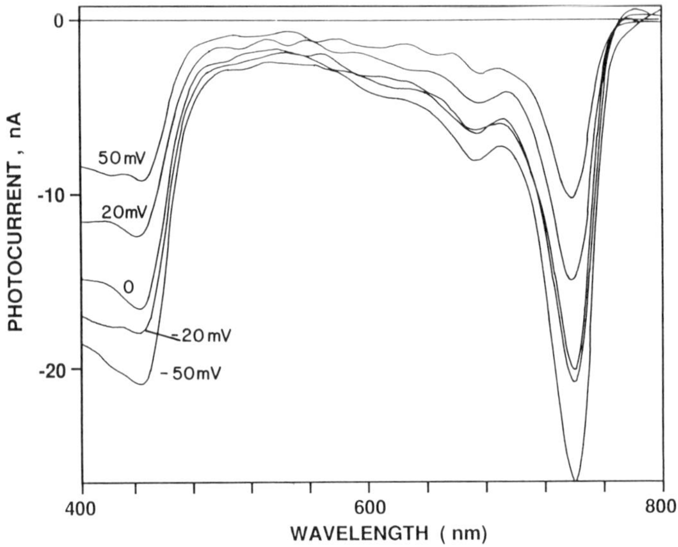

FIGURE 5. Action spectra of wet chlorophyll-a monolayer on SnO_2-OTE. Medium: 0.1 M KCl and 0.025 M NaH_2PO_4 at pH 6.9. Electrode potential: (1) +50 mV, (2) +20 mV, (3) 0 mV, (4) -20 mV, (5) -50 mV (the aqueous solution was initially saturated by O_2).

Under illumination of the wet chlorophyll monolayer at SnO_2-OTE in the aqueous solution in the presence of hydroquinone (QH_2), there arises an anodic photocurrent under polarization between +200 and -50 mV (Fig.4, curves 1 through 5). The action spectra under these conditions coincide with the absorption spectrum of dry chlorophyll monolayer. Under more negative polarization of the electrode, cathodic photocurrents are observed. The action spectra of these cathodic photocurrents correspond to the absorption spectrum of the hydrated oligomer of chlorophyll-a. Under polarization in the range of -50 to -150 mV, both cathodic and anodic photocurrents caused by wet and dry chlorophyll-a apear in the action spectra. When QH_2 is absent from the aqueous solution (initially saturated with O_2 before illumination) only cathodic photocurrent appear(Fig.5).

The cathodic photocurrent depends on the concentration of oxygen in the aqueous solution (Fig.6).

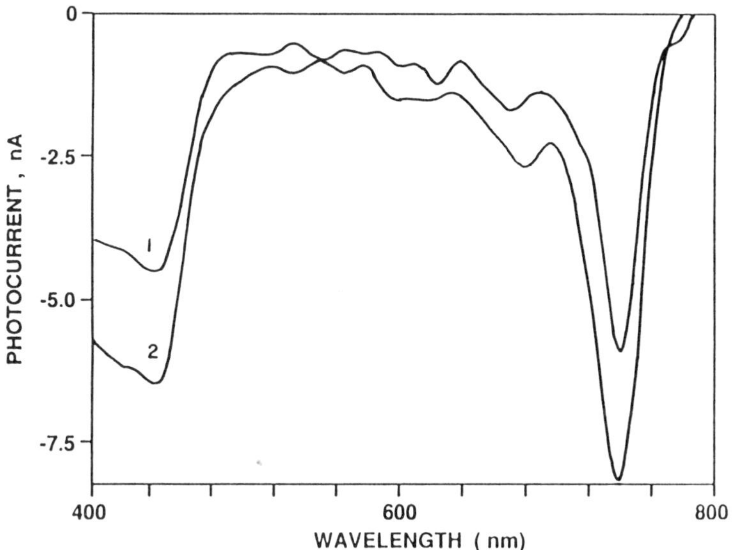

FIGURE 6. Action spectra of wet chlorophyll-a monolayer on SnO_2-OTE. Medium: 0.1 M KCl and 0.025 M NaH_2PO_4 at pH 6.9. Electrode potential $+50$ mV. The aqueous solution was initialy saturated by argon (1) or by air (2).

The cathodic photocurrent is caused by the oxygen uptake of the hydrated oligomer of chlorophyll-a monolayer and strongly depends on the concentration of oxygen and the electrode potential. Quantum efficiency ϕ_λ for photocurrent generation at wavelength λ was calculated by the following equation:

$$\phi_\lambda = I_\lambda / qn(1 - 10^{-A_\lambda})$$

where I_λ is the photocurrent density, q is the elementary charge, n is the number of incident photons, A_λ is the absorbance of chlorophyll monolayer in contact with aqueous solution. Using this relation, the maximum quantum yield of photorespiration at pH 6.9 and an electrode potential of -50 mV was found to be 0.45 $\pm$ 0.05%.

ACKNOWLEDGEMENT

This work was financially supported by the Natural Sciences and Engineering Research Council of Canada (NSERC) and Fonds pour la Formation de Chercheurs et l'Aide a la Recherche (FCAR). The authors are grateful to Prof. G.Babcock for critical reading of the manuscript and Mr. A. Tessier for supervision of the instrumentation.

REFERENCES.

Blanchet PF, Tessier A, Paquette A, Cote D and Leblanc RM (1989): Simple computer- supported photoelectrochemical instrument for studies of ultrathin films on transparent electrodes. *Rev Sci Instrum* 60: 2750 - 2755.

Tang CW and Albrecht AC (1975): Chlorophyll-a photovoltaic cell. *Nature* 254: 507- 509.

Volkov AG (1984): A possible mechanism of the photooxidation of water sensitized by chlorophyll adsorbed at the interface. *Bioelectrochem Bioenerg* 12: 15-24.

Applications of Bioelectrochemical Technology

THE COAXIAL-PORE MECHANISM OF CELL MEMBRANE ELECTROFUSION:
THEORY AND EXPERIMENT

Iziaslav G. Abidor
Department of Bioelectrochemistry
A.N. Frumkin Institute of Electrochemistry
Academy of Sciences of the USSR, Moscow

Arthur E. Sowers
Department of Biophysics
School of Medicine
University of Maryland
Baltimore, Maryland

INTRODUCTION

Cell membrane fusion starts with two membranes in close
contact which become reorganized and combined in a
topographically continuous structure. Membrane fusion is
followed by other complicated processes that may end, for
example, in the appearance of a new hybrid cell (for reviews
see e.g. Sowers, 1987a). Cell electrofusion is an
artificially induced phenomenon. Its significance is
discussed and reviewed elsewhere in this volume (Sowers and
Abidor). This chapter is limited to a discussion of the
coaxial-pore mechanism of electrofusion.

Since electrofusion can be applied to many different
kinds of cells it is expected to be a physical process highly
independent of the nature of cell membranes or metabolism
(Zimmermann, 1982; Kuzmin et al., 1988; Abidor et.al, 1989).
Because the lipid bilayer is similar in all cell membranes,
membrane proteins are assumed to play a minor role in
electrofusion.

Lipid bilayers are commonly understood to be held appart
by strong electric and hydration forces (for review see e.g.:
Blumenthal, 1987; Wilschut and Hoekstra, 1991). Membrane

fusion, however, is known to be a local event (Hui et al., 1981; Sowers, 1989; Kuzmin et al, 1988; Coackley and Gallez, 1989). Melikyan and Chernomordik (1989) have proposed that membrane electrofusion starts from stalks or thin lipid bridges which are formed between lipid bilayers in close contact (Markin et al., 1984; Chernomordik et al., 1985). While the stalks transform into so-called trilaminar structures (Melikyan et al., 1982), the associated cytoplasmic compartments are still separated by contact (central) bilayers formed by outer lipid monolayers from each of the pre-fusion cell membranes. Rupture of the contact bilayer due to irreversible electrical breakdown results in "completed" fusion, i.e. mixing of cell membranes and cytoplasms. This mechanism, however, is unlikely for cell electrofusion because electrofusion is essentially a nonspecific phenomenon while the tendency to form stalks is associated only with those lipids having a negative spontaneous curvature. Furthermore, the trilaminar structures have not been observed in EM pictures of electrofusion products (Stenger and Hui, 1986).

It has been reported (Sowers, 1986, 1987b, 1989; Teissie and Rols, 1986) that some cells can be fused even if the electric pulse is applied before, rather than after, the cell membranes are brought into contact. Hence, it was concluded that electric pulse-treated cells can fuse even if contact is induced many minutes after the pulse treatment. Up to now this long-lived fusogenic state has not been well studied. Some authors consider this state to be caused by electropermeabilization of cell membranes and suggest that this state could be the prime factor in the electrofusion process (Rols and Teissie, 1990; Teissie et al., 1989). In an alternate viewpoint (Sowers, 1989) electric field pulses may induce two independent processes: i) fusion of cells already in contact (i.e. normal electrofusion) and ii) transition of membranes into a fusogenic state as has been found when cells are treated by chemical fusogens (Wilschut and Hoekstra, 1991). The first process is found to be more effective and the importance of membrane contacts in the electrofusion process has been repeatedly stressed (Sowers, 1989; Abidor et al., 1989; Coackley and Galez, 1989; Sukharev et al., 1990). A dramatic increase in membrane permeability induced by electric fields is commonly attributed to reversible electric breakdown, as caused by the induction of a large number of minute pores in the lipid bilayers (Benz et al., 1979; Kinosita and Tsong, 1977; Chizmadzhev and Abidor, 1980). It is possible, however, that electroporation and electrofusion are unrelated with each other. However many authors (Zimmermann, 1982; Dimitrov and Jain, 1986; Kuzmin et al., 1988; Zhelev et al., 1988) accept pores as a necessary condition for electrofusion.

The possibility that electropores are an intermediate stage of cell electrofusion was first proposed by Pilwat et al. (1980). According to these authors electrofusion starts with the formation of pairs of coaxial pores filled by water and randomly positioned lipid molecules. The last ones form bridges between the pore edges as minute "tubes" combining adjacent cell membranes into a topologically single structures. Afterwards, the membrane tubes expand, thus leading to mixing of cell contents, which completes the fusion process.

This mechanism has at least two unclear points. It is not clear: i) how lipid molecules can be pulled out of the lipid bilayer, and ii) why they do not diffuse away instead of forming bridges. However the idea of membrane tube formation by combining coaxial pores was attractive and was further developed, mainly on a theoretical level (Dimitrov and Jain, 1984; Sugar et al., 1987; Kuzmin et al., 1988). The most detailed theoretical analysis of the coaxial-pores mechanism (Kuzmin et al., 1988) suggests that the presence of the electric field may be important at all stages of the fusion process. However, our understanding of cell electrofusion is still limited to a qualitative or semi-quantitative scheme. This is because most experimental studies on electrofusion are descriptive in character.

This chapter reviews the theory of the coaxial pore mechanism of membrane electrofusion (Kuzmin et al., 1988) and the experimental efforts (Abidor and Sowers, 1992) to verify this mechanism using erythrocyte ghosts as a model system. The original paper (Kuzmin et al., 1988) on the coaxial-pore mechanism of cell membrane electrofusion was published in Russian and thereby almost unknown in the West.

1. A THEORETICAL ANALYSIS OF THE COAXIAL-PORE MECHANISM OF ELECTROFUSION

The emphasis of the Kuzmin et al. (1988) analysis includes effects which would be expected when an electric field pulse is applied perpendicular to two membranes in close contact. These effects include: the drawing of adjacent membranes together, the formation and mutual attraction and merging of coaxial pores by electric current, and some effects impeding membrane fusion.

1.1. Electrocompression of Membrane Contacts

Fig. 1 shows a model of two parallel membranes in close contact (the contact zone). Here (and below) δ is the membrane thickness and $h=2d$ is the intermembrane distance. Membrane contact areas are assumed to be protein-free. Membrane surfaces carry a fixed charge of density q. Interaction between membranes is expected to depend on different physical forces: i) the Van-der-Waals attraction of membranes, ii) the mutual Coulomb repulsion of the double electric layers created by membrane surface charges, iii) the

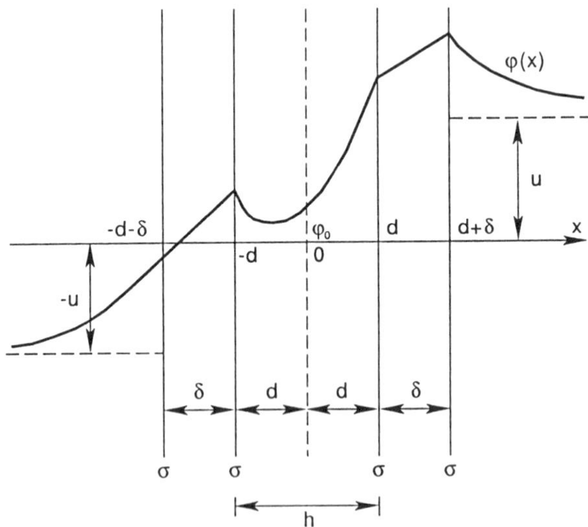

Fig. 1. Contact of two charged membranes in an external electric field. A constant voltage $2u$ is applied to the system.

osmotic pressure, iv) hydration repulsion of the adjacent surfaces, and v) mechanical forces promoting formation of the contact zone. The hydration forces decrease very quickly with distance between membranes and therefore may be neglected for sufficient large distances. Here the primary factor which prevents membranes from becoming closer is the electrostatic repulsion involving the double electric layers. Naturally a question arises: can the electric field which induces fusion of membranes in contact compensate or at least decrease their electrostatic repulsion?

It is assumed that an electric pulse of voltage $2u$ and duration τ is applied to the membrane system and the membranes become charged. This also affects the diffuse double layers in the gap between membranes. A charge relaxation time τ_c in the system is estimated to be $C_m R/G$ where C_m is the specific membrane capacitance, G is the specific electroconductivity of the electrolyte solution, and R is a characteristic distance (in the case of the cell contacts R is the cell radius). For physiological conditions $G \sim 1$ Mho/m, and assuming values of $C_m \sim 1$ μF/cm^2 and $R \sim 5$ μm, we have $\tau_c \ll \tau \sim 5$ μs. Thus it may be may be assumed that a constant voltage 2U is applied to the bulk solutions on both sides of the contact zone and the potential distribution in the system is constant.

To calculate the force affecting the membrane surface, we assume that the force consists of two components: i) osmotic, and ii) electric. The first is due to the electric field-induced shift in electrolyte concentrations in the membrane diffuse double layers. The second component is the difference between Maxwell's tensions on both sides of the membrane. The osmotic pressure in the electrolyte solution is defined by the electric potential at a given point while the Maxwell tension depends on the electric field strength. A linear approximation using the Laplace equation is used to calculate both forces in the course of deriving the potential distribution $\varphi(x)$ in the system:

$$\Delta\varphi = 0 \qquad (d<|x|<d+\delta) \qquad (1.1)$$

for the regions inside membrane, and the Poisson–Bolzmann equation:

$$\Delta\varphi = \kappa_i^2\varphi \qquad \left(\begin{matrix} 0<|x|<d,\ i=1 \\ d+\delta<|x|,\ i=2 \end{matrix} \right) \qquad (1.2)$$

for the space between membranes and outside, in the bulk solutions. Here κ_i is the reciprocal Debye atmosphere thickness, $i=1$ for the intermembrane gap, $i=2$ for the bulk solutions. Additional conditions for Eqs. 1.1 and 1.2 are i) potential continuity, ii) presence of fixed charges q on the membrane surfaces, and iii)

$$\varphi\,(\pm\infty)\ =\ \pm u \tag{1.3}$$

The osmotic pressure and Maxwell tensions in the system can be found as follows. A plane is chosen in outer solutions outside of the adjacent diffuse double layer. Note that in the bulk solutions ($x\to\pm\infty$) the osmotic pressure as well as the Maxwell tension equal zero. In the case of small potentials ($e\varphi/kT\ll1$) the osmotic pressure in the gap may be written as:

$$F_r = \left(\frac{\kappa_1 e}{kT}\right)^2 \frac{\varepsilon_w}{2}\,\varphi_0^{\,2} \tag{1.4}$$

where e is the unit charge, k is the Boltzmann constant, T is temperature ($^\circ K$), φ_0 is electric potential at $x=0$. Solving Eqs. 1.1 and 1.2, one obtain for φ_0:

$$\varphi_0 \ = \ \frac{1+2\gamma_2}{\gamma_1\cosh(\kappa_1 d)+(1+\gamma_2)\sinh(\kappa_1 d)}\cdot\frac{q}{\varepsilon_w\kappa_1} \tag{1.5}$$

where

$$\gamma_i \ = \ \frac{\varepsilon_m}{\varepsilon_w\delta}\cdot\frac{1}{\kappa_i} \tag{1.6}$$

and ε_m and ε_w are membrane and water dielectric permeabilities respectively.

Since the Maxwell tension in the bulk solutions is zero, in the intermembrane space Maxwell tension equals the pressure drawing the membranes together. Here

$$F_a \ = \ -\frac{\varepsilon_w}{2}\left(\frac{d\varphi}{dx}\right)^2_{x=0} \tag{1.7}$$

Substituting solutions of Eqs. 1.1 and 1.2 in Eq. 1.7 gives:

$$F_a \ = \ -\left(\frac{\kappa_1 e}{kT}\right)^2 \frac{\varepsilon_w}{2}\left[\frac{u\gamma_1}{(1+\gamma_2)\cosh(\kappa_1 d)+\gamma_1\sinh(\kappa_1 d)}\right]^2 \tag{1.8}$$

Eqs. 1.4 and 1.7 give opposite signs for F_a and F_r, therefore these forces have opposite directions. Thus in an electric field the mutual repulsion of the membranes becomes weaker. If the contact zone is in mechanical balance then:

$$F_a = F_r \qquad (1.9)$$

(mechanical, hydration, and Van-der-Waals forces are assumed to be negligible). Hence at $\kappa_1=\kappa_2=\kappa$:

$$(2\gamma+1)\coth\left[\kappa d+\frac{1}{2}\ln(2\gamma+1)\right] = \frac{u}{u_*} \qquad (1.10)$$

where u_* is a voltage under which the membrane charge induced by the external field becomes equal to the surface charge q:

$$u_* \equiv q\delta/\varepsilon_m \equiv q/C_m \qquad (1.11)$$

The intermembrane distance $h=2d$ can be found as a function of the applied voltage u from Eq. 1.10. The dependance $h(u)$ plotted in Fig. 2 shows that when u slightly exceeds some critical value $u_1=(2\gamma+1)u$ the distance between membranes drastically decreases to the value of the order of $2x$ Debye κ^{-1}. If the voltage increases to the value of $u_2=(2\gamma+1)u_*/\gamma=u_1/\gamma$ then h is zero.

For lipid bilayers in solutions having physiological ion compositions, the parameter γ is ~1/200, i.e. $\gamma \ll 1$. In this

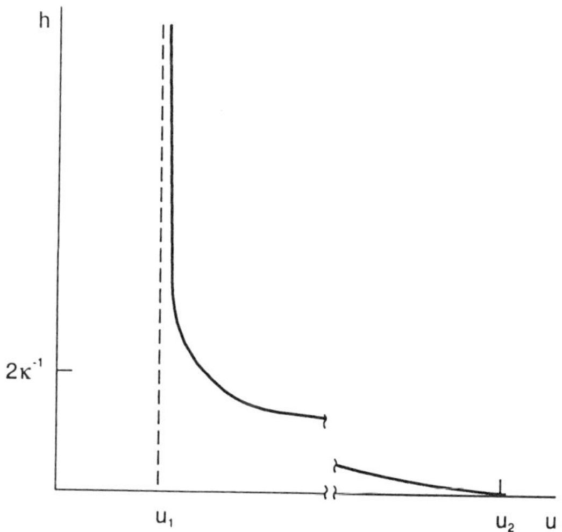

Fig. 2. The intermembrane distance as a function of the transmembrane voltage u.

case Eq. 1.10 is simplified to:

$$\coth(\kappa d) = u/u_* \qquad (1.12)$$

For typical values of $q \sim 1$ μCoul./cm^2 and $C_m \sim 1$ μF/cm^2 we have $u_1 \sim u_* \sim 1$ V. Voltages of this order are, indeed, used for cell electrofusion. For the same conditions u_2 is ~ 200 V and therefore out of physical range of voltages for biomembranes. This is why membranes cannot be drawn together by electric fields to very small distances where hydration and Van-der-Waals forces come into play.

Thus at voltages of order of 1 V applied to membranes in contact the membranes may be drawn together to distances approximately corresponding to two thicknesses of the double electric layer at the membrane surfaces. It can be expected that membranes drawn to such closer distances would further promote their fusion.

1.2. Formation, Interaction and Merging of Pores in Membrane Contacts

When $u \sim 1$ V is applied to the contact zone, electrical breakdown occurs with the formation of a large number of pores in both membranes. Since the lateral resistance of the intermembrane gap is very high (Abidor et al., 1987) the appearance of a conducting pore in one membrane would be expected to cause the voltage drop across the second membrane in the pore vicinity to be almost doubled. This highly increases the probability of the appearance of a coaxial pore in the second membrane. Therefore at least part of pores in the contact zone would be expected to be coaxial or almost coaxial. A distribution of electric field related with current through the coaxial pores is such that a pressure, which can be estimated using Fig. 3, will appear that draws the membranes together.

Let the contact zone contain a pair of coaxial pores of radius r_0 (Fig. 3). In cylindrical coordinates, the polar axis z coincides with the pore axis and the plane $z=0$ coincides with the left membrane surface. Since the membrane is actually an insulator, the current density j and related electric field $E=qj$ are both tangential to the membrane surface. This serves as an additional condition to the Laplace equation for the potential distribution in the system:

$$\Delta\varphi = 0 \tag{1.13}$$

To calculate the forces acting, for example, on the right membrane it is necessary to find the potential on the membrane surface at $r>r_0$. For simplification it is assumed that the current is uniformly distributed over the membrane surface, as restricted to the solution of Eq. 1.13 in the ranges of $0<z<h$ and $z>h+\delta$ with the boundary conditions:

$$\frac{\partial\varphi}{\partial z} = -\frac{I}{\pi r_0^2} \equiv -E_0 \qquad (z=h, \delta+h; r<r_0)$$

$$\tag{1.14}$$

$$\frac{\partial\varphi}{\partial z} = 0 \qquad (z=h, \delta+h; r>r_0)$$

where I is the current going through the pore and E_0 is the field strength in the pore itself. It is assumed that current flows to the right, i.e. along z-axis.

Solutions of Eqs. 1.13 and 1.14 for $r>r_0$ are:
in the intermembrane gap $(0<z<h)$

$$\varphi(r,z) = E_0\frac{4r_0}{h}\sum_{n=0}^{\infty} K_0(\lambda_n r)\, I_1(\lambda_n r_0)\, \frac{\cos\lambda_n z}{\lambda_n} \tag{1.15}$$

and in outer solution $(z>h+\delta)$

$$\varphi(z,0) = E_0 r_0\sum_{n=0}^{\infty} \left(\frac{r_0}{\rho}\right)^{n+1} \frac{P_n(0)\, P_n(\cos\theta)}{n+2} \tag{1.16}$$

Here $\lambda_n=(2n+1)\pi/h$; $\rho^2=r^2+(z-h-\delta)^2$; θ is the angle between the

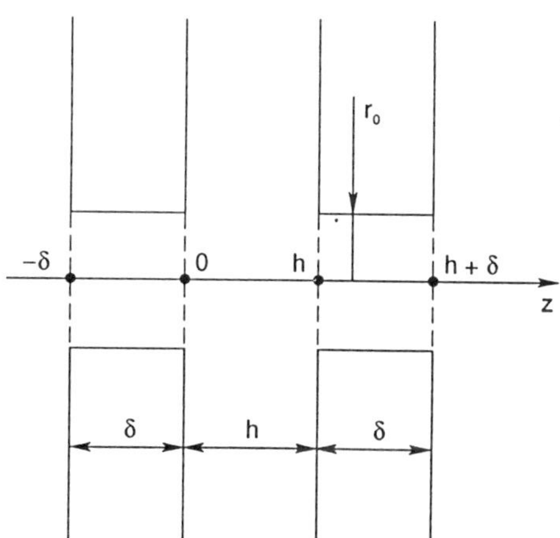

Fig. 3. A pair of coaxial pores in a membrane contact.

polar axis z and a radius-vector drawn from the point ($z=h+\delta$, $r=0$) to a given point; K_m is the MacDonald function of the order of m; I_m is the modified Bessel function of the order of m; P_m are the Legendre polynomials of the order of m.

Differentiation of Eqs. 1.15 and 1.16 with respect to r at corresponding values of z gives the field strength on both membrane surfaces:

for the surface $z=h$

$$E(r,h) \equiv E_1 = -E_0 \frac{4r_0}{h} \sum_{n=0}^{\infty} K_1(\lambda_n r)\, I_1(\lambda_n r_0) \qquad (1.17)$$

and for the surface $z=h+\delta$

$$E(r,h+\delta) \equiv E_2 = E_0 \sum_{n=0}^{\infty} \frac{n+1}{n+2} \left(\frac{r_0}{r}\right)^{n+2} P_n^2(0) \qquad (1.18)$$

The density of forces acting on the membrane surface normally equals the Maxwell tension difference for both sides of the membrane:

$$p(r) = \varepsilon_w (E_2^2 - E_1^2)/2 \qquad (1.19)$$

The sign of Eq. 1.19 is chosen so that the pores attract each other at $p(r)>0$.

Fig. 4 shows $p(r)$-curves for different values of the ratio h/r_0 at a constant value of the parameter $p_0=\varepsilon E_0^2/8$.

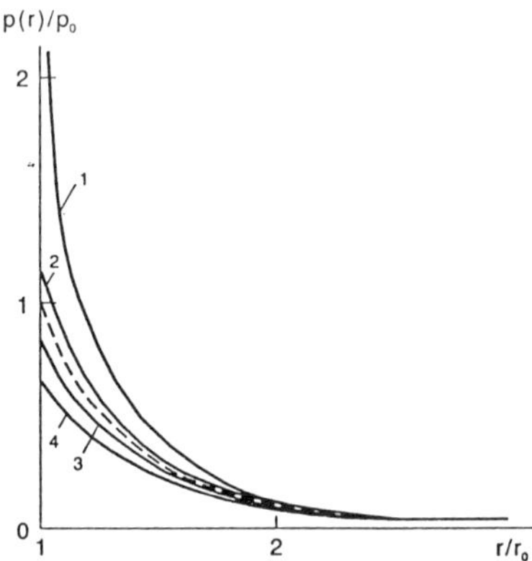

Fig.4. Distributions of the pressure pushing pore edges toward each other as functions of the normalized distance from the pore axis. The dashed curve was obtained by Eq. 20., while the solid curves, 1-4, were calculated using Eq. 19 with h/r_0 = 1.0, 1.5, 2.0, and 2.5, respectively.

Since $p(r)$ is always >0 there is mutual attraction of coaxialpores in the contact zone. The electrical forces drawing pores together are concentrated nearer the pore edge and drop rapidly with distance from the pore axis. Therefore one may suppose the compressing force is applied to the pore edge. To find this force Eq. 1.19 should be integrated over the membrane surface while taking into account Eqs. 1.17 and 1.18. Since a simple analytical solution could not be found it will be assumed for the estimation that the current flows along the pore axis. In this approximation the field strength on the inner surfaces of the membranes equals zero and:

$$p(r) = p_0(r_0/r)^4 \qquad (1.20)$$

A curve calculated using Eq. 1.20 is shown by the dotted line in Fig. 4. As can be seen for sufficiently small distances between membranes ($h<1.5r_0$) Eq. 1.20 gives a lower limit of $p(r)$ since the dotted curve lies under other curves calculated numerically using Eqs. 1.17-1.19.

The expression for p_0 includes current $I=u/R_p$ where R_p is the pore resistance which can be estimated as:

$$R_p = \frac{\delta + \pi r_0/4}{\pi r_0^2 G} \qquad (1.21)$$

Hence for p_0 we have:

$$p_0 = \frac{\varepsilon_w}{8}\left(\frac{u}{\delta}\right)^2 \Big/ \left(1 + \frac{\pi r_0}{4\delta}\right)^2 \qquad (1.22)$$

which gives $p_0 \sim 5 \cdot 10^6$ Pa (≈ 50 atm) for typical values of $u \sim 1$ V, $\delta \sim 50$ Å, $h \sim 100$ Å, $\varepsilon = 80\varepsilon_0$ (ε_0 is the dielectric permeability of vacuum).

Integrating of Eq. 20 gives a resultant force acting on the pore edge:

$$F_t = p_0 \pi r_0^2 \sim 2 \cdot 10^{-11} \text{ H} \qquad (1.23)$$

Fig. 5 shows the dependance of the force F on the distance between membranes h. The curve is computed by numeric integration of Eq. 1.19. As can be seen, the attraction between pores becomes weaker when the gap width h increases. The pore attractive force estimated by Eq. 1.23 is given in the figure by a line parallel to the abscissa. It confirms that Eq. 1.23 is a fairly good approximation for the attractive force in relatively small gaps ($h \leq 1.5r_0$).

1.3. Impeding Factors in Membrane Electrofusion

Mutual electrostatic repulsion of equally charged membrane surfaces, membrane resistance to bending deformation, and hydration repulsion due to the hydrophilic edges are all expected to impede fusion of membranes with each other. Neglecting the charge screen effect, the upper limit of electrostatic repulsion forces (per unit surface area) can be estimated as:

$$F_r \sim q^2/2\varepsilon_w$$

The surface charge of cell membranes density usually does not exceed 2-3 μCoul./cm^2 (the absolute value). Hence, F_r can reach several atmospheres which is much less than typical values $p(r_0)$ developed when current goes through coaxial pores.

It is necessary to compare the electric field work W_e needed to draw pore edges together with the membrane bending elastic energy W_m.

W_e is estimated as

$$W_e \sim F_t h \tag{1.24}$$

where F_t is the total force acting on the pore edge (Eq. 23). With h in the range from 10 to 100 Å W_e is in the range of $(2-20) \cdot 10^{-20}$ J or 5-50 kT.

To calculate W_m it will be assumed that the curved part

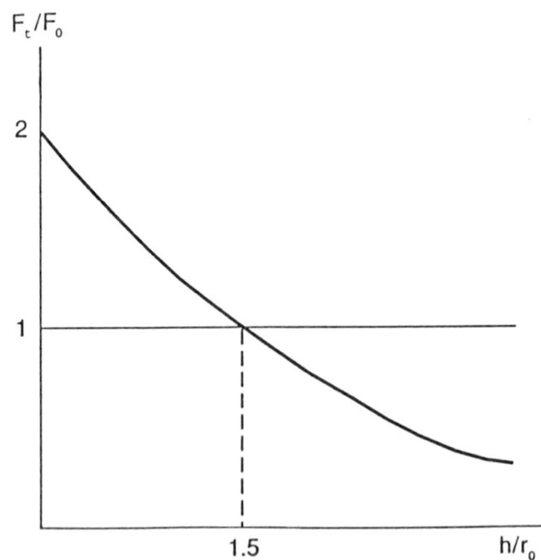

Fig. 5. Normalized force F_t/F_0 acting on the coaxial pore edges as a function of the relative intermembrane distance h/r_0.

of the membrane has a "fluted" shape with respect to the pore axis (Fig. 6). The membrane elastic energy is written as an integral over the curved surface (Evans and Skalak, 1980):

$$W_m = \frac{B_m}{2} \int \left(\frac{1}{R_1} + \frac{1}{R_2} \right)^2 dS \qquad (1.25)$$

where R_1, R_2 are the main primary curvature radii, B_m is the membrane bending modules. After substitution of the corresponding expressions for R_1 and R_2 in Eq. 25 we have:

$$W_m = 8\pi B_m \left[\left(\frac{\alpha+\beta}{1+\beta^2} \right)^2 \int_0^{\theta_0} \frac{d\theta}{\gamma - \sin\theta} - \frac{2}{1+\beta^2} \right] \qquad (1.26)$$

where $\alpha = 2r_0/h$, $\beta = 2L/h$, $\sin\theta_0 = 2\beta/(1+\beta^2)$, $\gamma = 2(\alpha+\beta)/(1+\beta^2)$, L is the curved membrane region width.

Using Eq. 1.26 to calculate the dependence of the curved membrane elastic energy on β for different values of α the results in Fig. 7 are obtained. The membrane bending modules B_m is assumed to be equal to $3 \cdot 10^{-20}$ J (Evans and Skalak, 1980). The line parallel to the abscissa corresponds to W_e for $h=50$ Å. It is seen that at large ratios of L/h the external electric field work W_e is more than the bending energy W_m. Thus the electric forces are large enough to overcome the elastic resistance of the membrane and to bring edges of coaxial pores into contact.

Up to this point the nature or structure of the pore edge has not been considered. It is understood (Abidor et. al., 1979; Chizmadzhev and Abidor, 1980; Glaser et al., 1980) that during electric breakdown hydrophobic pores are first formed. It is conceivable that such pore edges could fuse spontaneously due to hydrophobic attraction and form connecting bridges between two membranes. Since the direct contact of hydrophobic pore edges with water is energetically

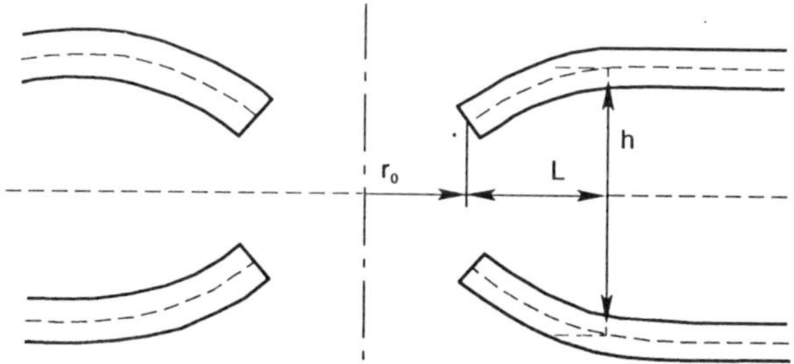

Fig. 6. Deformation of membranes in the vicinity of the coaxial pores.

unfavorable, the pore edges are expected to rapidly convert to the inverted (hydrophilic) configuration. Electrofusion could thus occur with the participation of *both* hydrophobic and hydrophilic pores, or be limited to one of the two classes of pore. In the case where hydrophilic pores are brought in contact during fusion, their coming together however would also be impeded by hydration repulsion between polar heads of lipid molecules. However, the hydration repulsion between hydrophilic pores may be expected to be weaker than between flat bilayers because the interaction is localized at "points" rather than spread uniformly over a surface. Moreover in this system hydration interaction may be weaker because the pores may be only partially hydrophilized before merging (Sugar et al., 1987).

The question of whether it is the hydrophobic or hydrophilic pores that are merging might be considered by comparing times required for pore hydrophilization (τ_h) and for drawing pore edges together (τ_d). To estimate these times it is necessary to estimate the driving forces and resisting factors for both processes.

The driving forces for the pore edge hydrophilization and merging processes are expected to be different: electrical

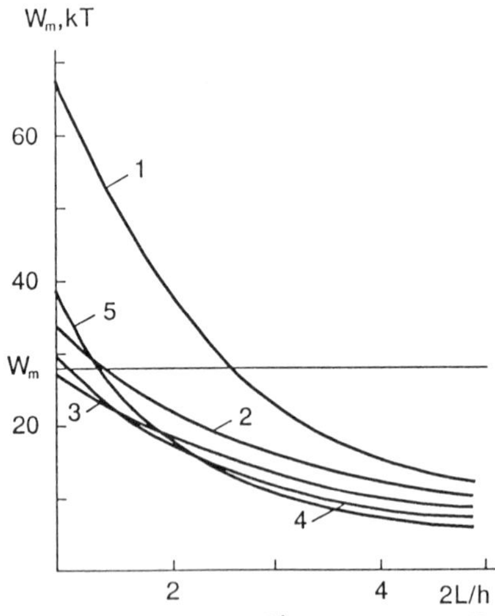

Fig. 7. Dependence of the bending energy *W*ₘ on the relative size of deformed membrane area *L/h*, in curves 1-4, for values of $2r_0/h$ = 0.2, 0.4, 0.6, 0.8, and 1.0, respectively. The line parallel to the abscissa axis corresponds to the value calculated by means Eq. 20 at *h*=50 Å.

forces acting on pore edges during merging should be proportional to the square of the transmembrane voltage while the forces involved in pore hydrophilization do not depend on this voltage. Thus it may be expected that the probability of the hydrophobic pore fusion would increase with transmembrane voltage.

The characteristic times of hydrophilization, τ_h, and the drawing of pores together, τ_a, are estimated by assuming that the first process depends mainly on surface tension, σ, and viscosity, η_m, of the lipid bilayer and the second process depends on water viscosity, η_w.

The gradient of the velocity, v, for bilayer movement of the pore edges towards each other is of the order of v/r_0. Therefore the viscous resistance force is $\sim \eta v/r_0$. This should be equal to the force σ/δ due to change of the linear tension in the hydrophilization process. Hence, $v = \sigma r_0/\delta\eta_m$, and:

$$\tau_h = \delta/v = \eta_m\delta^2/\sigma r_0 \qquad (1.27)$$

For $\eta = 1$ Poise, $\delta = 50$ Å, $\sigma = 0.5$ erg/cm^2 and $r_0 = 10$ Å we obtain $\tau_h = 5 \cdot 10^{-8}$ s.

The time for removing water while the pores are being drawn together with the velocity v_a is estimated as follows. The force, acting on the pore edge, is $p_0\pi r_0^2$. This force is balanced by the force of water friction on the pore wall which is $(v_a/r_0)\eta_w \cdot 2\pi r_0\delta = 2\pi\eta_w\delta v_a$. Thus the velocity of edges as they are drawn together is $v_a = r_0^2 p_0/2\eta_w\delta$ and the characteristic time of this process equals:

$$\tau_a = d/v_a = 2\eta_w\delta d/r_0^2 p_0 \qquad (1.28)$$

Hence, $\tau_a = 2 \cdot 10_{-9}$ s for $\eta_w = 10^{-2}$ Poise, $p_0 = 5 \cdot 10^6$ Pa, $r_0 = 10$ Å, $\delta = 50$ Å, $d = 10$ Å.

As can be seen, the time for drawing the pore edges together is about an order of magnitude less than the time needed for their hydrophobization. This would suggest that fusion is more likely to occur through hydrophobic pores. Since the difference between the estimated values of τ_h and τ_a is less than one order of magnitude, a clear choice between the hydrophobic and hydrophobic pores as being involved in the fusion mechanism is not obvious.

1.4. Summary of the Coaxial-Pore Mechanism

The above discussion suggests that the presence of the electric field may be important at all stages of electrofusion. The electric field can i) draw close-spaced membranes together, ii) promote pore formation in the membranes and iii) induce forces (due to an electric current through coaxial pores) strong enough to overcome impeding factors such as electrostatic and hydration repulsions, mechanical resistance to membrane bending. Pushing the pore edges toward each other electric forces facilitate merging of pores. Thus cell electrofusion may be directly related with coaxial pores induced in membrane contact zones due to reversible electric breakdown.

2. KINETICS AND MECHANISM OF CELL MEMBRANE ELECTROFUSION

The efficiency of electrofusion can be characterized by the fusion yield (i.e. a percentage of all cells became fused). However the study of the mechanism of electrofusion using fusion yield as a variable is hindered because it is a semi-quantitative parameter dependent on many other variables. The labeled erythrocyte ghost system extensively characterized by Sowers (Sowers, 1989) allows a new quantitative approach to study cell electrofusion to be developed (Abidor and Sowers, 1992). The principles of this approach are reviewed here.

2.1. Erythrocyte Ghosts as a Model Electrofusion System

A suspension of erythrocyte ghosts is a very simple and convenient system for the study of electrofusion. Erythrocyte ghosts are held in close contact by dielectrophoresis and then are treated with one fusogenic electric pulse. Individual fusion events are followed by lateral diffusion of the fluorescent lipid analog DiI from originally labeled to originally unlabeled adjacent ghosts. The alternating electric field which causes ghosts to come into contact through "pearl chain", the fusogenic electric pulse and the aqueous medium can all be manipulated independently of one another. It is important also that fusion of the erythrocyte ghosts does not depend on viability of cells because the ghosts are free of cytoplasmic elements. As will be seen below, the presence of a lipid analog fluorescent label in a fraction of all ghosts facilitates fusion yield measurement in a rigorous way.

Fig. 8 illustrates the ghost electrofusion system. In a florescence microscope a suspension of erythrocyte ghosts, some of which are DiI-labeled, only fluorescent ghosts are visible as randomized individual bright spheres against a dark background. If ghosts are brought into close contact to form "pearl chains" by means dielectrophoresis, then the application of a fusogenic electric pulse will induce a fraction of the labeled ghosts to fuse with neighbors. After fusion occurs, some of the labeled ghosts become part of labeled chains as DiI laterally moves into adjacent but originally unlabeled ghosts.

Theoretical approaches to any system are essentially dependant on the nature of the phenomenon under study. Generally, we can conceive three possible mechanisms of electrofusion with a different role for the electric field in each.

1. A "trigger" mechanism: the fusion process is initiated by the electric field pulse but then proceeds without any further involvement of the field.

2. A "force-based" mechanism: the fusion process is continuously "propelled" by the fusogenic electric field.

3. A "stochastic" mechanism: the electric field increases the probability of membrane fusion.

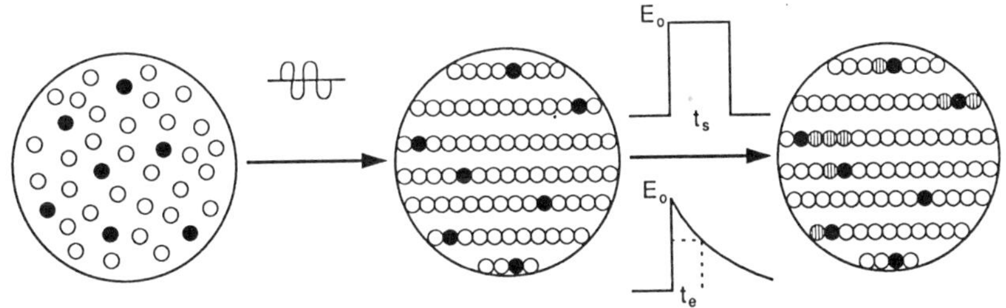

Fig. 8. Appearance of erythrocyte ghosts in suspension (left), in dielectrophoretic chains (middle) and after electrofusion (right). Most ghosts are originally unlabeled (open circles) while a few ghosts are labeled with a fluorescent membrane dye (solid circles). When a labeled ghost fuses with unlabeled neighbors the dye spreads over ghost membranes (dashed circles) thus causing fluorescence to appear in all ghosts that fuse.

These three mechanisms would be expected to show essentially different phenomenology.

<u>The trigger mechanism</u> is associated with the induction of the so-called long-lived fusogenic state in the membranes (Sowers, 1986; 1987b; 1989; Teissie and Rols, 1986; Teissie et al., 1989; Tsoneva et al., 1988). Generally, such mechanism would be expected when a given phenomenon occurs as an "all or nothing" type process initiated only when a stimulating factor (e.g. electric field) exceed some "critical value" or a field threshold. Meanwhile all earlier data (Abidor et.al, 1989; Sowers, 1989), as well as data presented below, show that the same values of fusion yield could be obtained both with strong short and weaker but longer pulses. Since there is no *distinctive* "field threshold" in the electrofusion process the pure trigger mechanism is rather unlikely.

<u>The force-based electrofusion mechanism</u> proposes that the fusion process caused by an electric driving force F_e involves movement against friction in a viscous medium (the external water solution or the membrane itself as characterized by the viscosity η). This process requires time $t_f \sim \eta/F_e$ for completion. If the membrane system is homogeneous then all fusion events would be synchronized by the fusogenic pulse. Thus short pulses ($\leq t_f$) would result in no fusion while longer pulses would result in 100% fusion. An increase or decrease of the field strength could be expected to lead to a "parallel shift" of this step function. In the case of a heterogeneous system the *dependence of the fusion yield on the pulse duration* should clearly be wider but with the same "parallel shift". However an analysis of previous experimental results (Abidor et.al, 1989; Sowers, 1989) shows that there are no signs of the "time threshold" or the "parallel shift" therefore the forced movement is unlikely to be the primary factor of electrofusion.

An erythrocyte ghost suspension with a fraction of the ghosts carrying a membrane label is a very convenient system to study <u>electrofusion as a stochastic process</u>. In this system fusion of labeled ghosts with adjacent neighbors is identified by labeled chains independent of each other. That the electrofusion is a stochastic process seems to be reasonable at least because of an unpredictable behavior of individual cells in electrofusion experiments: after the fusogenic electric pulse some cells form fused chains of different length while other ones stay unfused. Moreover, if electrofusion does relate to electroporation and if electroporation is a limiting stage of the whole electrofusion process, electrofusion could be expected to have a stochastic nature too. The stochastic character of electrofusion is presumed, below, to be the case. In principle, there are two possible descriptions of stochastic processes based on either statistics or kinetics.

2.2. Statistics of Cell Electrofusion

When originally DiI-labeled ghosts fuse with adjacent unlabeled ghosts, labeled chains of different length are formed. In this system the number of originally labeled ghosts is relatively small and any labeled chains formed in the electrofusion process are assumed to contain only one originally labeled ghost.

In long "pearl chains" each ghost has two neighbors (i.e. there are two membrane contacts per triplet of ghosts). If p is the probability that at time t a given membrane contact has fused, then the probability of finding the contact unfused is $1-p$. The probability P_n that a given labeled (fused) chain selected random from fusion products contains exactly n ghosts from which only one is originally labeled, can be find as follows. The chance that a given labeled ghost remains unfused after application the fusogenic electric pulse is the product of the probabilities that the ghost does not fuse with either of its two neighbors in the same "pearl chain", that is:

$$P_1 = (1 - p)^2 \qquad\qquad (2.1)$$

The probability that a given labeled ghost will fuse with a given neighbor is $p(1-p)^2$ because the fusion process involves simultaneously three events: one namely fusion at the contact between the labeled and unlabeled ghosts with the probability of p and two other are the absence of fusion at contacts of these ghosts with other ones with the probabilities $1-p$. The probability of forming chains containing two ghosts is actually 2× higher because to form a two-ghost chain a labeled ghost can fuse with either of its two neighbors, thus

$$P_2 = 2p(1-p)^2 \qquad (2.2)$$

The formation of a labeled three-ghost chain with a labeled ghost in a specific initial position involves two fusion events and two nonfusions on the chain ends with a chance of $p^2(1-p)^2$. Since in a three-ghost chain the originally labeled ghost can occupy three different positions (i.e. either of ends or the center), the probability finding such a chain is:

$$P_3 = 3p^2(1-p)^2 \qquad (2.3)$$

Generally, for chains containing n fused ghosts from which only one was originally labeled, the probability is:

$$P_n = np^{n-1}(1-p)^2 \qquad (2.4)$$

This satisfies the obvious condition:

$$\sum_{n=1}^{\infty} P_n = 1 \qquad (2.5)$$

From Eq. 2.4 it also follows:

$$p = (n-1)P_n/nP_{n-1} \qquad (2.6)$$

The Eqs. 2.4 and 2.5 give the length-distribution of labeled chains for the system with a few originally labeled ghosts when all events involved in electrofusion process are independent of each other. It should be also noted that this description is not valid for fusion of unlabeled ghosts, as observed, for example, in a phase-contrast microscope which is a typical case for the classical binomial law that under some conditions can be reduced to the Gaussian or Poisson distributions.

2.3. Cell Electrofusion as a First-Order Rate Process

Conversion of labeled ghosts from the unfused (G_{uf}) to the fused (G_f) state, as marked by appearance labeled chains, can be formally described as a first-order rate reaction:

$$G_{uf} \rightarrow G_f$$

If, in any given "pearl chain", any fusion event is

independent of other fusion events and if it is governed only by a pure chance, then the number of the labelled ghosts which remain unfused, n_{uf}, decreases with the rate given by:

$$\frac{dn_{uf}}{dt} = -k_f n_{uf} \qquad (2.7)$$

where t is time of the process and k_f is the fusion rate constant. Equations of this type are widely applied to describe various processes such as radioactive decay, bacterial growth, different chemical reactions (see e.g.: Marshall, 1978). In these applications equations of this type postulate that transitions of "individual but identical particles" (e.g. labeled ghosts) into a new state will occur with a probability independent of not only the other particles but also the prehistory of the population. Solving Eq. 2.7 for membrane electrofusion requires two considerations: i) the fusion rate constant should be a function of the electric field, E, since the fusion yield, F, is highly dependant on the pulse amplitude E_0, i.e. $k_f=k_f(E)$; ii) the fusion assay involves counting fused or unfused cells at least 1.5 min after the electric pulse. Thus it is important to define t, i.e. to find out at what stage (during pulse treatment or after it) electrofusion occurs. If the electrofusion process proceeds only during the electric pulse of duration t_p, then $0 \leq t \leq t_p$ and Eq. 2.7 can be transformed to:

$$\int_{N_0}^{N_{uf}} \frac{dn_{uf}}{n_{uf}} = \int_0^t k_f(E)\, dt \qquad (2.8)$$

or

$$\ln\frac{N_{uf}}{N_0} = \int_0^t k_f(E)\, dt \qquad (2.9)$$

where N_0 is the total number of labeled cells before fusion, N_{uf} is the number of unfused ghosts after electrofusion. Taking into account the definition of fusion yield, Eq. 2.9 can be rewritten as:

$$\ln(100 - F) = \int_0^t k_f(E)\, dt \qquad (2.10)$$

Further solution of this equation depends on the waveform used for electrofusion (square or exponential).

For <u>square pulses</u> $t_p=t_s$ and, as long as E does not change during the pulse and equals E_0, k_f is a constant. Hence:

$$\log(100 - F) = 2 - 0.43 k_f t_s \qquad (2.11)$$

Thus $\log(100-F)$ is a linear function of t_s with the slope proportional to k_f.

For <u>exponential pulses</u> E decreases from E_0 to 0 in the range $0 \leq t \leq t_p = \infty$ as

$$E = E_0 \exp\left(-\frac{\ln 2 \; t}{t_e}\right) \qquad (2.12)$$

In this case Eq. 2.10 may be transformed to:

$$\log(100 - F) = 2 - 0.63 k_{ef} t_e \qquad (2.13)$$

where

$$k_{ef} = \int_0^\infty k_f(E_0 \exp(-x)) \, dx \qquad (2.14)$$

and $y = \ln 2 \cdot t/(t_e)$. Eq. 2.13 is similar to Eq. 2.11, however Eq. 2.13 includes the effective constant k_{ef} and the pulse half-time t_e instead of k_f and t_s. It should be noted that, since Eq. 2.14 is integrated over $0 \leq y \leq \infty$, k_{ef} does not depend on t_e and is defined only by E_0.

2.4. Fusion Rate Constant and Related Characteristics

Depending on experimental and theoretical approaches, first-order rate processes can be described in different ways. The rate constant, k, is commonly used as a parameter in chemical kinetics. For fast relaxation processes the parameter of choice is usually the mean lifetime t_m. For example, in the case of irreversible electric breakdown of the bilayer lipid membranes (Abidor et al., 1979) this is a time interval during which a given electric field does not induce, on the average, an irreversible damage or rupture of the membranes. In a process such as radioactive decay the half-time t_h is in common use. All these parameters are related as follows (e.g. Marshal, 1978):

$$k = \frac{1}{t_f} = \frac{\ln 2}{t_h} \qquad (2.15)$$

The fusion rate constant, which can be found directly from experimental data, seems a more convenient characteristic of electrofusion. However because electrofusion is induced by short electric pulses and the most important part of this process may proceed only in the presence of electric field, the mean fusion lifetimes t_f (i.e. the time interval during which ghosts remain unfused in the fusogenic electric field) may be of interest also.

The fusion rate constants are also related to the probability that a pair of membranes in contact is fused during the fusogenic pulse. According to Eqs. 2.1, 2.11 and 2.13:

$$p = 1 - \exp(-k_f t_s/2) \tag{2.16}$$

A similar expression can be obtained for the exponentially-decaying waveform pulses. Note that Eq. 2.16 includes the product of the pulse width, t_s, and the fusion rate constant, k_f. The fusion rate constant, in turn, is a function of the field strength. Thus there is reciprocity between t_s and E_o. This means that the same values of p, and thereby the same result of electrofusion (including not only fusion yield but also the length-distribution of fusion products!), can be obtained with square (or exponential-decaying) waveform pulses having different parameters.

3. EXPERIMENTAL VERIFICATION OF THE ELECTROFUSION MECHANISM

The purpose of this Section is to show that the coaxial-pore mechanism of electrofusion is consistent with experimental data on statistics and kinetics of electrofusion.

3.1. Samples, Protocols and Primary Data Transformatin

All procedures are described in Abidor and Sowers (1992).

3.2. Experimental Results on Kinetics and Statistics of Rabbit Erythrocyte Ghost Electrofusion

Dependence of the fusion yield on the fusogenic pulse. The fusion yield, F, as function of the strength, E_o, width, t_s, and total duration, Σt_s, of the square waveform pulses for two different REG preparations is shown in Figs. 9a. As can be seen, F increases with both E_o and t_s. It should be noted that at lower electric fields ($E_o = 1.7$ and 2.0 kV/cm) the initial parts of the $F(t_s)$-curves have a distinctive "toe", i.e. a relatively slow increase of fusion yield with t_s. This "toe" becomes reduced at $E_0 = 2.3$ kV/cm and actually disappears at 3.0 kV/cm. Note that within the experimental error, application of several (2-4) pulses with a total duration Σt_s gives the same result as one pulse of this duration.

Results obtained for fusion of the REG using exponentially-decaying pulses (Fig. 10a) are qualitatively similar to those for square waveform pulses (Fig. 9a). The quantitative difference is that a square waveform pulse produces a higher fusion yield than an exponential one with corresponding parameters (i.e. with the same E_0 and $t_s=t_e$).

Figs. 9b and 10b show data for REG electrofusion plotted as log (100-F) vs. t_s or t_e. The straight lines were fitted to the experimental points by linear regression (in all cases the coefficient of variation was > 0.97). Within experimental error all of these plots are linear. As follows from Eqs. 2.11 and 2.13, slopes of these plots are proportional to the true (k_f) or effective (k_{ef}) fusion rate constants.

Effect of experimental conditions on the fusion rate constant. Fig. 11 shows the true, k_f, and effective, k_{ef}, fusion rate constants obtained from the slopes of log(100-F) vs. t_s or t_e - plots for REG treated with square and exponentially-decaying pulses respectively. It can be seen that both constants increase considerably with E_0 though effective constants are clearly much less than true constants.

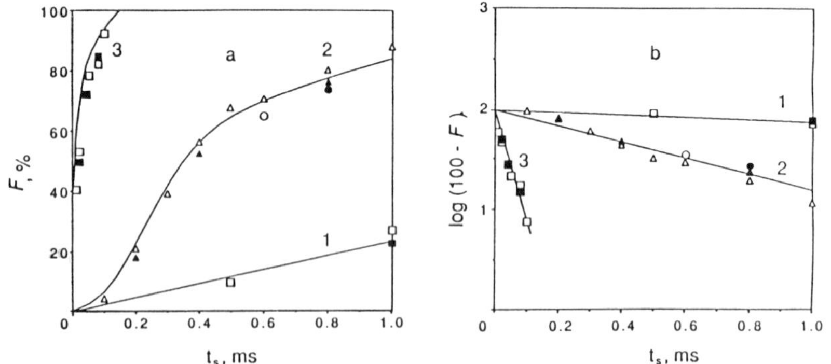

Fig. 9. (a) REG fusion yield a function of the total duration of square pulses Σt_s for E_0 = 1.7 (curve 1: □ - 1, ■ - 2 pulses), 2.0 (curve 2: △ - 1, ▲ - 2, ○ - 3, ◉ - 4 pulses) and 3.0 kV/cm (curve 3: □ - 1, ■ - 2 pulses). (b) The same data in a linear form.

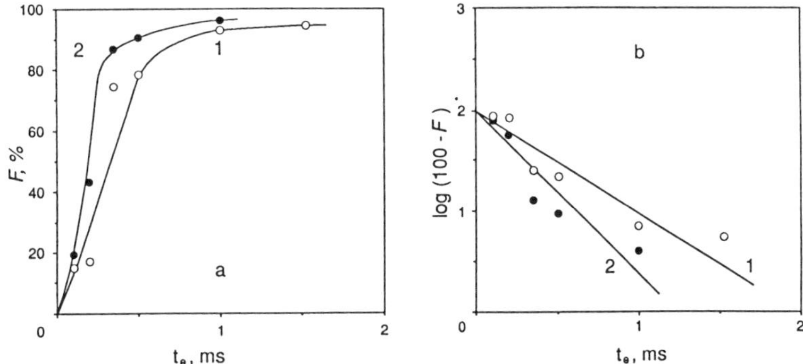

Fig. 10. (a) REG fusion yield as a function of the decay half-time of exponential pulses t_e for E_0 = 3.5 (curve 1), 4.0 (curve 2). (b) The same data in a linear form.

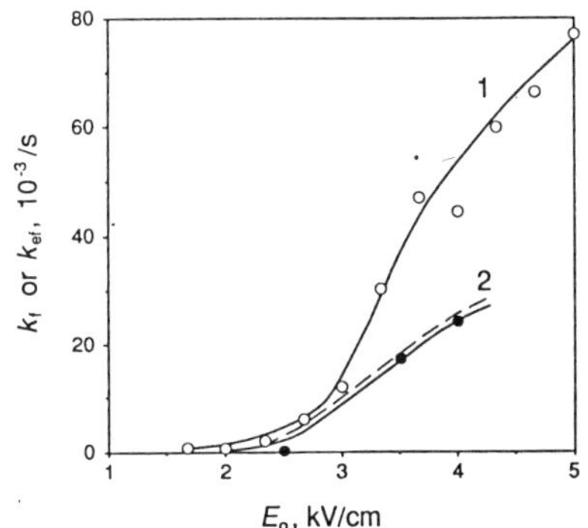

Fig. 11. REG fusion rate constants, k_f. as a function of the square (curves 1) and exponentially-decaying (curve 2) waveform pulse field strength, E_0. Dashed line - the dependence of k_{ef} on t_e as calculated from curve 1 by numeric integration.

Modification of the REG by heating at 42 °C for 10 min (to disrupt the spectrin network) or exposure to 0.1 - 1.0 mM uranyl-ions (to stabilize membrane lipid matrix) does not alter the shapes of the $F(t_s)$-curves and in fact results only in a decrease in k_f (Fig. 12).

Fig. 13 shows the temperature dependence of the REG fusion rate constants in the form of Arrhenius plots ($\ln k_f$ vs. reciprocal temperature $1/T$, °K) for E_0 = 2.3, 3.0, and 5.0 kV/cm. As can be seen, all of these plots are linear which is typical for activated processes described by the first-order rate kinetics. Slopes of these plots give the values of the activation energy E_a: 9.6, 6.8, and 9.7 kT, respectively.

Length-distribution of fusion products. Fig. 14 shows length-distributions of labeled fusion products obtained for REG with fusogenic square pulses of the same field strength (3.0 kV/cm) but different duration. It can be seen that an increase in the pulse width and the field strength (data are not shown here; for more details see Abidor and Sowers, 1992) leads to a widening of the distribution and a decrease in the yield, P_n, for short labeled chains including unfused labeled ghosts (n=1). Values of P_n (solid columns) calculated with mean p, estimated form experimental data using Eq. A-14,

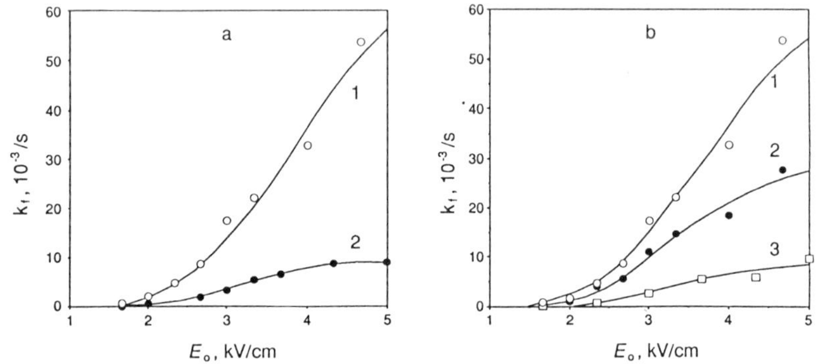

Fig. 12. Effect of (*a*) heating (42 °C, 10 min) and (*b*) uranyl-ions on the field strength dependence of the REG fusion yield -: (*a*) 1 - control, 2 - heating; (*b*) 1 - 0 (control), 2 - 10^{-4}, 3 - $3 \cdot 10^{-4}$ M uranyl acetate.

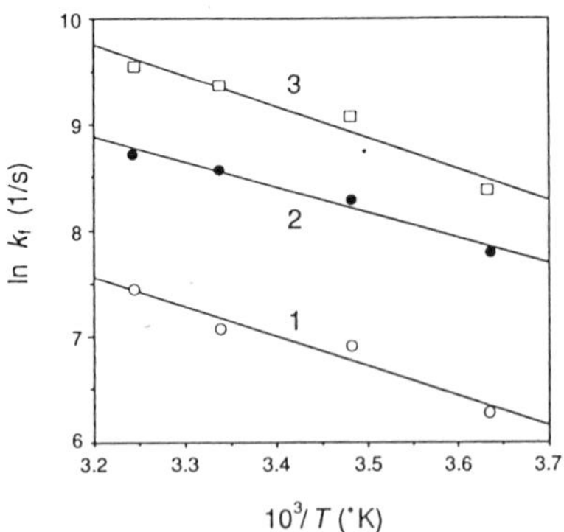

Fig. 13. Arrhenius plots for the REG fusion rate constants at E_o = 2.3 (curve 1), 3.0 (2) and 5.0 kV/cm (3).

(Abidor and Sowers, 1992) actually coincides with experimental distributions.

3.3. Discussion of Experimental Results

Statistics and kinetics of electrofusion. For statistic or kinetic analysis of the erythrocyte ghost fusion system unlabeled cells may be considered as *a background* which fusion causes single labeled ghosts to be replaced by fluorescent labeled "pearl chains". Since a very simple statistical

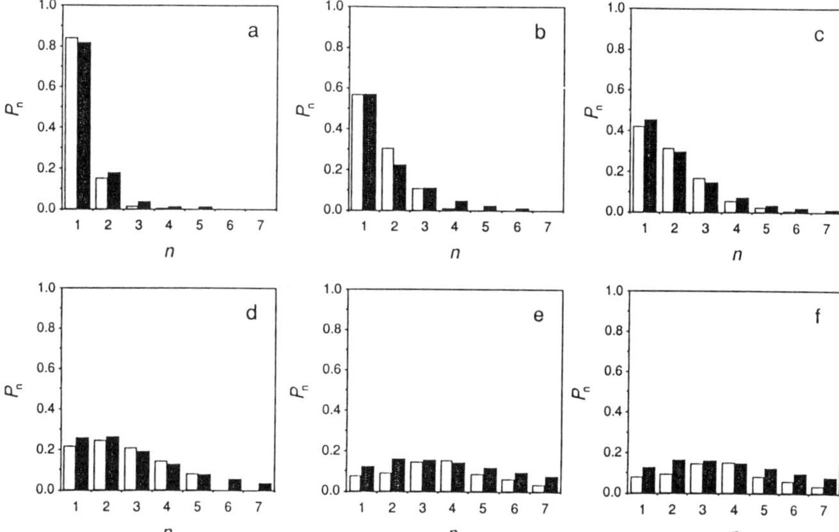

Fig. 14. Statistics of labeled fusion products for the REG treated with 3.0-kV/cm square pulses of different t_s: light columns -experimental data, solid columns -values calculated as in Fig. 9.

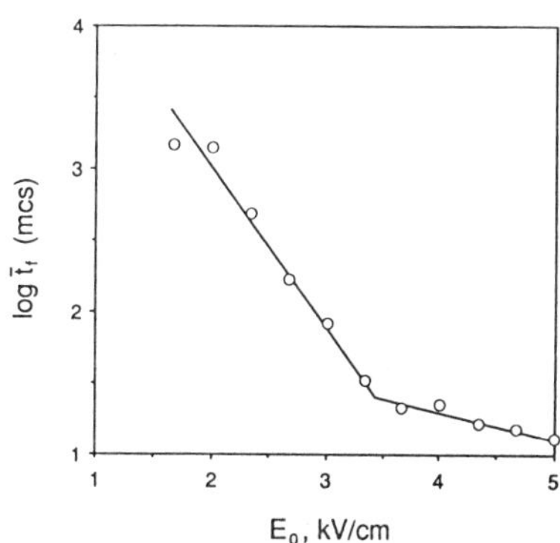

Fig. 15. Fusion mean lifetime t_f of the REG as function of E_0.

description based on an assumption of independence of every fusion event on other ones fits experimental data very well one may conclude that electrofusion or at least its limiting stage is a stochastic process. Under such conditions electrofusion may be formally described by the first-order kinetics (see Section 2.3), to which a variety of processes (e.g. radioactive decay, bacterial growth or different chemical reactions) may be reduced. Although the mechanisms of these processes are quite different the feature that they all share in common is that the behavior of individual particles is probabilistic and independent of other identical particles or time.

As can be seen from Figs. 9b and 10b, fusion yield data for the REG fall on straight lines which are based on the first-order rate kinetics (Eqs. 2.11 and 2.13). Some deviations from these kinetics were observed as the "toes" for short duration low-field pulses. This shows that in fact electrofusion is a more complicated process and may include more than one step. However the first-order rate kinetics always predominates under our experimental conditions.

The agreement of Eq. 2.13 with experimental data for the exponentially-decaying pulses gives an additional support to the first-order rate kinetics. To derive this equation required the assumption the fusion rate constant be dependent on electric field. Although this dependence is unknown, it was possible to obtain Eq. 2.13 in a form similar to Eq. 2.11 and thus to relate the square waveform pulse data with the exponential waveform pulse data. As can be seen from Fig. 6, values of k_{ef}, calculated from k_f (dash line) and found directly from experimental data for the exponentially-decaying pulses (solid line, curve 2), are very close.

Since the variable parameter in this kinetic description is the duration (i.e. t_s or t_e) of the fusogenic pulse, the agreement between experiment and theory actually indicates that ghosts fuse (or rather, start to fuse) only during the pulse rather than after it. Actually the fusion process itself may occur very fast while the fusogenic pulse should be much longer to initiate the membrane fusion process. Generally this is typical for stochastic phenomena. The mean time required for initiation of fusion or the mean fusion lifetime

t_f (i.e. the time intervals during which ghosts remain unfused in the fusogenic electric field) can be calculated as a reciprocal of the fusion rate constant (Eq. 2.15). For the REG this time is in the range from tens μs to several ms and drastically decreases with E_o (Fig. 15).

The fusion process, of course, does not stop occurring immediately after the pulse. However later events, e.g. expansion of fusion lumens (Chernomordik and Sowers, 1991) or lateral diffusion of DiI, do not affect the fusion yield. Therefore the long-lived fusogenic state is not likely to be involved in this process.

A correlation between fusion and pore formation rate constants as a clue to the mechanism of cell membrane electrofusion. Rate constants are commonly used as a fundamental characteristic in studies of different processes because in many cases they can be estimated on a model basis or compared with corresponding values for related processes. In the following discussion, it will be assumed that the fusion rate constant, k_f, depends on the number of pores, m, created in each pair of membranes. Actually k_f is the probability of transforming a pair of adjacent ghosts from the unfused state to the fused state. If fusion is a result of merging pores randomly scattered on both membranes in contact (i.e. two independent pores are involved in every fusion event), k_f should be proportional to m^2. However pore formation in close-spaced membranes may be strongly correlated (see Section 1.2) and a notable part of pores in the contact zone can be coaxial from the beginning. The number of coaxial pores should be proportional to the total number of pores m. Therefore, in this case it is more likely that $k_f \sim m$.

On the other hand, the number of pores m created during reversible electric breakdown should be proportional to the pore formation rate constant k_p. Hence:

$$k_f \sim k_p \tag{3.1}$$

The dependence of k_p on electric field can be expressed (Glaser et al., 1988) as:

$$k_p = k_p(0)\exp(Bu^2) \tag{3.2}$$

or

$$lnk_p = A_p + Bu^2 \qquad (3.3)$$

where u is the voltage drop across the membranes or the transmembrane potential; $k_p(0) = k_p$ at $u=0$; $A_p=lnk_p(0)$; B is a constant (of order of ~ 4.8 V^2) given by the equation:

$$B = \frac{\pi r_*^2 \varepsilon_m}{2\delta kT} \qquad (3.4)$$

where $r_* \approx 4$ Å is the critical radius after which originally hydrophobic pores become hydrophilic. Combining Eqs. 3.1 and 3.3 gives:

$$lnk_f = A_f + Bu^2 \qquad (3.5)$$

Thus we have found a functional dependence of the fusion rate constant on the fusogenic electric field. It shows that, in terms of the coaxial pore mechanism, the slope of the lnk_f vs. u^2 - curves should equal B. Demonstrating this experimentally requires an estimation of the membrane potentials, u, induced by the electric field pulse.

The transmembrane potential induced by an external electric field at the poles of suspended spherical cells of the radius R can be estimated (e.g. Kinosita and Tsong, 1977) as:

$$u = \frac{3}{2} ER \qquad (3.6)$$

However the distribution of an electric field around the "pearl chains" is expected to differ from that for ghosts in suspensions. In chains the ghosts are pressed toward each other by electric forces and tight contact zones are formed at the ghost poles. Such contacts can be treated as if they were a series of membrane capacitors. Thus if a unit of chain length (1 cm) contains N_c ghosts, φ equals:

$$u = \frac{E}{2N_c} = ER \qquad (3.7)$$

This is 1.5× less than for suspended ghosts. Actual values of φ should be between these two limits, and are more likely to be closer to the lower limit. Thus in the following discussion values of φ were calculated using Eq. 3.7.

As can be seen from Fig. 16, the initial part of the lnk_f vs. φ^2 - dependence for the REG is linear. Its slope equals 4.2 V^{-2}, i.e. close to $B = 4.8$ V^{-2} found both theoretically and experimentally for reversible electrical breakdown (Glaser et al., 1988). Moreover, as seen in Fig. 16, the initial slope of the $\ln k_f$ vs. u^2 - plot actually does not change after disrupting the membrane spectrin network by heating (42 $^\circ$C, 10 min) or altering the membrane lipid matrix by uranyl ions. This result is reasonable because B is expected to have a weak dependance on experimental conditions since Eq. 3.4 has no parameters that could significantly change B. On the other hand, the number of potential fusogenic sites (e.g. defects in the lipid bilayers) can be changed. This is why both heating and exposure to uranyl-ions decrease the fusion yields without changing B. In conclusion, it should be noted that the same slope of u^2 - dependence for fusion rate and pore formation constants actually means that a limiting stage of electrofusion is pore formation.

CONCLUDING COMMENTS

Thus our theoretical and experimental studies indicate that: i) the pulse electric field may be important at all stages of the electrofusion process - from drawing membranes in closer contact to formation and merging of coaxial pores; ii) electrofusion is a stochastic process proceeding only when a fusogenic electric field is applied to the membrane system; iii) this process may, indeed, relate to coaxial (more likely

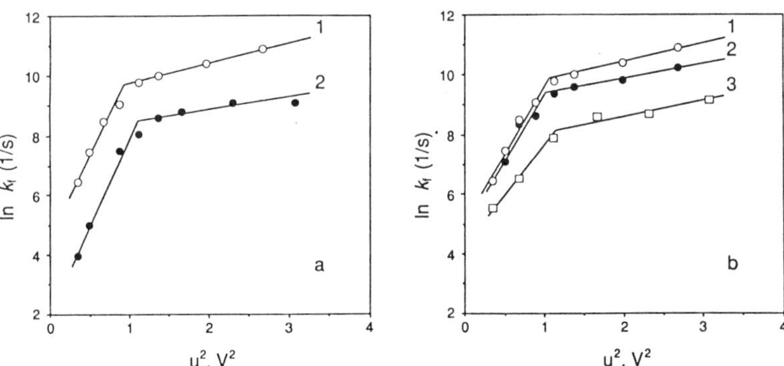

Fig. 16. Plots of $\ln k_f$ vs. φ^2 for the rabbit erythrocyte ghosts for different experimental conditions. (a) The ghosts were fused in normal conditions (curve 1) and after heating at 42 $^\circ$C (curve 2). (b) Electrofusion without (curve 1) and in the presence of 10^{-4} (curve 2) and $3 \cdot 10^{-4}$ (curve 2) M uranyl acetate.

hydrophobic) pores induced in the membranes due to reversible electric breakdown. The authors realize, however, that unequivocal evidence has not yet been obtained and for a complete understanding of cell electrofusion further theoretical and experimental studies are necessary.

ACKNOWLEDGEMENT

This work was supported by ONR grant N00014-89-J-1715 to AES.

REFERENCES

Abidor IG and Sowers AE (1992): Kinetics and mechanism of cell membrane electrofusion. *Biophys. J.* (accepted)

Abidor IG, Arakelyan VB, Chernomordik LV, Chizmadzhev YuA, Pastushenko VF and Tarasevich MR (1979): Electrical breakdown of bilayer lipid membranes. *Bioelectrochem. Bioenerg.* 6: 37-52.

Abidor IG, Pastushenko VF, Osipova EM, Melikyan GB, Kuzmin PI and Fedotov SV (1987): Relaxation studies of the plane contact of bilayer lipid membranes. *Biol. membrany (Russ.).* 4: 67-76.

Abidor IG, Barbul AI, Zhelev D, Sukharev SI, Kuzmin PI, Pastushenko VF and Zelenin AV (1989): Electrofusion and electrical properties of cell pellets in centrifuge. *Biol. membrany (Russ.).*12: 212-224.

Benz RF, Beckers F and Zimmermann U (1979): Reversible electrical breakdown of lipid bilayer membranes: a charge-pulse relaxation study. *J. Membr. Biol.* 48: 181-204.

Blumenthal R (1987): Membrane fusion. *Curr. Top. Membr. Transp.* 29: 203-254.

Chernomordik LV and Sowers AE (1991): Evidence that the spectrin network and a non-osmotic force control the fusion product morphology in electrofused erythrocyte ghosts. *Biophys. J.* (in press)

Chernomordik LV, Kozlov MM, Melikyan GB, Abidor IG, Markin VS and Chizmadzhev YuA (1985): The shape of lipid molecules and monolayer membrane fusion. *Biochim. Biophys. Acta.* 812: 643-655.

Chizmadzhev YuA and Abidor IG (1980): Membranes in strong electric fields. *Bioelectrochem. Bioenerg.* 7: 83-100.

Coackley WT and Gallez D (1989): Membrane-membrane contact: involvement of interfacial instability in the generation of discrete contacts. *Bioscience Reports.* 9: 675-691.

Dimitrov DS and Jain RK (1984): Membrane stability. *Biochim. Biophys. Acta.* 779: 437-468.

Evans EA and Skalak R (1980): Mechanics and Thermodynamics of Biomembranes. Boca Raton: CRC Press.

Glaser R, Leikin SL, Chernomordik LV, Sokirko AV and Pastushenko VF (1988): Reversible electrical breakdown of lipid bilayers: formation and evolution of pores. *Biochim. Biophys. Acta.* 940: 275-287.

Hui SW, Stewart TP, Boni LT and Yeagle PL (1981): Membrane fusion through point defects in bilayers. *Science.* 212: 921-922.

Kinosita K and Tsong TY (1977): Voltage-induced pore-formation and hemolysis of human erythrocytes. *Biochim. Biophys. Acta.* 471: 227-242.

Kuzmin PI, Pastushenko VF, Abidor IG, Sukharev SI, Barbul AI and Chizmadzhev YuA (1988): Theoretical analysis of a cell electrofusion mechanism. *Biol. membrany (Russ.).* 5: 600-612.

Markin V S, Kozlov MM and Borovjagin VL (1984): On the theory of membrane fusion: the stalk mechanism. *J. Physiol. Biophys.* 5: 361-377.

Marshall AG (1978): Biophysical Chemistry: Principles, Techniques, and Applications. New York: J Wiley & Sons.

Melikyan GB, Abidor IG, Chernomordik LV and Chailahyan LM (1982): Electrostimulated fusion and fission of bilayer lipid membranes. *Doklady AN SSSR (Russ.).* 263: 1009-1013

Melikyan GB and Chernomordik LV (1989): Electrofusion of lipid bilayers. In: Electroporation and Electrofusion in Cell Biology. Neumann E, Sowers AE and Jordan C, eds. New York: Plenum Press.

Pilwat G, Richter H-P and Zimmermann U (1981): Giant culture cells by electric-field induced fusion. *FEBS Lett.* 133: 169-174.

Rols MP and Teissie J (1990): Electropermeabilization of mammalian cells-quantitative analysis of the phenomenon. *Biophys. J.* 58: 1089-1098.

Sowers AE (1986): A long-lived fusogenic state is induced in erythrocyte ghosts by electric pulses. J Cell Biol 102: 1358-1362.

Sowers AE, ed. (1987a): Cell Fusion. New York: Plenum Press.

Sowers AE (1987b): The long-lived fusogenic state induced in erythrocyte ghosts by electric pulses is not lateral mobile. *Biophys. J.* 52: 1015-1020.

Sowers AE (1988): Fusion events and nonfusion contents mixing events induced in erythrocyte ghosts by an electric pulse. *Biophys. J.* 54: 619-625.

Sowers AE (1989): The mechanism of electroporation and electrofusion in erythrocyte membranes. In: Electroporation and Electrofusion in Cell Biology. Neumann E, Sowers AE, and Jordan C, eds. New York: Plenum Press.

Stenger DA and Hui SW (1986): Kinetics of ultrastructural changes during electrically-induced fusion of human erythrocytes. *J. Membrane Biol.* 93: 43-53.

Sugar IP, Forster W and Neumann E (1987): Model of cell electrofusion: membrane electroporation, pore coalescence and percolation. *Biophys. Chem.* 26: 321-335.

Sukharev SI, Bandrina IN, Barbul AI, Abidor IG and Zelenin AV (1990): Electrofusion of fibroblasts on porous membrane. *Biochim. Biophys. Acta.* 1034: 125-131.

Tsoneva I, Panova I, Doinov P, Dimitrov DS and Strahilov D (1988): Hybridoma production by electrofusion: monoclonal antibodies against the Hc antigene of Salmonella. *Stud. Biophys.* 125: 31-35.

Teissie J and Rols MP (1986): Fusion of mammalian cells in culture is obtained by creating the contact between cells after electropermeabilization. *Biochem. Biophys. Res. Comm.* 140: 258-266.

Teissie J, Rols MP and Blangero C (1989): Electrofusion of mammalian cells and giant unilamellar vesicles. In: Electroporation and Electrofusion in Cell Biology. Neumann E, Sowers AE and Jordan C, eds. New York: Plenum Press.

Wilschut J and Hoekstra D, eds (1991): Membrane Fusion. New York: Marcel Decker.

Zhelev DV, Dimitrov DS and Doinov P (1988): Correlation between physical parameters in electrofusion and electroporation of protoplasts. *Bioelectrochem. Bioenerg.* 20: 155-167.

Zimmermann U (1982): Electric field-mediated fusion and related electrical phenomena. *Biochim. Biophys. Acta.* 694: 227-277.

EFFECT OF PARAMAGNETIC LANTHANIDE(III) COMPLEXES OF A SIX-NITROGEN MACROCYCLIC LIGAND ON THE AQUEOUS NMR SPECTRA OF AMINO ACIDS

K. K. Fonda, J. Kroll, D. D. Shillady and L. M. Vallarino
Department of Chemistry, Virginia Commonwealth University,
Richmond, Virginia

INTRODUCTION

The lanthanide(III) complexes of formula $\{ML\}X_3 \cdot nH_2O$, in which M is a trivalent lanthanide, L is the macrocyclic ligand $C_{22}H_{26}N_6$ (Fig. 1a), and X is acetate and/or chloride, possess a unique combination of properties and structure (De Cola et al., 1986; Benetollo et al., 1990; Bombieri et al., 1989). They are soluble in water as well as in organic solvents. The metal-macrocycle entities retain their integrity in solution even in the presence of acids, bases, and competing ligands; the exocyclic counterions, in contrast, are labile and exchangeable. This solution behavior is consistent with the crystal structure established by X-ray analysis for the europium diacetate chloride analog (Fig. 1b). In this structure the Eu(III) center is linked to the six nitrogen atoms of the macrocyclic ligand in a quasi-planar arrangement that results in a highly inert {EuL} moiety. At the same time, the Eu(III) center remains easily accessible and can coordinate even fairly bulky exocyclic ligands on the two opposite open faces of the {EuL} macrocycle. Finally, complexes of the lanthanides with f^1 to f^{13} electronic configurations are paramagnetic.

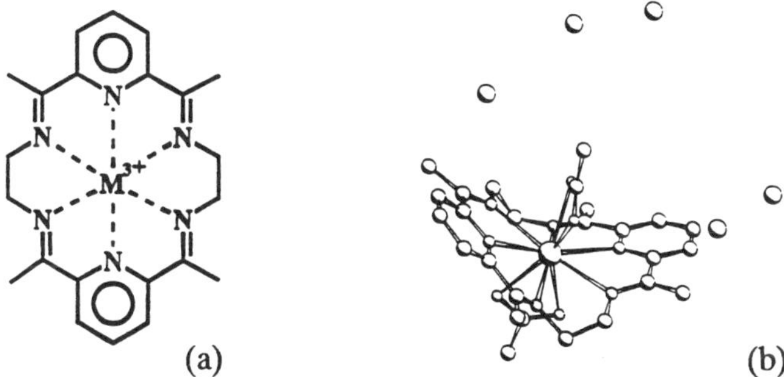

FIGURE 1. (a) Schematic of the {ML} moiety, $L = C_{22}H_{26}N_6$. (b) X-Ray crystal structure of $[Eu(CH_3COO)_2L]Cl \cdot 4H_2O$.

These complexes, therefore, have potential application as NMR shift reagents for the study of water-soluble biomolecules over a range of pH values. In the present study, the ability of $\{EuL\}Cl_3$ to act as a shift reagent in aqueous solution was investigated using as substrates a representative series of amino acids, including alanine, valine, leucine, phenylalanine, and histidine. A parallel study was carried out using [EuL(CH$_3$COO)$_2$L]Cl·4H$_2$O and [PrL(CH$_3$COO)$_2$L]Cl·4H$_2$O as the reagents with phenylalanine as the substrate. This study has focused on the complexes of Eu(III), f^6, and Pr(III), f^2, because these ions are known to cause considerable shifts in the resonances of organic substrates without excessive line broadening (Wenzel, 1987).

EXPERIMENTAL

<u>Starting Materials</u>. The [M(CH$_3$COO)$_2$L]Cl·4H$_2$O complexes, where M is Eu(III) and Pr(III), were synthesized from 2,6-diacetylpyridine and 1,2-diaminoethane by metal-templated Schiff base condensation, as described in ref 1. The Eu-trichloride analog was prepared by the method of Fonda and Vallarino (1991).

<u>NMR Spectral Measurements</u>. Experiments were carried out in D$_2$O at ambient temperature and natural pH on a GE QE-300 Fourier transform NMR spectrometer. Since DSS, (CH$_3$)$_3$SiCH$_2$CH$_2$CH$_2$SO$_3$Na, is not a suitable reference for these paramagnetic complexes, spectra were referenced to the residual HOD peak (4.80 ppm), which remains unshifted by the paramagnetic metal (Fonda and Vallarino, 1991). For each experiment, the ^{1}H NMR spectrum of the 0.03 M substrate alone was measured first. Spectra of solutions containing the same substrate concentration with increasing metal to substrate ratio (0.33:1, 0.66:1, and 1:1) were also recorded.

RESULTS AND DISCUSSION

The aqueous ^{1}H NMR spectra of [Eu(CH$_3$COO)$_2$L]Cl·4H$_2$O (Benetollo, 1990) and {EuL}Cl$_3$·nH$_2$O (Fonda and Vallarino, 1991) have been reported; that of [Pr(CH$_3$COO)$_2$L]Cl·4H$_2$O is reported here: 16.25 s, 4H (ß py); 14.13 s, 2H (γ py); 5.44 s, 12H (macrocycle CH$_3$); 4.02 s, 8H (CH$_2$); -4.69 s, 6H (acetate CH$_3$). It was noted that the resonances of the macrocyclic ligand L differ markedly, as expected, for the two different paramagnetic centers and are also somewhat sensitive to changes in the counterions (Fonda and Vallarino, 1991). The macrocycle resonances, however, are always readily identifiable and remain constant over the concentration range investigated in this study.

The results obtained in the ^{1}H NMR titrations of selected amino acid substrates with the {EuL} and {PrL} complexes, as well as with EuCl$_3$, are summarized in the Table. In every case the proton signals of the amino acid substrate are considerably shifted downfield by the {EuL} reagent; the magnitude of the shift increases with the Eu-to-substrate mole ratio and decreases with the distance of the observed proton from the -COO$^-$ metal-binding site. There was a progressive loss of resolution as the Eu(III) concentration of the solution was increased; in each case, however,

Sample in D_2O	α H (ppm)	ß H (ppm)	Other (ppm)
0.03 M D,L-alanine	3.76	1.46	--
+ 0.01 M {EuL}Cl$_3$	4.38	1.77	--
+ 0.03 M {EuL}Cl$_3$	5.47	2.30	--
0.03 M L-valine	3.58	2.28	γCH$_3$ 1.03, 1.01
+ 0.01 M {EuL}Cl$_3$	4.36	2.72	γCH$_3$ 1.27, 1.25
+ 0.03 M {EuL}Cl$_3$	5.58	3.43	γCH$_3$ 1.74, 1.63
0.03 M L-leucine	3.73	1.73	γH 1.71 δCH$_3$ 0.98, 0.96
+ 0.01 M {EuL}Cl$_3$	4.80[a]	2.46, 2.31	γH 2.16 δCH$_3$ 1.17, 1.14
+ 0.02 M {EuL}Cl$_3$	5.76	2.98, 2.74	γH 2.50 δCH$_3$ 1.30, 1.27
+ 0.03 M {EuL}Cl$_3$	6.68	3.50, 3.16	γH 2.80 δCH$_3$ 1.44, 1.39
0.03 M L-histidine	3.96	3.18, 3.11	ring H2 7.73 H4 7.03
+ 0.01 M {EuL}Cl$_3$	4.74[a]	3.71, 3.60	ring H2 7.86 H4 7.29
+ 0.02 M {EuL}Cl$_3$	5.44	4.00, 3.88	ring H2 7.92 H4 7.39
+ 0.03 M {EuL}Cl$_3$	5.96	4.29, 4.14	ring H2 7.94 H4 7.48
0.03 M L-phenylalanine	3.96	3.25, 3.14	ring H's 7.35
+ 0.01 M {EuL}Cl$_3$	4.53	3.56, 3.45	ring H's 7.44
+ 0.02 M {EuL}Cl$_3$	4.99	3.81, 3.70	ring H2,6 7.54 H3,5 H4 7.44
+ 0.03 M {EuL}Cl$_3$	5.42	4.07, 3.94	ring H2,6 7.64 H3,5 7.46 H4 7.38
+ 0.03 M EuCl$_3$	3.77	3.06, 2.94	ring H's 7.32
+ 0.01 M {EuL}Ac$_2$Cl	4.16	3.30, 3.17	ring H's 7.36
+ 0.01 M {PrL}Ac$_2$Cl	3.64	3.12, 3.05	ring H's 7.32

[a] Resonance partially obscured by solvent.

TABLE. ^{1}H NMR data for amino acids with {EuL}Cl$_3$, EuCl$_3$, {EuL}Ac$_2$Cl, and {PrL}Ac$_2$Cl.

assignments remain unambiguous. Fig. 2 illustrates schematically how the spectrum of phenylalanine changes upon addition of equimolar quantities of {EuL}Cl₃ or EuCl₃.

The {EuL} complexes acted as downfield shift reagents, while the {PrL} complex was an upfield shift reagent. In comparative experiments with phenylalanine, the {EuL} diacetate chloride reagent caused greater shifts than the corresponding {PrL} reagent. Also, {EuL}Cl₃·nH₂O caused greater paramagnetic shifts than the corresponding diacetate chloride, [Eu(CH₃COO)₂L]Cl·4H₂O. This result is consistent with previous work,

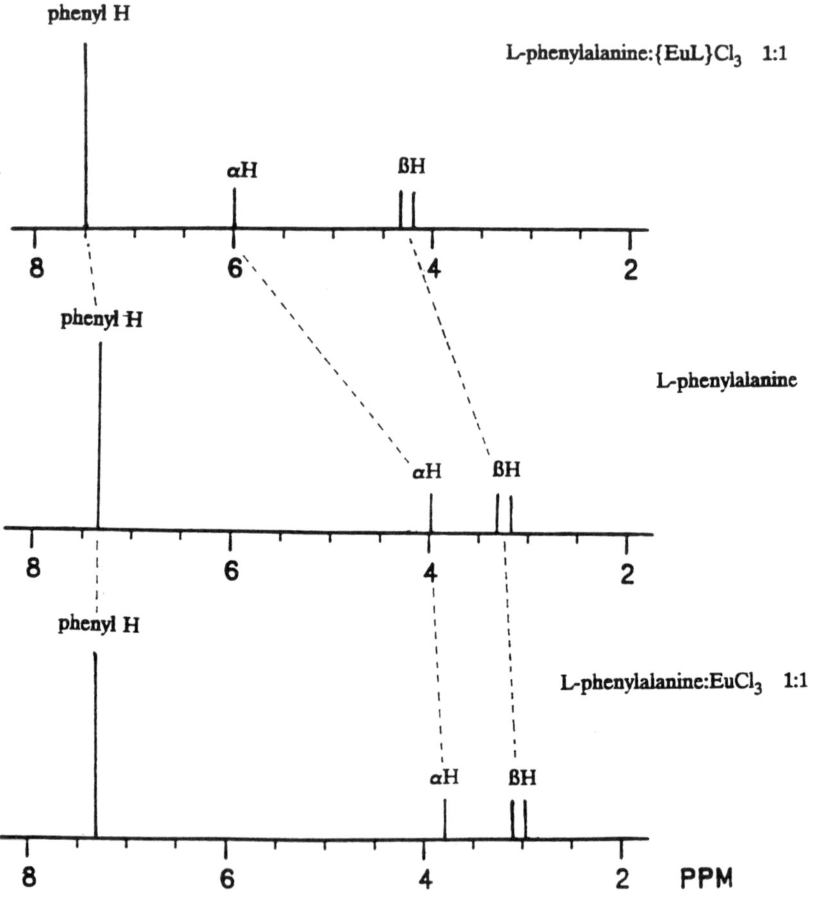

FIGURE 2. Schematic ¹H NMR spectra in D₂O of phenylalanine (0.03 M) alone and with equimolar amounts of {EuL}Cl₃·nH₂O or EuCl₃.

which had shown that the acetate groups remain at least partially coordinated to the metal even in aqueous solution (Fonda and Vallarino, 1991); thus the presence of the acetate groups limits access of the amino acid substrate to the paramagnetic center.

Additional experiments using phenylalanine as the substrate and $EuCl_3$ as the shift reagent (Fig. 2) were conducted in order to provide a comparison between the effects of the macrocyclic complexes and those of a simple salt of the same metal ion. The shifts caused by the trichloride complex {EuL}Cl$_3$ were greater in magnitude than those induced by the simple salt $EuCl_3$. Even more significantly, the macrocyclic complex caused a downfield shift of the substrate resonances, instead of the upfield shift produced by $EuCl_3$. This is a distinct advantage for any complex to be employed as a shift reagent because downfield shifts usually result in less crowded spectra.

CONCLUSIONS

This work has shown that the {ML}X$_3$•nH$_2$O complexes containing paramagnetic lanthanides may fulfill the need for a class of NMR shift reagents which retain their identity in water over a wide pH range and are suitable for the study of amino acids and other biologically related anionic species.

ACKNOWLEDGEMENTS

This work has been supported in part by Coulter Electronics, Hialeah, Florida, by a Virginia Commonwealth University Grant-in-Aid, and by N.A.T.O. Bilateral Project No. 184-185.

REFERENCES

Benetollo F, Polo A, Bombieri G, Fonda KK and Vallarino LM (1990): X-Ray Crystal Structures and Nuclear Magnetic Resonance Spectra of Macrocyclic Complexes of Neodymium(III) and Europium(III). *Polyhedron* 9: 1411.

Bombieri G, Benetollo F, Hawkins WT, Polo A and Vallarino LM (1989): Synthesis, Characterization and X-Ray Crystal Structure of Yttrium(III) Macrocyclic Complexes with a Six-Nitrogen-Donor Cavity. *Polyhedron* 8: 1923.

De Cola L, Smailes DL and Vallarino LM (1986): Hexaaza Macrocyclic Complexes of the Lanthanides. *Inorg. Chem.* 25: 1729.

Fonda KK and Vallarino LM (1991): NMR Properties of Lanthanide(III) Complexes of the Macrocyclic Ligand $C_{22}H_{26}N_6$. Submitted for publication in *Magn. Res. Chem.*

Wenzel TJ (1987): *NMR Shift Reagents*, Boca Raton, Florida: CRC Press.

MODEL SYSTEM FOR THE STUDY OF GONORRHEA CREATED BY CELL-TISSUE ELECTROFUSION.

Richard Heller
Departments of Surgery and Medical Microbiology
University of South Florida, College of Medicine

Richard Gilbert
Department of Chemical Engineering
University of South Florida, College of Engineering

INTRODUCTION

An increased understanding of receptor mediated processes of species specific pathogens can be obtained by creating novel animal models which possess distinctive features that differ fundamentally from those biological properties displayed naturally by the unaltered animal species. For example, the pathogenesis of certain infectious diseases of importance to human and veterinary medicine is initiated by the binding of the etiological agents directly to host or tissue specific microbial receptors located on cell surfaces. Therefore, biologically relevant animal models for these particular infectious diseases could theoretically be established by transferring cells that contain the appropriate microbial attachment receptors to selected tissues of common laboratory animals. The histologically modified animals should now become susceptible to infection by those host and tissue restricted microbial pathogens to which the unaltered animals are naturally resistant.

Most microorganisms which are intracellular pathogens utilize host cellular membrane components as attachment sites (receptors) to begin the process of infection. The specificity of the receptor is a key factor in determining the type of cell and intracellular environment that the organism infects. Some bacteria can enter eucaryotic cells of a diverse nature; e.g., *Salmonella* and *Yersinia* can enter embryonic *Drosophila* cell lines in addition to human intestinal cells (Finlay, 1988). Complement receptors are used by some bacteria, viruses and parasites to mediate their uptake, e.g., *Legionella pneumophila* (Payne, 1987), *Mycobacterium tuberculosis* (Finlay, 1988), *Leishmania donovani* (Blackwell, 1985), and Epstein Barr virus (Nemerow, 1986). Human immunodeficiency virus (HIV) binds to the human specific marker CD4 (Dalgleish, 1984 and Klatzman, 1984). The cellular receptor for *N. gonorrhoeae* has yet to be fully characterized. However, due to the species specific property of the gonococci, this receptor, like the receptor for HIV, most likely is unique to human cells.

Currently, there are no practical, suitable, biologically relevant animal models for establishing reproducible gonococcal lesions or for testing potential vaccines. Many investigators have tried to infect animals with this obligate human pathogen since Neisser first described the organism in 1879. Early attempts at developing a model were reviewed by Hill (1944) and subsequently by Kraus (1977). Recent attempts at creating laboratory models for experimental gonococcal infections were reviewed by Arko (1989).

The interspecies transfer of membrane surface components can be accomplished by the formation of somatic cell hybrids. Electrical fields have been shown to be effective in creating these hybrid cells. (Bates, 1987; Finaz, 1984; Foung, 1989; Glassy, 1988; Hewish, 1989; Lo, 1984; Sowers, 1989; Sukharev, 1990; and Zimmermann, 1986) Cell-tissue electrofusion (CTE) is one technique available for the transfer of surface membrane components from susceptible cells to intact tissue of naturally resistant laboratory animals. CTE was successfully demonstrated by Grasso et. al (1989) and Heller and Grasso (1990).

This manuscript will focus on the development of a model system for studing the pathogenic mechanisms of *N. gonorrhoeae* by using cell-tissue electrofusion. To establish the validity of this CTE created model it was necessary to accomplish the following:

(a)   to demonstrate that individual human cells containing gonococcal attachment receptors can be electrofused to intact tissue;

(b)   to demonstrate that there is receptor mediated attachment of the gonococci to the newly formed somatic cell hybrids;

(c)   to demonstrate that cell tissue electrofusion techniques are safe and cause no ill-effects to an animal during *in vivo* transfers;

(d)   to demonstrate that *N. gonorrhoeae* does cause an infection in a naturally resistant animal; and

(e)   to demonstrate the differences between piliated and nonpiliated bacterial attachment.

ELECTROFUSION TO INTACT TISSUE

The first step in obtaining this goal is the establishment of a system which would allow for the transfer of membrane surface components from susceptible cells to intact tissue of laboratory animals. Individual cells which contain the appropriate receptors can be incorporated into excised tissue which is naturally resistant to attachment by host or tissue specific microbial pathogens. The details of the transfer procedure and the usefulness of the system can be tested and characterized in vitro before attempting to establish an animal model.

Procedure

Cell-tissue electrofusion can be accomplished *in vitro* with commercially available cell fusion instruments. For the *N. gonorrhoeae* model system specially designed chambers and electrodes were developed to facilitate the fusion of individual cells to intact tissue. The new design utilized mechanical pressure to juxtaposition individual cells with intact tissue.

The *in vitro* chamber designed for the electrofusion of pressure aligned cells consists of six wells. Each insulated well has a flat round 10 mm diameter platinum electrode at the bottom. The chamber can be attached to a water bath to maintain the desired thermal environment. The second electrode is a collection of six individual electrode heads of various shapes and sizes that screw into a common insulated handle. The appropriate tip is selected to ensure maximum cell-tissue contact during the fusion process. The electrode tip shapes included a concave electrode which was machined to reflect the curvature of the rabbit cornea as well as 2 mm and 6 mm diameter polished titanium disks.

The procedure utilizing this CTE equipment is as follows. Briefly, a 10 mm corneal button is excised and placed in one of the insulated wells. A suspension of human lymphoma cells (U937 or HL60) were placed in contact with the excised tissue. The electrode contact with the cell tissue system is controlled via a micromanipulator. Pressure is exerted through the electrode until desired force is reached (600-700 gm/cm^2). The force can be estimated by placing the entire apparatus on a torsion beam balance (or similar scale). The pressure is maintained for 3 min and then one to three 20 μsec, 0.20 kV/cm^2 square pulses are delivered 1 sec apart to initiate the fusion process. The disc electrode is raised and the cornea removed from well. A vigorous wash and fixation in 2.5% glutaraldehyde follows. After the fixation step, the fusion results can be examined by scanning electron microscopy (SEM).

Results

The above protocol was used to demonstrate the fusion of the two lymphoma cell lines to the rabbit corneal epithelium. The U937 cells were clearly visible on the surface of the rabbit corneal epithelium (Fig 1) after the DC pulses were delivered. The U937 cells were easily distinguishable from the rabbit epithelial cells because of their spherical shape and smooth membranes that lack microvilli. In contrast, rabbit corneal epithelial cells were larger, flat and have a centrally located nucleus and rough surface due to the presence of microvilli. Although the human cells appeared to be lying on the surface of the epithelium (Fig 1A), when observed under higher magnification and at different angles (Fig 1B-D), several fusion points between the two cells were clearly discernable. At a different orientation the two membranes appeared to be flowing together. From these observations, it was apparent that individual cells could be fused to intact tissue.

RECEPTOR MEDIATED ATTACHMENT

N. gonorrhoeae strain Pgh 3-2 was utilized to determine if gonococcal receptors present on human cells were present on the surface of these newly formed human-rabbit somatic cell hybrids. Initial experiments were carried out *in situ* on freshly killed rabbbits. These fusion experiments were performed in combination with bacterial adherence assays. Results from these experiments were also important in determining if the procedure could be performed on an intact animal.

Procedure

A complete description of the *in situ* procedures are described in Grasso, et. al. (1989) and Heller and Grasso (1990). Briefly, human and nonhuman cells are washed by centrifugation in PBS and layered onto Millipore filters by centrifugation. Supernatant fluids are discarded and the filters containing about 10^7 cells are removed with forceps and placed cell side down on the PBS washed corneal surfaces of rabbits. Mechanical pressure plus three 20 μsec, 20 V square pulses at a 1 second repetition rate were applied simultaneously to a concave titanium electrode shaped to the curvature of the rabbit eye. Following the administration of the electrical pulses all eyes are washed with PBS to remove unfused cells. After electrofusing human

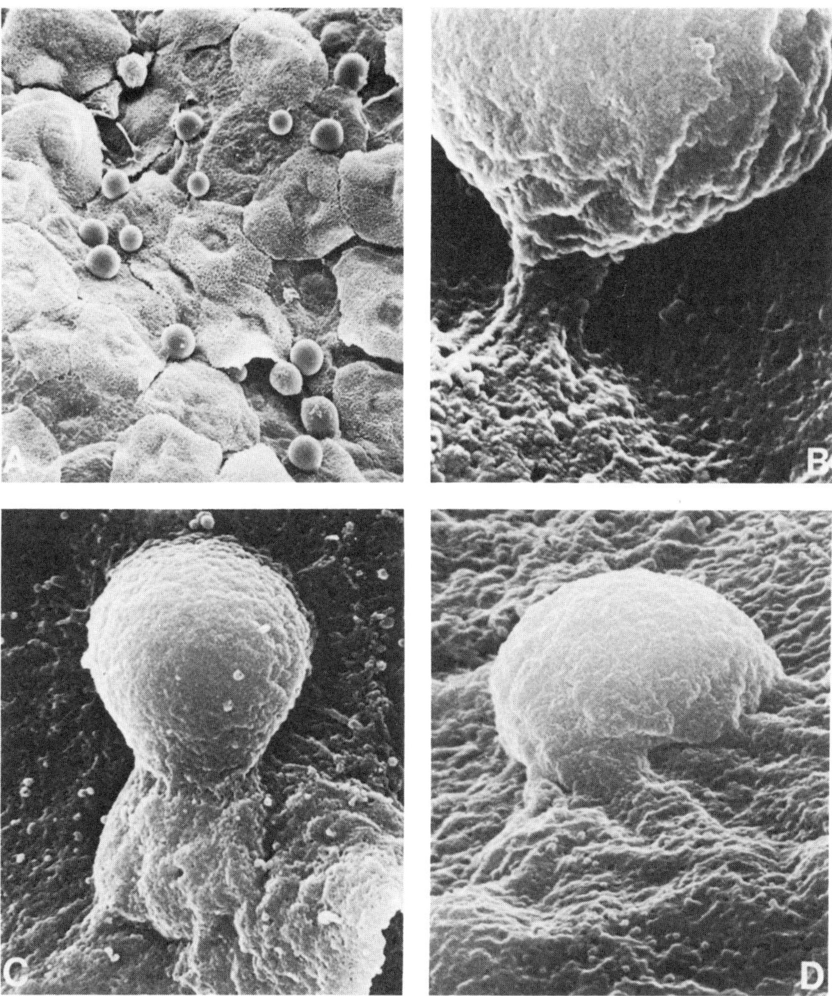

FIGURE 1. Electrofusion of human U937 and HL60 lymphoma cells directly to rabbit corneal epithelium. Panel A: Corneal surface (1300X) showing a representative distribution spherically shaped human cells electrofused to rabbit epithelium. Panel B: Plasma membrane "stalk" that forms shortly after electrofusion between the upper smoother human cell and the lower microvilli covered rabbit cell surface (15,000X). Panels C (6000X) and D (8600X) plasma membranes of human-rabbit somatic cell hybrids beyond the initial stages of their formation in the intact corneal epithelium. Reprinted with permission of Elsevier Science Publishers from Grasso, et.al., 1989: Biochim. et Biophys. Acta, vol 980, p 9-14.

HL60 or U937 lymphoma cells, rabbit skin cells or Vero cells directly to the rabbit tissue, the corneas were excised and employed in qualitative gonococcal adherence assays and then examined by SEM.

<u>Results</u>

Control experiments were performed using nonhuman cells as well as corneas not subjected to electrofusion. No attachment was observed after fusion of nonhuman cells. In addition, no attachment was observed on a untreated rabbit cornea. The gonococci attached only to those corneas which were fused with human cells (Table 1). It is evident that the administration of an electrical field is not sufficient to modify the rabbit corneal epithelium to allow the attachment of this obligate human pathogen to the rabbit tissue. Furthermore, the incorporation of nonhuman cells into the intact tissue did not allow attachment of the organism. Successful attachment of this diplococcal organism only occurs if the human membrane components containing the gonococcal receptors are present on the histologically modified rabbit corneas. Figure 2 illustrates that the human and rabbit plasma membranes have coalesced completely due to the extended incubation times (60 min) required to perform the adherence assays. In contrast, figure 1 shows the early stages of the fusion process.

CORNEA NO.	CELL TYPE ELECTROFUSED TO RABBIT CORNEAS	GONOCOCCI ATTACHED TO CORNEAL EPITHELIUM
1	None	No
2	Rabbit Skin Cells	No
3	Monkey Vero Kidney Cells	No
4	Murine WEHI-3 Cells	No
5	Human U937 Lymphoma Cells	Yes
6	Human HL60 Lymphoma Cells	Yes

TABLE 1. Specific adherence of *N. Gonorrhoeae* Pgh 3-2 to rabbit corneas after electrofusion of human cells *in situ* and observed by scanning electron microscopy.

In Vivo TRANSFERS

Human HL60 cells were electrofused to the corneas of anesthetized rabbits to determine if cell-tissue electrofusion could eventually be utilized to develop an animal model for the study of *N. gonorrhoeae*. It was vital to

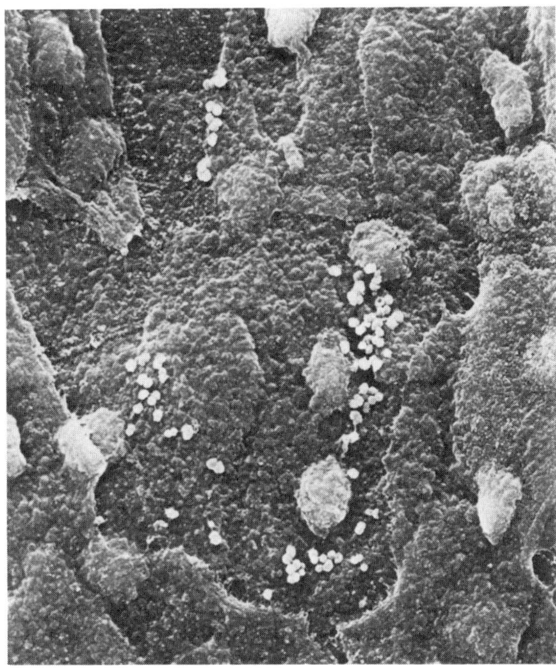

FIGURE 2. Receptor-mediated attachment of *N. gonorrhoeae* Pgh 3-2 to somatic human-rabbit cell hybrids formed within rabbit corneal epithelium by electrofusion in situ. After human HL60 cells were electrofused to corneal epithelium, the corneas were excised, employed in gonococcal adherence assays, and examined by SEM (2000 X). Reprinted with permission of Elsevier Science Publishers from Heller and Grasso, 1990: Biochim Biophys Acta, vol 1024, p 185-188.

ascertain if the CTE procedure caused an inflammatory response in the rabbits. In addition, it was necessary to examine if the procedure produced any long term ill effects on the rabbits.

<u>Results</u>

The electrofusion procedure utilized was the same as that performed *in situ* except the animals were anesthetized. Two rabbits were utilized in this protocol. Table 2 shows the experimental design and results from this experiment.

Rabbit #1 served as the electrofusion and human cell control. The left eye of this animal was subjected to DC fusion pulses in the absence of cells, the right eye was subjected to human cells in the absence of the pulses. These treatments failed to produce ocular inflammation over the 24 hr test period after a 5 min exposure to the pathogen.

EXPERIMENTAL DESIGN[a]				
TREATMENT OF CORNEAS	RABBIT #1		RABBIT #2	
	LEFT	RIGHT	LEFT	RIGHT
Human HL60 cells	-	+	+	+
In vivo Electrofusion	+	-	+	+
N. gonorrhoeae Pgh 3-2	+	+	-	+
Clinical Evaluation[b] 6 hr Postinfection				
Injection	-	-	+/-	++
Chemosis	-	-	-	+
Exudate	-	-	-	+++

TABLE 2. Incorporation of human HL60 cells into the corneal epithelium of anesthetized rabbits.

Note:

a   "Experimental Design" - performing (+) or not performing (-) the indicated treatments on the respective left and right corneas of the two rabbits.

b   Ocular symptoms of injection (reddening of the sclera), chemosis (conjunctival edema) and exudate (extravascular cells and fluid) were evaluated for their severity with the following grading scheme: - = no symptoms; +/- = trace; + = slight; ++ = mild; +++ = moderate and ++++ = severe.

Rabbit #2 had HL60 cells fused to the corneas of both eyes. Only the right eye of this animal was exposed to the pathogen. The left eye showed that the presence of the fused HL60 cells in the absence of the organism did not cause an inflammatory response. Ocular inflammation was observed in the right eye within 2 hrs following exposure to the gonococcal organism.

Neisseria gonorrhoeae INFECTION

To have a viable model system it is necessary to demonstrate that a human specific pathogen can cause an infection in a naturally resistant animal. In this case the goal is to elicit an inflammatory response in the rabbit against the gonococci. In addition, it is important that CTE does not interfere with the specificity of the invading organism.

<u>Results</u>

Four rabbits were utilized in each of the infectivity experiments. *In situ* procedures were followed except that the animals were anesthetized. Table 3 presents the experimental design and results from a representative experiment of 4 replicate experiments that were performed.

Rabbit #1 served as the electrofusion and human cell control. The left and right eyes of this rabbit were set up exactly as in the initial *in vivo* experiment described above. In all experiments, there were no observable signs of ocular inflammation in these eyes during the 24 hr period after exposure to the pathogen.

Both eyes of rabbits 2, 3, and 4 were electrofused with rabbit, murine, and human cells, respectively, as usual the left eye of each animal served as the inflammation control and was not exposed to gonococci. When the right eyes of Rabbits 2 and 3, bearing electrofused non-human cells, were exposed to the bacteria neither inflammatory reactions nor ocular lesions developed.

A purulent keratoconjunctivitis developed in the right eye of Rabbit 4 due to the presence of electrofused human cells. Clinical symptoms of these ocular lesions appeared approximately 2 hr after infection and peaked between 8 and 12 hr (Fig. 3). Inflammation was resolved after 24 hr. Microscopic examination of the exudate revealed diplococcal organisms engulfed by polymorphonuclear leukocytes. The rabbits were repeatedly observed for seven days following the electrofusion procedure. No inflammation was observed during this time period other than that described above.

BACTERIAL ADHERENCE: PILIATED VS. NONPILIATED

The important assessment of the CTE created model for gonococcal infectivity is whether the model can be used to test the adherence capabilities of different gonococcal strains. Attachment of the organism is the first step in gonococcal infectivity. The presence or absence of pili plays an important role in the ability of the organism to adhere. Piliated strains of *N. gonorrhoeae* have been shown by many investigators to bind more readily than nonpiliated strains (Cooper, 1984; Easmon, 1987; Mardh, 1979; and McGee, 1981).

426

EXPERIMENTAL DESIGN[a]								
Treatment of Corneas	Rabbit 1		Rabbit 2		Rabbit 3		Rabbit 4	
	left	right	left	right	left	right	left	right
Rabbit Skin Cells	-	-	+	+	-	-	-	-
Mouse WEHI-3 Cells	-	-	-	-	+	+	-	-
Human HL60 Cells	-	+	-	-	-	-	+	+
dc Fusion pulses	+	-	+	+	+	+	+	+
N. gonorrhoeae	+	+	-	+	-	+	-	+
Clinical Evaluation[b] 3 h postinfection								
Injection	-	-	-	-	-	-	-	+
Chemosis	-	-	-	-	-	-	-	+ +
Exudate	-	-	-	-	-	-	-	+
Clinical Evaluation[b] 6 h postinfection								
Injection	-	-	-	-	-	-	-	+ +
Chemosis	-	+ /-	-	-	-	-	-	+ +
Exudate	-	+ /-	-	-	-	-	-	+ +

TABLE 3. Purulent gonococcal keratoconjunctivitis mediated by human receptors in rabbits infected with N*N. gonorrhoeae* Pgh 3-2. Reprinted with permission of Elsevier Science Publishers, vol 1024, p185-188.

Note:

a   "Experimental Design" - performing (+) or not performing (-) the indicated treatments on the respective left and right corneas of the two rabbits.

b   Ocular symptoms of injection (reddening of the sclera), chemosis (conjunctival edema) and exudate (extravascular cells and fluid) were evaluated for their severity with the following grading scheme: - = no symptoms; +/- = trace; + = slight; + + = mild; + + + = moderate and + + + + = severe.

Procedure

Several piliated and nonpiliated strains were examined for their ability to bind to a rabbit cornea fused with human cells *in vitro*. Following fusion, gonococcal adherence assays were performed on rabbit corneas in the presence or absence of fused human ME-180 cervical carcinoma cells. Both piliated (p^+) and nonpiliated (p^-) strains of *N. gonorrhoeae* were utilized. Corneas were fixed with 2.5% glutaraldehyde following the 30 min exposure to the gonococci and examined by SEM.

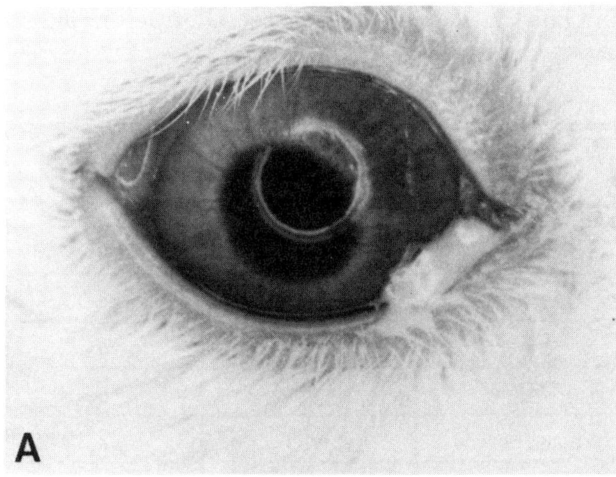

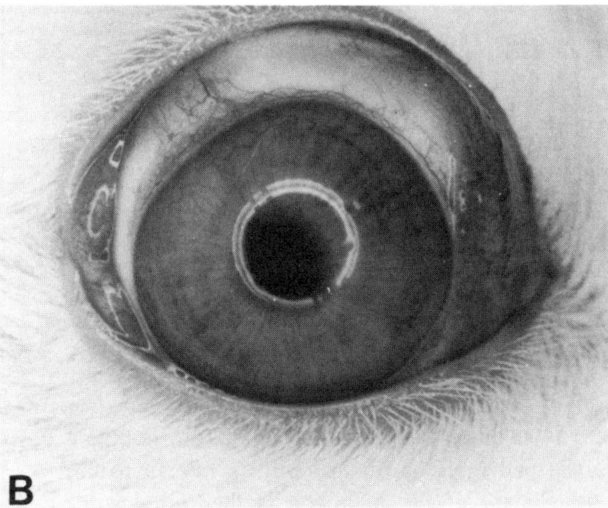

FIGURE 3. Purulent gonococcal conjunctivitis in rabbit cornea.
A: Photograph of a mild gonococcal lesion that occurred in the right eye of Rabbit #4 (Table 3) only 4 hours after initiating the bacterial infection.
B: Photograph of the right eye of Rabbit #3 (Table 3), that had been similarly infected 4 hours earlier. Reprinted with permission of Elsevier Science Publishers from Heller and Grasso, 1990: Biochim Biophys Acta, vol 1024, p 185-188.

Results

The relative attachment was determined by examining several SEM fields. One field was defined as an area visible at a SEM magnification of 200X (0.25 mm^2). A field was considered positive if at least one gonoccal organism was present. The greater the number of positive SEM fields the more adherent the gonococcal strain. Strains MS11 (p$^+$) and C30 (p$^+$) showed an increased ability to attach to corneas containing fused human cells than strains MS11 (p$^-$) and B2 (p$^-$) (Table 4). Nonpiliated gonococci showed the expected low attachment rate. In addition, the gonococcal organisms did not attach to unfused corneas.

CORNEA	GONOCOCCAL STRAIN	ELECTROFUSION WITH HUMAN ME-180 CELLS	NO. OF POSITIVE SEM FIELDS[a]
1	MS11 P^+	No	0
2	MS11 P^-	No	0
3	C30 P^+	No	0
4	B2 P^-	No	0
5	MS11 P^+	Yes	4
6	MS11 P^-	Yes	1
7	C30 P^+	Yes	3
8	B2 P^-	Yes	0

TABLE 4. Comparison of the binding ability of piliated (p^+) and nonpiliated (p^-) gonococcal strains to rabbit corneas following the electrofusion of human ME-180 cells.

Note
[a] one field is defined as the area visible at a magnification of 200x (0.25 mm^2)

FUTURE DIRECTIONS

The application of cell-tissue electrofusion in the creation of a laboratory animal model is an important first step in the development of an animal model to study the pathogenesis of the obligate human pathogen *N. gonorrhoeae*. Since attachment of the pathogen is the initial step in the infection process, it was an important to determine if the model could be used to test the adherence capability of different gonococcal strains. Several piliated and nonpiliated gonococcal strains were examined for their ability to bind to rabbit corneas fused with human cells. As seen in other systems attachment was enhanced by the presence of pili. However, in order to develop fully this ocular model, several questions need to be answered. How long do the transferred receptors remain fuctional on the corneal epithelium? Is this time frame the same *in vivo* as *in vitro*? Can this model be used to determine quantitatively the attachment capabilities of different gonococcal strains?   Is there an increase or decrease in the number of gonococcal attachment receptors over a period of time? What effect will varying the infectious dose of bacteria have on eliciting the observed inflammatory response? Can the infecting organism be recovered following the infection and will it express the same phenotype?

Once fully developed, this system could be employed to test virulence factors of a variety of gonococcal strains. The most important aspect of this model system is the correlation between *in vitro* and *in vivo* observations. Experimental protocols can be developed and tested *in vitro* before using live animals. However, the testing of vaccines could not be accomplished with this system as it exists. To utilize a model to test vaccines, a cervical model must be developed utilizing the procedures discussed above. A cervical model for gonorrhea would be more relevant and allow analysis of the immune response to the pathogen.

It is important to demonstrate that other human membrane components can be present on a CTE formed somatic cell hybrid. The ability to develop animal models for other species specific diseases may be dependent on the ability to transfer other human membrane components. For example, CTE treated rabbit corneas could be examined for the presence of human specific HLA antigens after the fusion of human cells to their surface. Then, to establish a model for HIV (AIDS) infectivity, H9 cells which contain CD4 on their surface could be fused to a selected tissue site to allow the binding of this pathogen.

In summary CTE has been used to show that: a) individual cells can be electrofused to intact tissue *in vitro, in situ* and *in vivo*; b) *in vivo* electrofusion can be performed without causing lethality or ocular inflammation to anesthetized animals; c) human gonococcal membrane receptors remain functional after their incorporation into human-rabbit somatic cell hybrids; d) an obligate human pathogen can produce a purulent inflammatory response mediated by transferred human microbial attachment receptors in a common laboratory animal; and e) the rabbit ocular model can be utilized to examine various gonococcal components for their potential role in aiding the organism to bind to the host cell. When these results are examined it is evident that a model system for studying the infectivity of species specific diseases can be developed by fusing susceptible cells to the tissue of naturally resistant animals by the process of cell-tissue electrofusion. This is the first major step in the development of animal models which could aid in the development of effective vaccines.

REFERENCES

Arko RJ (1989): Animal models for pathogenic *Neisseria* species. *Clin. Microbiol. Rev.* 2(Suppl):S56-S59.

Bates GW, Saunders JA, and Sowers AE (1987): Principles and applications of electrofusion. In *Cell Fusion* A. E. Sowers ed. Plenum Press, New York pp.367-395.

Blackwell JM, Ezekowitz RA, Roberts MB, Channon JY, Sim RB, and Gordon S (1985): Macrophage complement and lectin-like receptors bind *Leishmania* in the absence of serum. *J. Exp. Med.* 162:324-331.

Cooper MD, McGee ZA, Mulks MH, Koomey JM and Hindman TL (1984): Attachment to and invasion of human fallopian tube mucosa by an IgA1 protease deficient mutant of *N. gonorrhoeae* and wild type parent. *J. Infect. Dis.* 150:737-744.

Dalgleish AG, Beverley PCL, Clapham PR, Crawford DH, Greaves MF and Weiss RA (1984): The CD4 (T4) antigen is an essential component of the receptor for the AIDS retrovirus. *Nature* 312:763-767.

Easmon CSF and Ison CA (1987): *N. gonorrhoeae*: a versatile pathogen. *J. Clin. Pathol.* 40:1088-1097.

Finaz C, Lefevre A and Teissie J (1984): A new, highly efficient technique for generating somatic cell hybrids. *Exp. Cell Res.* 150:477-482.

Finlay BB and Falkow S (1988): Comparison of the invasion strategies used by *Salmonella cholera-suis*, *Shigella flexneri* and *Yersinia enterocolitica* to enter cultured animal cells: endosome acidification is not required for bacterial invasion or intracellular replication. *Biochimie* 70:1089-1099.

Foung SKH and Perkins S (1989): Electric field-induced cell fusion and human monoclonal antibodies. *J. Immunolog. Meth.* 116:117-122.

Glassy M (1988): Creating hybridomas by electrofusion. *Nature* 333:579-580.

Grasso RJ, Heller R, Cooley JC and Haller EM (1989): Electrofusion of individual animal cells directly to intact corneal epithelial tissue. *Biochim. Biophys. Acta.* 980:9-14.

Heller R and Grasso RJ (1990): Transfer of human membrane surface components by incorporating human cells into intact animal tissue by cell-tissue electrofusion *in vivo*. *Biochim. Biophys. Acta.* 1024:185-188.

Hewish DR and Werkmeister JA (1989): The use of an electroporation apparatus for the production of murine hybridomas. *J. Imunolog. Meth* 120:285-289.

Hill JH (1944): Experimental infections with *N. gonorrhoeae*, II. Animal innoculations. *Amer. J. Syph. Gon. Vener. Dis.* 28:471-510.

Klatzmann D, Champagne E, Chimaret S, Gruest J, Guetard D, Hercend T, Gluckman J -C and Montagnier L (1984): T-Lymphocytes T4 molecule behaves as the receptor for human retrovirus LAV. *Nature* 312:767-768.

Kraus SJ (1977): Laboratory Models for *N. gonorrhoeae* infection. In *The Gonococcus*. R. B. Roberts, ed. John Wileys and Sons, New York pp. 415-431.

Lo MMS, Tsong TY, Conrad MK, Strittmatter SM, Hester LD and Snyder SH (1984): Monoclonal antibody production by receptor-mediated electrically induced cell fusion. *Nature* 310:792-794.

Mardh PA, Baldetorp B, Hakansson CH, Fritz H and Westrom L (1979): Studies of ciliated epithelia of the human genital tract. 3. Mucociliary wave activity in organ cultures of human fallopian tubes challenged with *N. gonorrhoeae* and gonococcal endotoxin. *Br. J. Vener. Dis.* 55:256-264.

McGee ZA, Johnson AP and Taylor-Robinson D (1981): Pathogenic mechanisms of *N. gonorrhoeae*: observations on damage to human fallopian tubes in organ culture by gonococci of colony type 1 or type 4. *J. Infect. Dis.* 143:413-422.

Nemerow GR, Slaw MFE, Cooper NR (1986): Purification of the Epstein-Barr virus/C3d complement receptor of human B lymphocytes by monoclonal antibody affinity chromatography , and assessment of its functional capacities. *J. Immunol. Methods* 58:709-712.

Payne NR and Horwitz MA (1987): Phagocytoses of *Legionella pneumophila* is mediated by human monocyte complement receptors. *J. Exp. Med.* 166:1377-1389.

Sowers AE (1989): Evidence that electrofusion yield is controlled by biologically relevant membrane factors. *Biochim. Biophys. Acta.* 985(3):334-338.

Sukharev SI, Bandrina IN, Barbul AI, Fedorova LI, Abidor IG, and Zelenin AV (1990): Electrofusion of fibroblasts on the porous membrane. *Biochim Biophys Acta* 1034:125-131.

Zimmermann U (1986): Electrical breakdown, electropermeabilization and electrofusion. *Rev. Physiol. Pharmacol.* 105:176-256.

NMR STUDIES OF THE INTERACTION OF CATECHOL AND ASCORBIC ACID WITH POLY(N-VINYLPYRROL-IDONE) POLYMER

George P. Kreishman, Helen J. Johnson, Toshihiko Imato[1] and William R. Heineman
Department of Chemistry, University of Cincinnati,
Cincinnati, OH 45221-0172 U.S.A.

INTRODUCTION

The catecholamines are an important class of neurotransmitters secreted in the regions of the brain that control locomotion. Many neurological and phychiatric disorders such as Parkinson's disease, schizophrenia, psychosis and cocaine addiction are the result of alterations in the levels of these neurotransmitters and/or their receptors in the brain. Since the measurement of endogenous levels of these neurotransmitters would be extremely useful in clinical and scientific studies of these disorders, the development of chemical sensors specific for catecholamines _in vivo_ has been the subject of extensive research (Adams, 1976).

One approach that has received attention is the use of microelectrodes which takes advantage of the presence of the electroactive hydroquinone group in these molecules. For _in vivo_ applications, these electrodes must be analyte specific because of the presence of interfering species in the brain tissue. Particularly, ascorbate causes large interferences because it has a reduction potential close to that of the catechols and is present at concentrations which are approximately ten times greater than that of the catechols. _In vitro_ studies have shown that the use of poly(N-vinylpyrrolidone) (PNVP) coated graphite electrodes can enhance the electrochemical signal due to the catechols by a factor of three while attenuating the signal from ascorbate by a factor of three. It has been postulated that the apparent preconcentration of the catechols at the electrode surface is the result of hydrogen bonding of the analyte to the polymer. The lack of hydrogen bonding ability of the ascorbate to the polymer remains unresolved (Coury et al., 1988).

---

[1]Present Address, Department of Chemical Science and Technology, Kyushu University, Hakozaki, Higashi-Ku, Fukuoka 812, Japan

In order to begin to resolve these questions, preliminary results of NMR spectroscopic studies on the mode of interaction of catechol and ascorbate with the PNVP polymer are presented in which the ability of the NMR method to differentiate hydrogen bonding from other modes of interaction is exploited (James, 1975).

EXPERIMENTAL

Materials: The PNVP polymer and D_2O (99.9% atom D) were obtained from Aldrich Chemical Co., Inc.. The catechol and ascorbic acid were obtained from Fisher Scientific. The d_6-DMSO (99.96% atom D) was obtained from Sigma Chemical Co. All the reagents were used without further purification.

Methods: The 400 MHz ^{1}H NMR spectra were obtained on a Bruker AMX-400 MHz multinuclear spectrometer using a 5 mm ^{1}H probe. The sample temperature was maintained at 303.1°K and a presaturation pulse sequence was used for solvent suppression. Chemical shifts are reported with respect to the internal standard, $(CH_3)_4NBr$, at 3.0 ppm.

Cyclic voltammograms were obtained on a BAS-100A electrochemical analyzer equipped with a Houston Instruments DMP-2 digital plotter. The standard three electrode experimental setup was used with a Pt auxiliary electrode and a BAS RE-1 Ag/AgCl (3M NaCl) reference electrode. The working electrode was fabricated from spectrographic rods (Type FX-365T) obtained from Poco Graphite Inc., Decatur, TX. The rods were initially coated with Krylon and the graphite surface on one end of the rod was exposed by abrasion with K-622 emery paper and polished with 600A emery paper. Several minutes of sonication was utilized to remove any residual polishing grit and loose graphite. For the polymer coated electrodes, a 10 µL aliquot of 8% (w/v) PNVP was delivered to the polished end and allowed to dry. The electrodes were then subjected to gamma irradiation (25 Mrad) in an argon atmosphere. The periphery of the bare and polymer coated electrodes were covered with Teflon film. The electrodes were allowed to equilibrate in the blank solvent for one hour prior to use. All solutions were deoxygenated with Ar.

RESULTS AND DISCUSSION

The ^{1}H NMR spectrum of PNVP in D_2O is shown in Figure 1a. The individual resonances of the ring and backbone protons can be resolved and have been previously assigned (Coury et al., 1989). Upon addition of catechol, all of the polymer resonances are to varying degrees shifted upfield (Figure 1b). With increasing catechol concentration, the chemical shift of the polymer resonances is increasingly upfield (Figure 2). Within experimental error, the chemical shifts and linewidths of the catechol resonances were not altered by the presence of the polymer. The concentration dependence of the polymer resonances is the result of fast exchange averaging of the chemical shifts of the polymer in the absence and presence of catechol binding. Since with increasing concentration of catechol, the polymer resonances are shifted upfield, the chemical shift of the polymer with bound catechol must be upfield from that of the polymer alone. On the average, the polymer protons must be situated in the shielding region of the catechol ring which is either above or below the aromatic ring.

If the mode of interaction between the catechol and the polymer is via hydrogen bonding, any interaction with the –OH groups along the edge of the catechol would require that at least some of the polymer protons would be

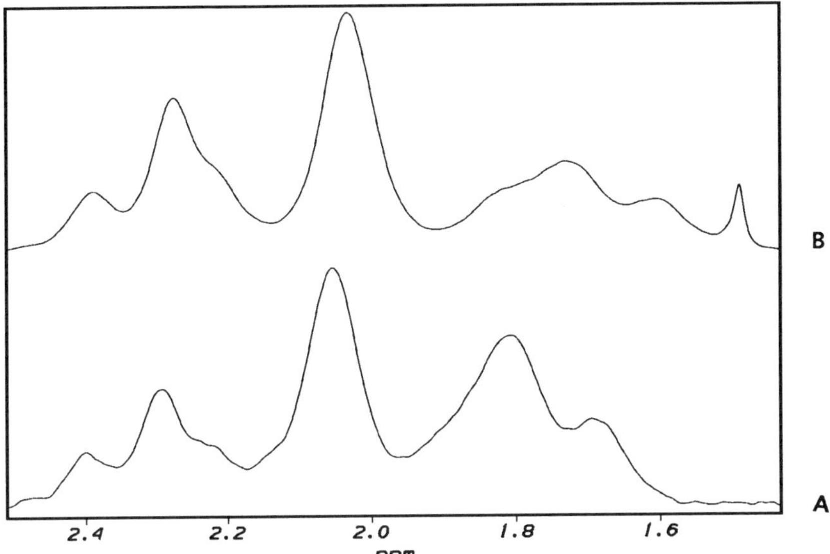

FIGURE I.   (A) 400 MHz spectrum of PNVP in D_2O at 30°C. (B) 400 MHz spectrum of [catechol]/[PNVP] = 6 in D_2O at 30°C.

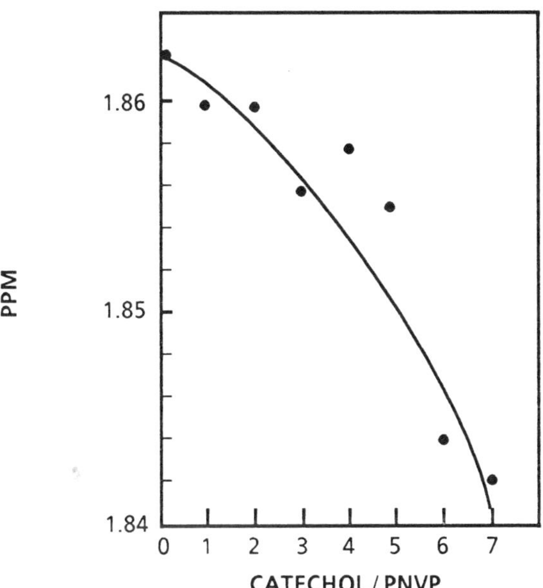

FIGURE II.   Chemical shift of the H_1 resonance vs. moles catechol/moles PNVP.

situated in this deshielding region of the catechol resulting in downfield shifts. Since the downfield shifts in the polymer resonances are not observed, an alternate mode of interaction must be considered. It is proposed that the hydrocarbon regions of the polymer are associating with the aromatic ring of the catechol and that this association is driven by the hydrophobic interaction. Since the hydrophobic interaction is the result of the hydrogen bond network in water, the observed interactions should not be observed in a non-hydrogen bonding solvent such as DMSO. As can be seen in Figure 3, no measurable shifts for the polymer resonances can be observed upon addition of catechol in d_6-DMSO.

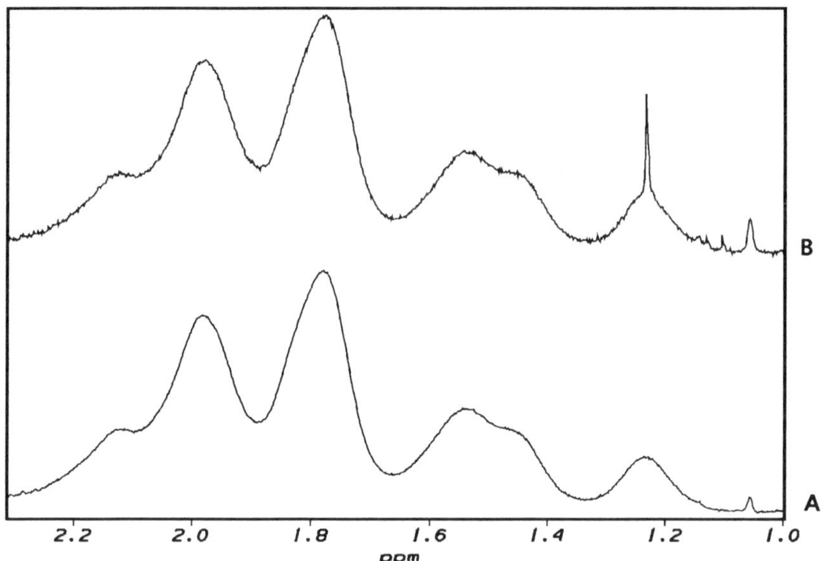

FIGURE III.   (A) 400 MHz spectrum of [catechol]/[PNVP] = 1 in DMSO at 30°C. (B) 400 MHz spectrum of [catechol]/[PNVP] = 5 in DMSO at 30°C.

In contrast to the induced upfield shifts of the polymer in the presence of catechol in water solution, no measurable chemical shifts in the polymer resonances could be observed even with the addition of large excesses of ascorbic acid in either water or DMSO solutions. No interaction between the polymer and ascorbic acid could be detected by the NMR method. The lack of interaction of the ascorbic acid with the polymer in water solution could be simply the result of its greater hydrophilicity as compared to the catechol.

In light of these results, the cyclic voltammetry of catechol on PNVP coated electrodes was repeated in 90% DMF/10% H_2O (v/v) with 0.5M LiCl supporting electrolyte. The added water was required so that the same electrochemical mechanism is detected at the electrode surface. As can be seen in Figure 4, the threefold enhancement found in water solutions is not observed in the non-hydrogen bonded solvent.

These preliminary results show that the enhanced electrochemical signal for catechol on PNVP coated carbon electrodes in water solution is primarily the result of a preconcentration of the analyte at the electrode surface that is driven by the hydrophobic interaction between the catechol

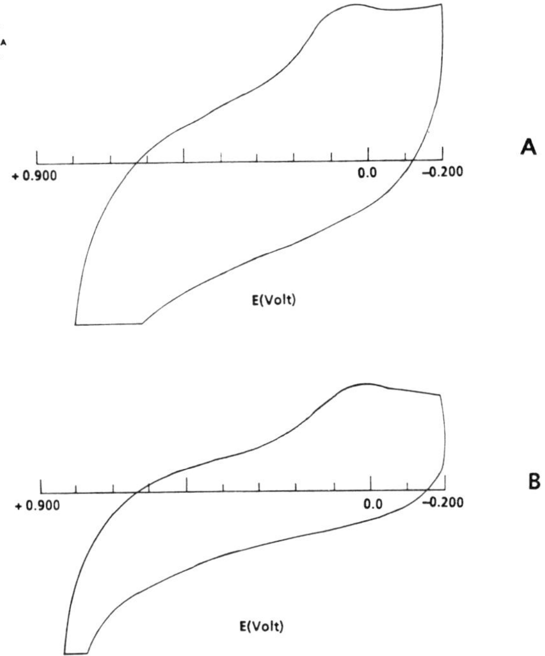

FIGURE IV. (A) Cyclic voltammogram of 1 mM catechol at bare graphite electrode. (B) Cyclic voltammogram of 1mM catechol at PNVP modified electrode.

and the polymer. The development of polymeric materials for use in more selective electrodes will require consideration of this type of interaction.

REFERENCES

Adams RN (1976): Probing brain chemistry with electroanalytical techniques. *Anal Chem* 48:1126A-1138A.

Coury LA Jr., Birch EM and Heineman WR (1988): γ-irradiated polymer-modified graphite electrodes with enhanced response to catechol. *Anal Chem* 60:553-560.

Coury LA Jr., Huber EW, Birch EM, Heineman WR (1989): Polymer-modified, carbon fiber microelectrodes as catechol sensors. *J. Electrochem Soc* 136:1044-1049.

James TL (1975): Nuclear Magnetic Resonance in Biochemistry, New York: Academic Press.

Wightman RM (1981): Microvoltammetric electrodes. *Anal Chem* 53:1125A-1134A.

FREQUENCY AND AMPLITUDE DEPENDENCE
OF THE EFFECT OF A WEAK OSCILLATING FIELD
ON BIOLOGICAL SYSTEMS

Baldwin Robertson and R. Dean Astumian

Biotechnology Division

National Institute of Standards and Technology

Gaithersburg, MD 20899

The frequency and amplitude dependences of the average rate of a reaction catalyzed by a membrane enzyme in a weak oscillating electric field are given. The average rate is a sum of Lorentz curves whose characteristic frequencies are the inverse relaxation times of the kinetic system. The amplitudes of the Lorentzians are quadratic in the field strength. Since the amplitudes can be either positive or negative, the average rate can have a variety of shapes including frequency windows. The theoretical expressions are compared with published experimental ion-transport rates catalyzed by membrane enzymes in erythrocytes and in plant protoplasts. This provides a powerful method for determining the kinetic parameters of an electrically-sensitive enzyme from measurements of the average rate of a catalyzed reaction——even when the relaxation times are much shorter than the time required for the measurements.

MEMBRANE POTENTIAL DUE TO APPLIED FIELD

When a biological system responds to an applied weak low-frequency oscillating electric field, the effect occurs in the plasma membrane of a cell.[1] At low frequencies the applied electric field is strongly concentrated in the plasma membrane and is nearly zero inside the cell. This occurs because the cytosol is a good ionic conductor while the membrane is not. When an electric field is applied across the cell, ion currents flow in the cytosol to cancel the field there. The ion currents cannot flow through the membrane, so charge builds up at the membrane surfaces producing a large field in the membrane. Thus the potential drop across a cell occurs almost entirely across the plasma membrane, and only molecules in the plasma membrane can be affected by it.

The membrane potential due to an electric field $\mathbf{E}$ applied to a cell is

$$\psi = 1.5 \, \mathbf{r} \cdot \mathbf{E}, \tag{1}$$

where $\mathbf{r}$ is the radius of the cell. The effect on the membrane potential is thus proportional to the diameter of the cell. The factor 1.5 arises because the equipotential surfaces outside the cell converge somewhat in the vicinity of the cell. An applied field of 20 V/cm changes the membrane potential of a 5 micron diameter cell by $1.5 \times (5 \text{ microns} / 2) \times (20 \text{ V/cm}) = 7.5$ mV. This approximately equals $RT/3F$, where R is the universal gas constant, T is the absolute temperature, and $F = 9.648 \times 10^4$ C/mol is the Faraday constant. The resulting electric field in a 75 Å bilayer membrane is 7.5 mv / 75 Å $= 10^6$ V/m, which is large enough to cause nonlinear effects.

ELECTROCONFORMATIONAL COUPLING

The applied field can affect a membrane protein because the conformational changes of many membrane proteins involve large displacement charge and because the membrane prevents the protein from rotating and thereby evading the effect of the field.[1] The protein in turn can affect a chemical reaction in the cell. Membrane enzymes undergo conformational changes in their catalytic cycle. The motion of charge during this conformational change provides a coupling between an electric field and the conformational state. This is the basis for the effect of an applied oscillating field on a biological system.

The effect of an applied potential on the conformational state of a protein can be described as follows. In zero potential, thermodynamic equilibrium of the conformational state is given by

$$[E^*]/[E] = K, \tag{2}$$

where $[E^*]$ and $[E]$ are concentrations of the enzyme states, and the equilibrium constant is

$$K = \exp(-\Delta G^0/RT). \tag{3}$$

When a membrane potential ψ is turned on, the energy of interaction with the conformational state is $q\psi$, where q is the net charge that moves across the potential ψ during the conformational change. It follows from thermodynamics that this interaction energy subtracts from ΔG^0 to give[2,3,4]

$$[E^*]/[E] = K \exp(q\psi/RT), \tag{4}$$

where K is the equilibrium constant (3) in zero potential. Equation (4) shows that an applied potential affects the conformational state of the enzyme.

This electroconformational coupling permits an oscillating membrane potential to affect the reactions catalyzed by the enzyme.[5] A qualitative mechanistic understanding of the effect of an oscillating potential can be obtained in the limit of large amplitude. This allows a thermodynamic derivation of the flux[6] and makes possible a physical picture of how the oscillating field causes the enzyme to cycle.[7]

FREQUENCY DEPENDENCE

When an enzyme system is suddenly disturbed by a small perturbation from a steady state, the concentration of any one of the enzyme states is a sum of exponentially decaying functions of time, as is well known from linear response theory.[8] The time constants of the exponentials are the relaxation times of the unperturbed enzyme system. When the enzyme system is perturbed sinusoidally in time, the linear response also varies sinusoidally, and it has an amplitude that is a sum of Lorentz curves. This Lorentzian frequency dependence is related mathematically to the exponential time dependence by a Laplace transformation. The time average of the sinusoidal linear response is of course zero, and so the linear response can be detected accurately only by using a rapid measuring technique.

Recently we have derived the relaxation kinetics for the effect of a weak oscillating potential on the average flux of the reaction catalyzed by the enzyme.[9] This flux is the time average of the sinusoidal linear response multiplied by a coefficient that is linear in the applied potential, and so the average flux is also a sum of Lorentz curves

$$\bar{V} = \bar{V}^{\infty} + \sum_{k=1}^{n-1} \frac{V_k}{1 + \tau_k^2 \omega^2}. \tag{5}$$

Here $\bar{V}^{\infty}$ is the average rate at infinite frequency, τ_k is the relaxation time of the kth normal mode of the unperturbed system, $\omega/2\pi$ is the frequency, V_k is the average rate for the kth normal mode, and n is the number of states of the enzyme system. The time constants τ_k are the same as for the linear response and can be calculated for a proposed mechanism.[9] Since the amplitudes V_k of the Lorentzians can be either positive or negative, frequency windows are possible as shown in Fig. 1.

The relaxation times of a plasma-membrane enzyme system can be determined from an experimentally-observed average catalyzed flux versus the frequency of an applied electric field. Tsong and coworkers[10] applied a 20 V/cm oscillating electric field to a suspension of erythrocytes and used

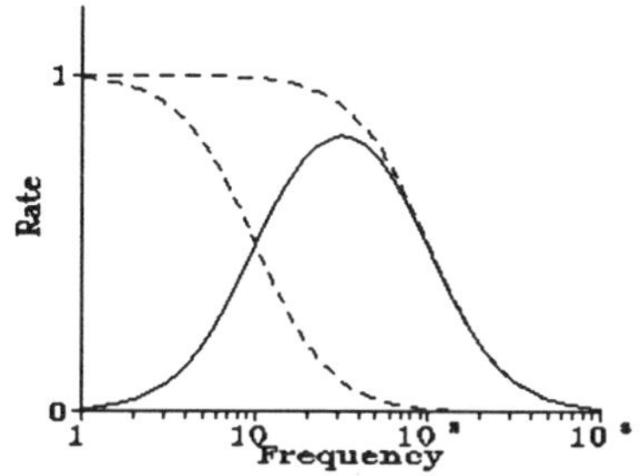

Figure **1**. Difference of two Lorentzians forming a window.

radiotracers to measure the average rates of Na^+ efflux and Rb^+ influx. The transport is catalyzed by Na^+-K^+-ATPase in the plasma membrane of the erythrocytes, and Rb^+ is an analog of K^+. A comparison of Eq. (5) with the data versus frequency is shown in Fig. 2. The upper curve is a sum of three Lorentzians, and the lower curve is a sum of two Lorentzians, both fit to the data. The error bars on the data points represent one standard deviation in the experimental uncertainty. There is only one point that deviates much from the theoretical curve, and it is within 1.7 standard deviation. The relaxation times for Rb^+ transport shown in the upper curve are 282, 265, and 0.855 μs, and the relaxation times for the Na^+ transport shown in the lower curve are 855 and 37 ns.[9]

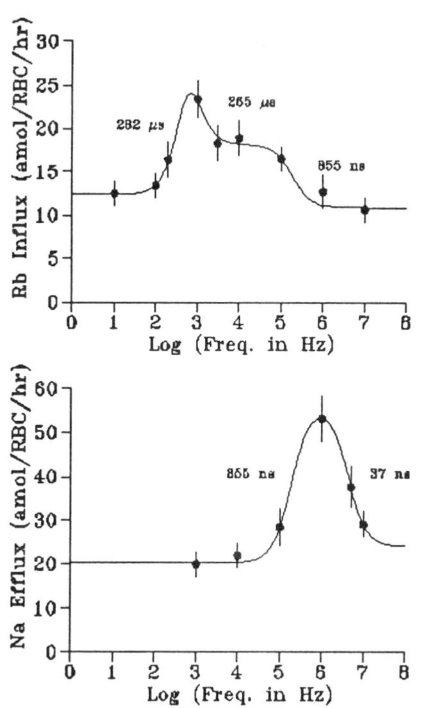

Figure **2**. Theoretical rate vs frequency compared with ion transport in erythrocytes.

Teissié and coworkers[11] used radioactive Ca^{2+} to measure the average rate of Ca^{2+} uptake by plant protoplasts in an aqueous suspension in an oscillating electric field. The upper figure in Fig. 3 shows the uptake versus frequency at constant amplitude of the oscillating field (25 V/cm). The curve is Eq. (5) with two Lorentzian terms in the sum fitted to the experimental data. The bars represent the one-standard-deviation uncertainty reported in the experimental paper. The fit gives 20 ms and 33 μs for the relaxation times of the membrane-enzyme system.

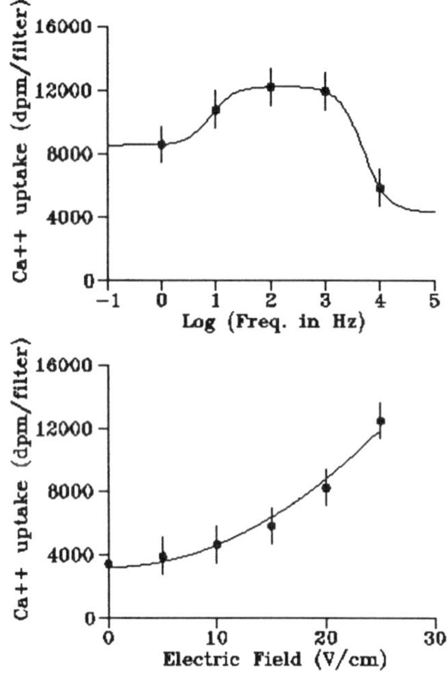

Figure **3**. Theoretical rate vs frequency and amplitude compared with ion uptake by plant protoplasts.

The agreement between Eq. (5) and the data of Fig. 3 is excellent even with only two terms in the sum. If the five data points were different, the fit might not have been as good——even though there are five adjustable parameters. One can imagine five data points that could not be fit at all well with just two Lorentzians, e.g. the five points 10, 1, 10, 1, and 10, at the five frequencies 10^0, 10^1, 10^2, 10^3, and 10^4, respectively. These five points would require the sum in Eq. (5) to contain at least four Lorentzian terms to fit them reasonably.

An enzyme system that has N distinct relaxation times requires a mechanism with at least N+1 kinetically-significant states. Thus the erythrocyte data require a mechanism with at least 5 states and the plant protoplast data require at least 3 states. By determining the relaxation times as a function of substrate and product concentrations, one can eliminate proposed mechanisms and find rate constants consistent with the data, just as in classical relaxation kinetics.[8]

AMPLITUDE DEPENDENCE

The amplitudes of the Lorentzians in Eq. (5) are each proportional to the square of the amplitude of the oscillating field. This prediction is compared with the experimental data of Teissié and coworkers[11] in the lower part of Fig. 3, which shows the effect of increasing field amplitude at constant frequency (100 Hz). The curve is a constant plus a term proportional to the square of the amplitude of the oscillating field fitted to the experimental data. The curve agrees with the data well within one standard deviation.

The fit parameters can be used to calculate the charge displacement of the plasma-membrane enzyme effected by the field. For weak fields the average rate at the peak frequency is[12]

$$\bar{V} = [1 + (zF\psi_1/RT)^2/4]\,k_{cat}[E_T], \tag{6}$$

where $k_{cat}[E_T]$ is the rate in the absence of the field, ψ_1 is the amplitude of the oscillating membrane potential, and z is the effective number of elementary charges displaced across the membrane by the conformational change. By averaging Eq. (6) over the spherical surface of the cell[13] and using the 21 μm diameter of the plant protoplasts, we get z = 4.7.

The average rate at the peak frequency also depends on substrate and product concentrations as described by a Michaelis-Menten equation.[14] This can be used to determine the kinetic constants in the same way as the usual Michaelis-Menten equation.

SUMMARY

We have described a powerful technique for determining the kinetic parameters of membrane enzymes by measuring the average rate of the catalyzed reaction in an oscillating field. This technique can be used even when the relaxation times of the enzyme system are rapid compared with the measurement time.

REFERENCES

1.   T. Y. Tsong and R. D. Astumian, Absorption and conversion of electric field energy by membrane bound ATPases. *Bioelectrochem. Bioenerg.* **15**, 457-476 (1986).

2.   J. Darnell, H. Lodish, and D. Baltimore, *Molecular Cell Biology* (W. H. Freeman, 1990) 2nd. ed., pp. 33, 539.

3.   L. Stryer, *Biochemistry* (W. H. Freeman, 1988) 3rd ed., pp. 182, 1015.

4.   D. Voet and J. G. Voet, *Biochemistry* (Wiley, 1990), pp. 49, 416.

5.   H. V. Westerhoff, T. Y. Tsong, P. B. Chock, Y.-D. Chen, and R. D. Astumian, How enzymes can capture and transmit free energy from an oscillating electric field. *Proc. Natl. Acad. Sci. USA* **83**, 4734-4738 (1986).

6.   V. S. Markin, T. Y. Tsong, R. D. Astumian, and B. Robertson, Energy transduction between a concentration gradient and an alternating electric field. *J. Chem. Phys.* **93**, 5062-5066 (1990).

7.   B. Robertson and R. D. Astumian, Kinetics of a multistate enzyme in a large oscillating field. *Biophys. J.*, **57**, 689-696 (1990).

8. M. Eigen and L. DeMaeyer, in *Techniques of Chemistry*, Vol. VI, Part II, *Investigation of Rates and Mechanisms of Reactions*, edited by G. G. Hammes (Wiley-Interscience, 1974), pp 63-146.

9. B. Robertson and R. D. Astumian, Frequency dependence of catalyzed reactions in a weak oscillating field. *J. Chem. Phys.* **94**, 7414-7419 (1991).

10. D.-S. Liu, R. D. Astumian, and T. Y. Tsong, Activation of Na^+ and K^+ pumping modes of (Na,K)-ATPase by an oscillating electric field. *J. Biol. Chem.* **265**, 7260-7267 (1990).

11. A. Graziana, R. Ranjeva, and J. Teissié, External electric fields stimulate the electrogenic calcium/sodium exchange in plant protoplasts. *Biochemistry* **29**, 8313-8318 (1990).

12. B. Robertson and R. D. Astumian, Interpretation of the effect of an oscillating electric field on membrane enzymes. Submitted to *Biochemistry*.

13. R. D. Astumian and B. Robertson, Nonlinear effect of an oscillating electric field on membrane proteins. *J. Chem. Phys.* **91**, 4891-4901 (1989).

14. B. Robertson and R. D. Astumian, Michaelis-Menten equation for an enzyme in an oscillating electric field. *Biophys. J.* **58**, 969-974 (1990).

LARGE VOLUME CELL ELECTROPERMEABILIZATION AND ELECTROFUSION BY A FLOW PROCESS

Justin Teissié,   Sophie Sixou and Marie Pierre Rols
Centre de Recherches de Biochimie et de Génétique Cellulaires du CNRS
Toulouse, France

INTRODUCTION

Since the pionneering work of Neumann and Rosenheck in 1972, cell electropulsation is a very efficient way to permeabilize the cell membrane while keeping the cell viability (Kinosita and Tsong, 1977; Zimmermann, 1982). The electric field when higher than a certain threshold allows the membrane to become permeable to external molecules. In 1982, Neumann et al. reported that plasmids can be integrated in the genome of the host cell when a cell-DNA mixture was pulsed (Neumann et al., 1982). This physical method, called "electrotransformation", is under the control of various parameters inherent to the technique and easily adjustable. Cell fusion is routinely used for the obtention of somatic hybrids. This process can be achieved by chemical or viral means but electrofusion, introduced in 1979 (Senda et al, 1979)) appears now as a more efficient technique. It is taken advantage of the observation that strong electric field pulses are able not only to reversibly permeate the cell membrane but, if cells are in close contact, the transient permeability is going to disappear in an intercellular process leading to the cell fusion. Different approaches have been described in order to create the cellular contact (Neumann, Sowers and Jordan, 1989). In our laboratory, we have been pionneering a technique where anchorage dependent cells were pulsed directly on the culture dish (Teissié et al, 1982). When cells are growing in suspension, it was postulated that cells must be in close contact when the field was applied in order to induce the fusion. The field was directly involved in the initial steps of the membrane fusion (Zimmermann, 1982). Events occured as described in Fig.1 A. This concept was proved not to be true (Sowers, 1986; Teissié and Rols, 1986). The electropermeabilization of the cell membrane induced a new fusogenic state of the cell. If the contact is created after the pulse, cell fusion was observed. The process occurred as described in Fig.1 B.

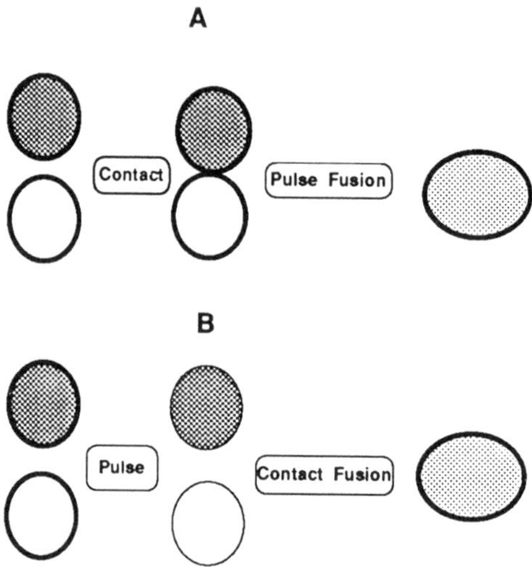

Figure1: Two operational pathways lead to electrofusion

A- In the classical approach, the cells must be brought into contact before the pulse in order to fuse. The perturbations present during the application of the field are the cause of the membrane fusion which gives the cell fusion.

B- In the new approach, the permeabilization induced by the field is associated to a long lived new organization of the membrane. When this new structure is present the repulsion between cells do not exist anymore and cells will fuse if they are brought into contact.

These different techniques from electropulsation are routinely used at the laboratory scale due to the availability of several commercial electropulsators. However, the electro-treated volume is generally small (several microliters) and the method has not until now been applicable to large volumes often required in biotechnology. This drawback is due to technical reasons. The power of the generator is the major limit.

The power P of a generator is given by :

$$P = U^2/R,$$

where U is the applied voltage and R , the resistance of the sample. As:

$$R = d / \lambda S$$

where λ is the conductivity of the medium, d is the distance between the electrodes and S their section (with the assumption of two parallel flat electrodes), that leads to :

$$P = U^2 \lambda S/d, \text{ and then :}$$

$$P = E^2 \lambda V,$$

where E is the electric field and V, the sample volume. As P , E and λ are all experimental constants, the experimental conditions impose the volume V which in the microliter scale in most cases. It appears necessary to develop another technology to treat large volumes of cell suspension. We have designed a flow method of electropulsation that allows to pulse continuously several milliliters of cells per minute.

After optimizing the methodology by a careful selection of the physical parameters, we are reporting in this communication that electropermeabilization and electrotransformation can be obtained by the use of this new flow technique. This technology was applied for cell fusion on large volumes. When working with anchorage dependent cells, the contact inhibition approach was followed by using cells grown on microcarriers. For cells in suspension, the post pulse contact methodology was proved to be efficient.

MATERIALS and METHODS

Cell culture and plasmid.

Chinese hamster ovary cells

We selected the wild-type Chinese hamster ovary (CHO) cell line (WTT clone) which is not strictly anchorage dependent. It has been adapted for suspension culture at 37°C under gentle agitation (100 rpm) in Eagle's minimum essential medium (MEM 0111, Eurobio, France) supplemented with 8% new-born-calf serum (Boehringer, Germany), penicillin (100 units/ml), streptomycin (100 mg/ml) and L-glutamine (0.584 mg/ml). Cells were maintained in the exponential growth phase ($4-10 \times 10^5$ cells/ml) by daily dilution of the suspension. Cells grown in suspension can be replated readily either on Petri dishes (35 mm diameter, Nunc, Denmark) and kept at 37°C in a 5% CO_2 incubator (Jouan, France) or on microcarriers (Biosilon, Nunc, Denmark) under gentle agitation (30 rpm) in a spinner flask (Tecne, U.K.) and kept at 37°C following the manufacturers instructions. For fusion, cells were grown during about 5 days in order to obtain a confluent culture on the beads with a lot of cellular contacts.

Blood cells

Blood samples were kindly provided by Dr Tarbes (CTS Rangueil, Toulouse, France). Erythrocytes were obtained by centrifuging blood samples (2000 g, 5 min., room temperature) and removing supernatant. Leucocytes were used as "leucocytes concentrates" : after elimination of most of the erythrocytes by centrifugation of blood samples (1100 g, 30 min.), a leucocytes and erythrocytes mixture in a 1/100 ratio was obtained.

Plasmid

Plasmid pUT531 used to transform CHO cells is a pBR322 shuttle vector that carries the SH-GAL fusion gene under the control of the SV40 early promoter and as such codes for the β-galactosidase activity. It was a gift from Prof. G. Tiraby (D. Stasi and G. Tiraby, unpublished). It is extracted by standard procedures (Birnboim and Doly, 1979).

Cell electropulsation.

Cells in suspension were centrifuged for 5 min at 350 g (1000 rpm, Jouan C500 centrifuge, France) and resuspended in the pulsing buffer PB (10 mM Phosphate buffer, pH 7.5, 1 mM $MgCl_2$, 250 mM sucrose) at a concentration of 10^7 cells per ml . Under our conditions, the cell density was similar for cells in suspension or cells on microcarriers. When large cell volumes were treated, a

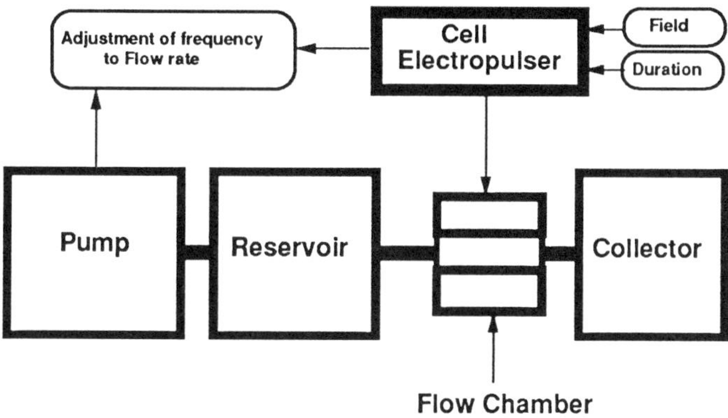

Figure 2: Flow system of the electrofuser

The flow is generated by the depression created by the pump, the tubing system between the reservoir and the pump being air-tight.

----

flow system was used (Figure 2). The flow was obtained from the depression created by a peristaltic pump (Gilson, France). Cells flowed between two parallel stainless-steel electrodes, 4 mm apart, which were the walls of a plexiglas flow-through chamber, the volume of which was 128 µl. The electrodes were connected to a voltage generator which gave square-wave electric pulses (CNRS cell electropulser, Jouan, France). By means of an electronic trigger, a repetitive train of electric pulses with a controlled duration in the micro-millisecond range were applied to the aqueous suspension. The shape of the pulses was a square wave. The voltage pulses applied to the cells were monitored with an oscilloscope incorporated into the cell electropulser. The frequency F of pulses (Hz) was chosen in order to apply a set number N of pulses to the cells according to the relationship :

$$\Phi = (F \times 60 \times V)/N$$

Where Φ was the flow of the pump (mL/min) and V the volume of the pulsation chamber (mL). Cells were pulsed at a 10 Hz frequency, at a flow of 7 ml/min that corresponded to 10 pulses applied to the cells during their residence in the pulsation chamber. The flow could be as large as 36 mL/min with our electropulsator. Sterility of the flow system was obtained by washing the system with a flow of bleach for 20 minutes and then rinsing with sterile PB. The field strength could be as high as 3 kV/cm in this configuration. The accumulation of heat on the sample due to Joule heating was calculated to be negligible. Electropulsing was carried out at 21°C under a laminar flow hood to work under sterile conditions.

Electropermeabilization procedure

Electropermeabilization of cells was quantified by penetration of unpermeant dye (Teissié and Rols, 1988). Penetration of trypan blue (T 0887, 4 mg/ml in the pulsing buffer;Sigma, USA) was used to monitor permeabilization. Cells were pulsed, incubated 5 min at room temperature and then observed under an inverted light microscope (Leitz, Germany). The percentage of permeabilized cells was 100 times the ratio of the number of blue stained cells to the total number of cells and was evaluated by observing 300-500 cells.

A slightly different procedure was followed with blood derivates. Experiments were performed at 21°C in the following low ionic content saline buffer : 125 mM sucrose, 69 mM KCl, 1 mM $MgCl_2$ in 10 mM phosphate buffer pH = 7.4. This "pulsing buffer" (PBB) was added after eliminating culture medium or plasma by centrifugation (100 g, 4 min. or 2000 g, 5 min. respectively at room temperature). Cells were washed twice in PBB. For experiments on blood samples, cells were directly pulsed in plasma. Permeabilization was assayed by Trypan Blue penetration in the case of leucocytes and by hemolysis for erythrocytes.

Electrotransformation procedure.

Cells were incubated at room temperature in pulsing buffer containing the plasmid (25 µg/ml) 5 minutes before electropulsation. Microcarriers were washed 3 times with PB and then resuspended in a volume of plasmid containing PB equal to the volume of beads. The surface (mm^2) to volume (ml) ratio was about 15.10^3 as compared to a value of 10^3 for a Petri dish. After a 5 minute incubation following the pulses, culture medium was added and cells were cultured for 24 hours in a 5% CO_2 incubator at 37°C for the expression of the electrotransfered activity. Cells were then stained in order to detect the expression of the electrotransfered β-Galactosidase activity. They were fixed in PBS buffer containing 2% formaldehyde and 0.2% glutaraldehyde 5 minutes at room temperature. Activity revealing solution was then added (1 mg/ml of X-Gal, the substrat of the enzyme, 4 mM potassium ferricyanide, 4 mM potassium ferrocyanide and 2 mM magnesium chloride in PBS buffer pH 7.2). Cells were incubated at 37°C and the cells expressing the β-Galactosidase activity appeared deep blue within one hour of incubation. Transformation efficiency is defined as the ratio of blue cells over the total number of surviving cells. Again 300-500 cells were observed under an inverted microscope.

Determination of electropulsed cell viability.

Cells were pulsed in the pulsing buffer in absence of dye under the same conditions as for permeabilization assays. They were kept for 5 minutes at room temperature and the pulsing buffer was discarded and replaced by 2 ml of culture medium. Viability was measured by observing the growth of cells over 48 h (about 2 generations) under a phase-contrast inverted light microscope. In the case of leucocytes, the membrane integrity was used as an assay.

All experiments were repeated at least three times at 2- or 3-day intervals in order to avoid possible fluctuations due to different physiological states of the cells. Reproducible results were obtained in all cases.

Electrofusion Procedure

Before electropulsation, the culture medium was substituted with a pulsing buffer PB (10 mM Phosphate buffer, pH 7.5, 1 mM MgCl2, 250 mM sucrose) (Teissié and Rols, 1988). This medium was proved to be very convenient for the electrofusion of plated CHO cells (Blangero and Teissié, 1985). This change is easily operated in the case of microcarriers by letting the beads settle by sedimentation

The pulsed cells were collected in a Nalgene container (Polylabo, France) which was cooled with a ice bucket. Immediately after the pulsation, the cells were centrifuged (25 g, 5 min., room temperature, Jouan C500, France) and the pellet was incubated (30 min., 37°C). It should be emphasized that the centrifugation speed must be carefully controlled in order to avoid lytic effects . The microcarriers were collected in a Nalgene container (Polylabo, France) and were incubated during 1 hour at 37°C. Cells were observed to be able to grow during several days without any problem after this treatment and we considered this observation as a good evidence that their viability was not affected by the electrical pulsations.

 Fusion and viability were controlled by plating the cells (after trypsinization in the case of microcarriers) and observing them under the microscope.

The extend of fusion was quantitated by the polynucleation index I:

$$I = \left(_{n=2}\Sigma^{n}C_n\right)/\left(_{n=1}\Sigma^{n}C_n\right)$$

where C_n was the number of cells where n nuclei were present in the cytoplasm. A more precise description was given by the histogram of polynucleation.

RESULTS

Electropermeabilization by a flow procedure :

CHO cells can be electropermeabilized either grown plated or in suspension (Teissié and Rols, 1988). More interestingly, this can be obtained on cells growing on microcarriers (Teissié and Conte, 1988). In these electropulsation experiments, the volume of treated cells is related to the size of the pulsation chamber (i.e. the geometry of the electrodes, the distance between them and their length). The physical (and economical) reason of this limit is the power of the pulse generator as explained in the introduction. Large volumes of cells can be treated by using the pulsation flow technique. A mixture of CHO cells and Trypan Blue was continuously pulsed. The extent of permeabilization (as assayed by the blue staining) as well as the post pulse viability (assayed by the ability to grow) are similar in the batch and in the flow procedures (Fig. 3. and Teissié and Rols, 1988).

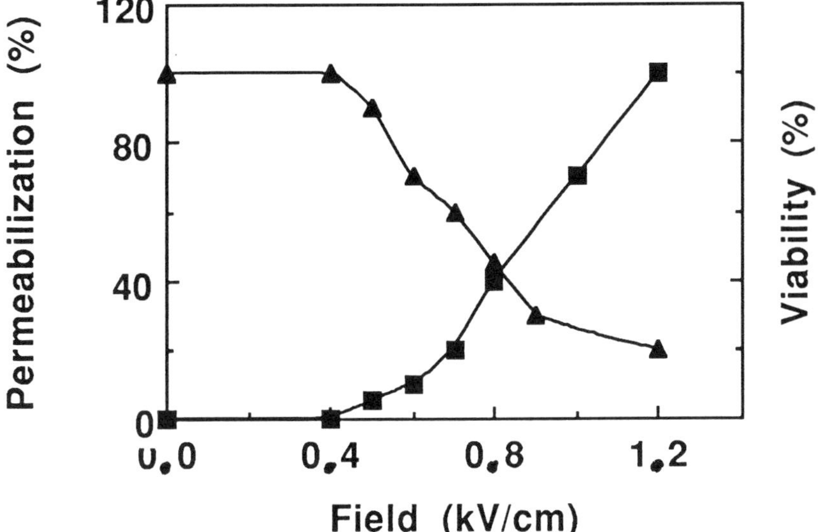

Figure 3: Effect of electric field strength on viability and permeabilization of CHO cells in suspension in flow.

Cells were pulsed 10 times, 5 ms duration. Viability (Δ) and permeabilization ( ) were measured as described in "Methods". Cells were pulsed in large volumes (i.e. several milliliters, flow method, flow rate : 7ml/min).

Cell electropermeabilization is physically size dependent. When working with a mixed cell population, it is possible to specifically permeabilize large cells while leaving the smaller ones unaffected (Sixou and Teissié, 1990). In the design of new drug vehicles, such a property would be of great value by opening the opportunity for targeting. The flow technique was checked to be prone to such applications.

In a first approach, a mixture of CHO cells and erythrocytes in PBB was treated by synchronously pulsing a flow of cells, pulses lasting 100 µs with a frequency of 10 Hz. The flow rate was 11 ml/min resulting in 7 pulses were applied for each cell. The size specific permeabilization occurs as in batch process. It was checked that the stress due to the flow was not lethal to pulsed cells by following the growth of cells (control and pulsed cells) for 24 hours. Viability was not affected up to 1.5 kV/cm. Consequently, by applying 10 pulses lasting 100 μs with a field intensity of 1.4 kV/cm, 100% of CHO cells are permeabilized without alteration of their viability and erythrocytes are not haemolysed.

But a more pharmacological assay is to pulse blood samples. The target cells would then be leucocytes. The procedure must be such as to leave the red blood cells unaffected. Cells were directly pulsed in plasma and permeabilization rates were recorded (figure 4).

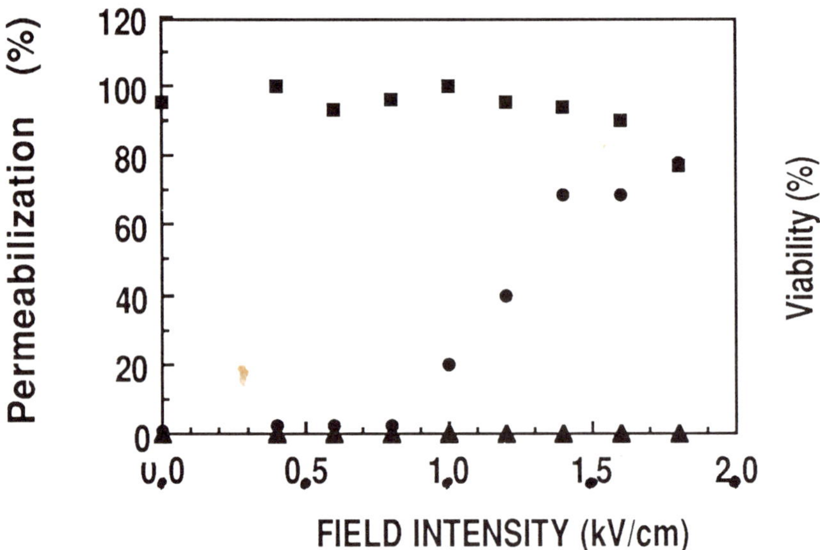

Figure 4: Permeabilization curves of leucocytes and erythrocytes. Leucocytes permeabilization (•) was quantitated by counting blue-stained cells and percentage of haemolysis (▲) was the permeabilization indicator for erythrocytes. Leucocytes viability (■) was measured after 24 hours by the trypan blue test.

Results obtained with the flow system were similar with those of batch experiments. Leucocyte viability decreased only above 1.6 kV/cm. At this field value, 70% of the leucocytes (polymorphs and monocytes) are permeabilized without alteration of viability. No hemolysis of erythrocytes was observed.

Flow electropulsation allows to transform large volumes of cells.

The flow technique can be used on a mixture of cells and plasmids. Transient expression of the coded activity was obtained in our approach. Under 10 pulses (average value) of 5 ms duration and 0.6 kV/cm intensity, (frequency 10Hz) i.e. for a flow of 7 milliliters per minute, we obtain a transformation efficiency of 25% while keeping the cell viability rather unaffected. Application of the flow technique to cells cultured on microcarriers also results in transformation of cells. Under the same electric parameters conditions, we observe that 20% of the cell population is transformed.

Under the best conditions, transformation efficiencies up to 60% for cells growing in suspension and in monolayers are obtained with up to 60-70% of cell viability. Under the same electric field parameter values, these efficiencies decrease only by a factor 2 when the pulsation is performed using the flow system, the viablity being affected to the same extend. This can be explained by the tumbling of the cells in the flow leading to application of the pulses under different orientations during the time of residence of the cells between the electrodes. Consequently, different parts of the cell surface are permeabilized because of the vectorial character of the effects of electric fields on cells (Bernhard and Pauly, 1973). As a consequence, the positive effect on gene transfer of the pulse accumulation at the same part of the cell surface is lost, giving a slight decrease in efficiency.

Electrofusion

1- Electrofusion of cells growing on microcarriers

Electropulsation of a flow of microcarriers triggered the fusion of cells attached on the beads. This is clearly shown by the direct observation of cells under a microscope and by the polynucleation histogram of the two strains.

A systematic investigation of the dependence of the electrofusion on the parameters of the electric pulsation was done. Fusion was observed only when the field intensity was large enough (Fig.5). The threshold values were specific of the strains; field intensities must be larger than 0.8 kV/cm for CHO . A plateau value of the fusion index was present for large field intensities (more than 1.5 kV/cm).

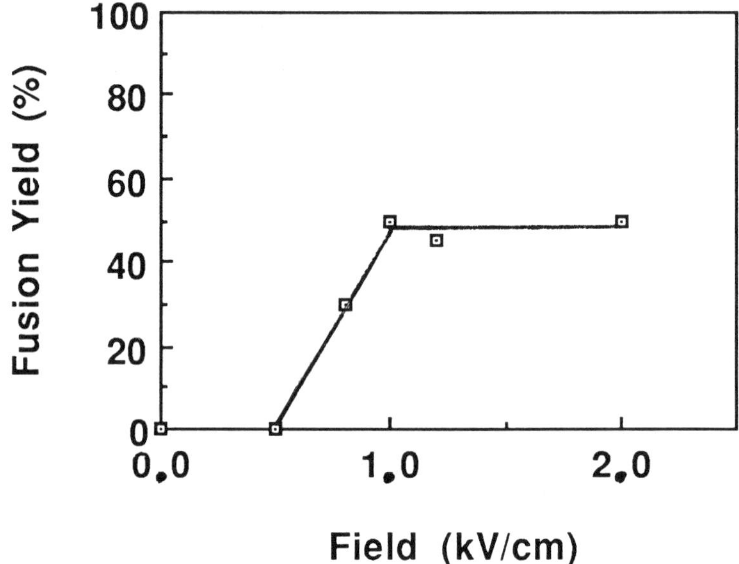

Figure 5: Fusion index of cells growing on microcarriers as a function of the field intensity

 Cells grown on Biosilon were pulsed at a frequency of 5 Hz (duration: 100 µs, flow rate: 5 mL/min)

The fusion index wa under the control of the pulse duration and very short pulses could not induce the fusion . The threshold value was clearly larger than in the case of cells growing on dishes.

The number of successive pulses which were applied on the cells was dependent on several parameters: the delay between the pulses, the flow rate and the volume of the electropulsation chamber. The frequency of the pulses was changed between 10 and 1 Hz to increase the number of applied pulses. The fusion yield was clearly increased by an accumulation of successive pulses in contradiction with cells growing on dishes (Blangero and Teissié, 1983), where 2 pulses were enough for the obtention of a high fusion index. In fact, we already showed that the fusion index is dependent on the relative geometry between the cells and the field (Teissié and Blangero, 1984). As the microcarriers were tumbling, the number of different orientations of cells relative to the field was going to increase with the number of pulses. This explanation was in agreement with the observation of highly polynucleated cells, which was predicted by the vectorial character of electrofusion (Teissié and Blangero, 1984). In some experiments, viable cells with 100 nuclei in their cytoplasm were observed. They were linked to the cooperative fusion of clusters of cells growing on the same microcarriers and pulsed successively under very different orientations.

2- Electrofusion of CHO in suspension by the flow procedure

A train of electric field pulses was applied on a flow of CHO cells. The parameters were choosen from the data observed in the batch experiments (field intensity: 2 kV/cm; pulse duration: 100 µS). The frequency of the pulses and the flow were adjusted in such a way that only 5 pulses were applied on each cell (frequency: 2 Hz; flow rate: 7.5 mL/min). A fusion index of 35% was routinely observed and the polynucleation histogram was indicative of the occurence of highly polynucleated cells. Taking into account the tumbling of the cells in the flow, the pulsations were applied under different orientations during the time of residence of the cells between the electrodes. Consequently, different parts of the cell surface were permeated because the electric field induced permeation loci were shown to be fixed in position on the surface (Sowers, 1986). In other words, the fusogenic part of the cell surface is increased. As these loci must face each other on two cells in contact to induce their fusion, these repetited permeabilizations of the cell surface appeared to facilitate the fusion of the cells during their centrifugation by increasing the probability of contact between the loci.

The stress due to the flow was only slightly lethal for the pulsed cells. More than 75 % of the pulsed cells were checked viable by observing their growth. Contacts which were needed for fusion were created only during the centrifugation step. In order to be definitively sure that no contact was present during the pulse, we observed that the fusion yield was not modified when using a dilute cell suspension (10^5 cells/mL). The post pulsation centrifugation step created the high density, a procedure avoiding repetitive centrifugations. Nevertheless, it was not possible to avoid a pre-washing of the cells because the fusion process appeared strongly modulated by the ionic content of the pulsation medium as with plated CHO cells (Blangero and Teissié, 1985). Pulsation media with different ionic contents were obtained by the use of mixtures between the culture medium and the pulsation medium. An increase of the ionic content in the pulsation medium decreased the observed fusion index. A low ionic content pulsing buffer was needed (Teissié and Rols, 1988).

CONCLUSIONS

Flow technology opens new perspectives for electropulsation in Biotechnology. Electropermeabilization allows the release of cytoplasmic molecules with limited loss in viability. Connecting flow electropulsing chambers to fermentors would permit an increase in the recovery of metabolites.

Electropermeabilization appears as a powerful tool in drug delivery. We describe electrical conditions suitable to allow the specific electroloading of leucocytes in a blood sample and treatment of large volumes of cells under conditions leaving them viable. Interest concerning the use of permeabilized cells as drug vehicles has up to now been focused on erythrocytes (Kinosita and Tsong, 1977) or platelets (Hughes and Crawford, 1989). It was needed to isolate cells from a blood sample, to load them and then to inject them back to the donor. New prospects are opened with the ability to specifically electropermeabilize (electroload or electrotransform) leucocytes directly in blood samples. The flow system can be adapted to an extracorporeal circulation in order to treat the whole blood volume. Our laboratory is currently studing the possibility of using leucocytes (polymorphs and monocytes) electroloaded with an antibiotic or an antiinflammatory drug as efficient delivery vehicles to sites of infection.

In the case of electrotransformation, the number of transformed cells is linearly related to both the number of pulsed cells and the concentration of plasmids (Förster and Neumann, 1989). Obtaining transformants with only one transfered copy requires a low plasmid concentration i.e. at the expense of the number of transformants. One obvious advantage of the flow technique is the increase in the quantity of pulsed cells. A compromise is provided by the flow technique where a large number of cells can be pulsed with a low plasmid concentration leading to a large number of transformed cells with a single plasmid copy.

Large cell volume electrofusion is obtained for anchorage dependent cells (use of microcarriers) as well as for cells in suspension. Taking into account the observed dependence of the fusion on the experimental parameters (strength, duration, number of pulses), it can be calculated that 10^8 cells can be treated per minute with the obtention of fusion index of 50%. This data proves that, with this approach, electrofusion can be used with cell volume on an industrial scale. These experiments with microcarriers took advantage of our knowledge on the electrofusion of cells growing on culture dishes (Teissié and Blangero, 1984). As predicted by the technology of microcarriers, electrofusion is obtained under conditions rather similar to the case of dishes (Blangero and Teissié, 1983). The main difference is the need for stronger fields in order to induce the fusion. This can be linked to small differences in the morphology of the cells when they are growing on the microcarriers. The dependence on the number of pulses is in fact a direct consequence of the vectorial character of the interaction between electric field and cells in electrofusion (Teissié and Blangero, 1984). The good condition of the fused cells was clearly established by the experimental observation that they were able to be treated by trypsin and then to be plated again.

The results with cells in suspension are one more evidence of the induction of a long lived electroinduced new organization of the cell membrane. It is associated with a permeabilization of the matrix but leads to a membrane fusion and to an associated cell fusion when the electropulsed cells are brought into contact. It can be called electrofusogenic.

A common feature of all results obtained by the flow approach when compared to the batch technique is the advantage provided by accumulation of pulses. When cells are fixed relative to the electrodes, increasing the number of pulses is of limited advantage (except for transformation). With the flow, as they are tumbling, different parts of the cell surface are exposed to the field effect even with reduced field strengths. This increase in the treated cell surface brings a lot of benefits in the practical results of electropulsation.

A definitive advantage of these flow methods is the good conditions of sterility. All operations are performed under a laminar flow hood.

Acknowledgments.

The authors are indebted to Dr D. Stasi and Prof G. Tiraby for providing the plasmid and wish to thank Mrs J. Zalta for her assistance in cell culture, Mr. Trotard and Greyze for their technical assistance, Ms Caparos and Mr D. Coulet for preliminary experiments This work was supported by the CNRS and by grants of the MRT ("Essor des Biotechnologies"), the ANVAR ("Aide à l'innovation"), the "Région Midi-Pyrénées" and the AFLM (to J. T.). Dr M.P. Rols and Dr S. Sixou were recipients of fellowships of the Fondation pour la Recherche Médicale and of the AFLM respectively.

REFERENCES

Bernhardt J and Pauly H (1973): On the generation of potential difference across the membranes of ellipsoidal cells in an alternating field. Biophysik (Berlin)10 : 89-98

Birnboim MC and Doly DJ (1979): A rapid alkaline extract procedure for screening recombinant plasmid DNA. Nucl. Acid. Res. 7:1513-1523.

Blangero C and Teissié J (1983): Homokaryon production by electrofusion: a convenient way to produce a large number of viable mammalian fused cells. Biochem.Biophys.Res.Comm. 114: 663-669

Blangero C and Teissié J (1985): Ionic modulation of electrically induced fusion of mammalian cells. J.Membr.Biol. 86: 247-253

Förster W and Neumann E (1989): Gene transfer by electroporation: A practical guide. *In: Electroporation and electrofusion in cell biology.* Neumann E, Sowers AE and Jordan CA (Eds). Plenum press, New York.

Hughes K and Crawford N (1989): Reversible electropermeabilization of human and rat blood platelets: evaluation of morphological and functional integrity in vitro and in vivo. Biochim.Biophys.Acta, 981: 277-287.

Kinosita K and Tsong TY (1977) : Hemolysis of human erythrocytes by transient electric fields. Proc.natl.Acad.Sci. USA 74 :1923-1927

Neumann E and Rosenheck K (1972) Permeability changes induced by electric impulses in vesicular membranes. J. Membr. Biol. 10: 279-290

Neumann E, Schaefer-Ridder M, Wang Y and Hofschneider PM (1982): Gene transfer into mouse lyoma cells by electroporation in high electric field. EMBO J. 1:841-845.

Neumann E, Sowers AE and Jordan C (1989) in "Electroporation and Electrofusion in Cell Biology", New York, Plenum Press

Senda M, Takeda J, Shunnosoke A and Nakamura T (1979): Induction of cell fusion of plant protoplasts by electrical stimulation . Plant cell Physiol. 20: 1441-1443

Sixou S and Teissié J (1990): Specific electropermeabilization of leucocytes in a blood sample and application to large volumes of cells. Biochim. Biophys. Acta. 1028:154-160.

Sowers A.E. (1986): A long lived fusogenic state is induced in erythrocytes ghosts by electric pulses. J.Cell Biol. 102: 1358-1362

Teissié J and Blangero C (1984): Direct experimental evidence of the vectorial character of the interaction between electric pulses and cells in cell electrofusion. Biochim.Biophys.Acta 775: 446-448

Teissié J and Rols MP (1986) : Fusion of mammalian cells is obtained by creating the contact between cells after their electropermeabilization. Biochem.Biophys.Res.Comm. 140: 258-266

Teissié J and Conte P (1988): Electrofusion of large volumes of cells in culture. Part I: anchorage-dependant strains. Bioelectrochem. Bioenerg. 19: 79-57.

Teissié J and Rols MP (1988): Electrofusion of large volumes of cells in culture. Part II: cells growing in suspension. Bioelectrochem. Bioenerg. 19: 59-66.

Teissié J and Rols MP (1988): Electropermeabilization and electrofusion of cells. In: Dynamics of membranes proteins and cellular energetics. Latruffe N, Gaudemer Y, Vignais P and Azzi A (Eds.). Springer verlag, Berlin, Germany.

Zimmermann U (1982): Electric field fusion and related electrical phenomena. Biochim.Biophys.Acta 694 : 227-277

ELECTRIC FIELD INDUCED ASYMMETRIC BREAKDOWN OF CELL MEMBRANES

Ephrem Tekle and P. Boon Chock
Section on Metabolic Regulation, Laboratory of Biochemistry,
National Heart, Lung and Blood Institute, NIH, Building 3,
Room 202, Bethesda, MD 20892.

R. Dean Astumian
Biotechnology Division, National Institute of Standard and
Technology, Building 230, Room 105, Gaithersburg, MD 20899.

ABSTRACT

Electric field of sufficient magnitude and duration
generates an induced local membrane potential, $V_m(R, E, \theta)$,
ultimately responsible for the formation of transient pores
in cell membranes (Chang and Reese, 1990). Earlier
experimental studies of this process revealed the *uptake* of
an indicator by plant cells occurs predominantly through the
membrane site facing the anode (Mehrle et al., 1985, 1989)
while in erythrocyte ghosts, the indicator probes *exit*
through the site facing the cathode (Sowers, 1988; Dimitrov
and Sowers, 1990). To reconcile these observations,
symmetrical pore formation and a mechanism of molecular
exchange by elecroosmosis has been proposed (Dimitrov and
Sowers, 1990). Here we present experimental evidence
supporting i) Membrane pores are formed either on one side or
asymmetrically when the perturbing field is unipolar. ii)
Symmetrical pores are only formed with bipolar fields. iii.)
Electroosmosis cannot be a major contributor to the molecular
exchange mechanism in electropermeabilized cells. In the
lower electric field regime of our study, the asymmetric pore
formation can be explained (both qualitatively and semi-

quantitatively) with the vectorial interaction of $V_m(R, E, \theta)$ with the resting (i.e intrinsic) membrane potential V_i. At excessively large electric fields, however, the observed asymmetry may derive from other factors (Tekle et al., 1990).

INTRODUCTION

Electropermeabilization of cell membranes has become a reliable and simple technique primarily for genetic transformation of a variety of mammalian, bacterial, yeast, and plat cells. In recent years, improvement in the efficiency of the method have been derived mainly based on manipulation of various experimental parameters (Chu et al., 1987; Potter et al., 1984; Yorifuji et al., 1989; Tatsuka et al., 1988), e.g. temperature, buffer system, plasmid type or electrical variables (Chang, 1989; Tekle et al., 1991) e.g. frequency, waveform. On the other hand, fundamental problems regarding the mechanism of electric field induced pore formation and the subsequent fate of these pores after the termination of the electric field are less well understood. In the present paper, we focus on some aspects of this general mechanistic problem, namely, the localization of membrane pores in a spherical cell subjected to a brief homogeneous electric field and the subsequent molecular exchange processes between intra and extra cellular compartments.

Figure I shows (after Neumann and Rosenheck, 1973) cross section of a spherical cell placed between two parallel electrodes (see Figure captions for meaning of symbols). In the presence of an external electric field, a cross membrane potential V_m is induced whose amplitude is dependent on the parameters given by

$$V_m = f(\lambda_i, R_i, d, E, \theta) \tag{1}$$

Considerable simplification to eq.(1) occurs with the assumptions that $\lambda_1 \approx \lambda_3 \gg \lambda_2$ and $d \ll R_1 \approx R_2$. These conditions are easily fulfilled in living cells and eq.(1) can thus be written as (for details see Neumann, 1989)

$$V_m \cong f(R, E, \theta) = 3/2\ R\ E\ \cos\theta \tag{2}$$

When the applied electric field is of a certain frequency, f, then the amplitude of V_m is attenuated by a factor $1/(1 + (2\pi f\tau)^2)^{1/2}$, where τ (of the order of 10^{-6} sec) is the charging time of the membrane (Schwan, 1983). According to eq.(2), the maximum amplitude of V_m occurs at the two poles of the membrane facing the electrodes (i.e. $\cos\theta=1$ for $\theta=0$ and -1 for $\theta=\pi$), thus symmetrical permeabilization of the membrane

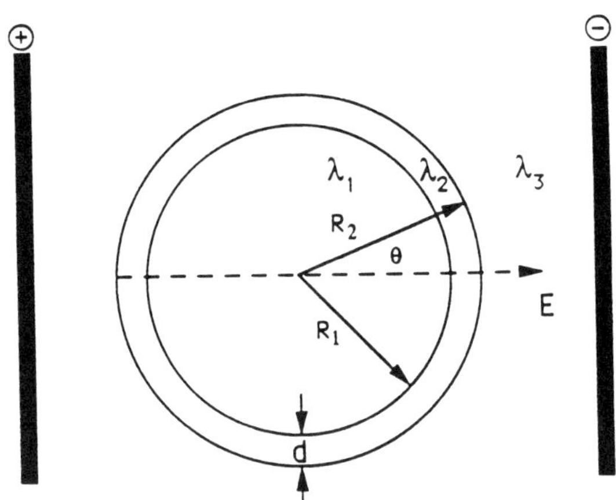

FIGURE I. Schematic illustration of the cross section of a model cell placed between two parallel electrodes. R_1, R_2 are the inner and outer radii respectively; θ, the angle between the electric field direction E (denoted by the dotted lines) and a point on the membrane; d, membrane thickness and λ_1, λ_2, λ_3, the conductivities of the intercellular, membrane and extracellular regions.

is expected. Here we demonstrate that in the low field regime (i.e. near critical membrane breakdown potentials, V_c), membrane pore is essentially asymmetric when perturbing field are unipolarly directed. Symmetric permeabilization occurs only with bipolar fields. These results indicate the resting membrane potential plays a role in electropermeabilization of cell membranes. Time resolved measurements further show that electroosmosis cannot be invoked to explain the dominant molecular exchange process in electropermeabilzed cells. At excessively large electric fields, the permeabilization pattern at the two membrane sites facing the electrodes appears asymmetric. Here, however, polarization of the positively charged probe and hence its depletion from the membrane site facing the negative electrode may be responsible.

MATERIALS AND METHODS

Electric pulse apparatus. For the purpose of this study, we have assembled an electric pulse apparatus capable of producing several type of waveforms with frequencies up to 1 MHz. The block diagram is shown in Figure II.

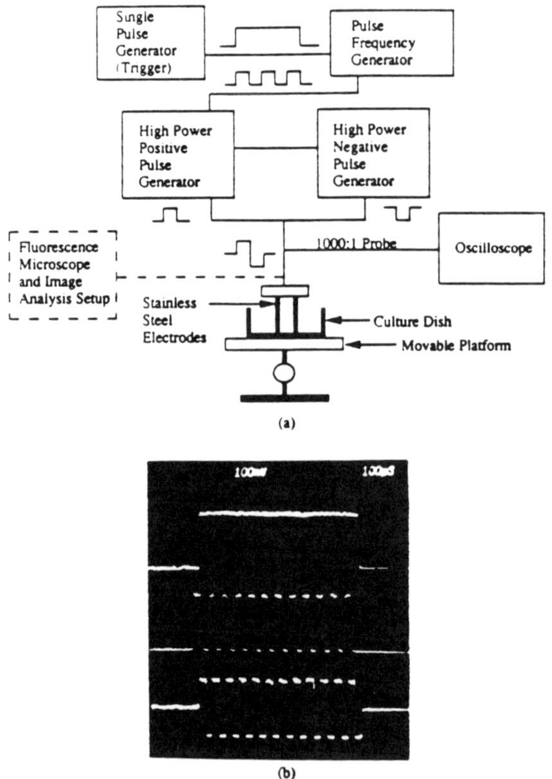

FIGURE II.   a) Block diagram of electropermeabilization apparatus, showing connections arranged to produce bipolar pulses. Connections for the other waveforms are explained in the text. The dotted block represents attachments of the instrument to a fluorescence microscope and image analysis setup with a separate electrode chamber. b) Oscilloscope traces of the different waveforms used measured with a 1000:1 probe across a 500 Ohm load resistor. Abscissa is 100 µs/division and ordinate is 100 mV/division.

The electric pulses were produced by serially gating several square wave generators. Duration of the bipolar square wave burst is controlled by a single square pulse generator (HP-8011A) of variable pulse width whose output gates the pulse frequency generator (Wavetek 183). The rising edge of each incoming single square pulse from the Wavetek triggers the high power positive pulse generator (Cober 606P). Amplitude and duration of the pulse from the Cober is independently adjustable from 0 to +2.5kV, and 50ns to 10ms, respectively.The falling edge of a synchronous 10V square pulse from the Cober is then allowed to trigger the second high power pulse generator (Cober 606P) with similar adjustable output (but with amplitude of 0 to -2.5kV). Combination of these pulses produces the desired bipolar waveform of peak-to-peak amplitude of up to 5kV. The maximum delay in the trigger mechanism between the two high power generators is <0.5µs. The magnitude of the electric pulses were measured across the electrodes with a 1000:1 probe (Tektronix P6015) and monitored on a storage oscilloscope (Tektronix 7704A). Single square pulses of either polarity were produced by disabling the pulse frequency generator and one of the high power pulse generators. Similarly, unipolar oscillating pulses could be obtained by disabling any one of the high power generators depending on the polarity desired. For transfection experiments two 50 mm long stainless steel electrodes (separation 1 or 2 mm) were used.

Fluorescence imaging. For single cell experiments an electrode chamber sandwiched between two microscope slides (electrode separation 0.75mm) with a capacity to hold 25µl of cell suspension was constructed. The chamber was mounted on a ZEISS (ICM405) inverted low light fluorescence microscope. The increase in fluorescence intensity (monitored at 610nm, excited at 520nm) resulting from the binding of ethidium bromide to cytosolic DNA and RNA was used as indicator of electropermeabilization. Real time images of these events were acquired with an image intensifier (Videoscope International, KS-1381) attached to a CCD camera (COHU, 4815) and recorded on a videotape (Sony, VO-5600). The time resolution of the videoscope recordings was 33ms. The recorded images were later transferred to a digital disk recorder (Panasonic, TQ2028F) and analyzed with a digital image processor (Recognition concepts, 55/48Q) using the RTIPS software library from TAU Inc.(Los Gatos, CA). Some of the video images in the time series were then displayed on a color monitor and photographed by transferring to a freeze frame recorder (Polaroid Corp.).

Cells and media. NIH 3T3 cells were grown to ≈50-70% confluency in Dulbecco's modified Eagle's medium (DMEM) in the presence of 10% calf serum. The average radii of the cells was 15 μm. The pulsing buffer used consisted of 250 mM sucrose, 10 mM phosphate, 1 mM $MgCl_2$, pH 7.2.

RESULTS AND DISCUSSION

Figure III summarizes results of our electro-permeabilization experiments using ethidium bromide as an indicator. The series of frames on the left column of the figure show a single cell subjected to a unipolar square wave pulse (frequency 250kHz; field strength 1.1kV/cm) for a total duration of 0.400ms. This field strength is close to the critical membrane breakdown potential (≈0.8-1 kV/cm). The data clearly show only the hemisphere of the cell membrane facing the positive electrode is permeabilized and with single square pulses of the same polarity and comparable amplitude, qualitatively similar results were obtained (data not shown). When the perturbing electric field is incrementally increased, often by ≈100-300 V/cm, the entry of the fluorescent probe at a much lower quantity is observed at the membrane site facing the negative electrode. These asymmetrical pore formations can be accounted for (at least in the low electric field regime) with the hypothesis that a negative resting membrane potential V_i interacts with the induced potential V_m. The equilibrium electric field in the membrane due to V_i is always directed inwards while that due to the external electric field is (+)outside, (−)inside on the hemisphere facing the negative electrode and (+)inside, (−)outside on the opposite hemisphere. This arrangement makes the membrane site facing the positive electrode to be first susceptible to electric breakdown. To test the self consistency of this hypothesis, we determined, from a number of cells of comparable size, the minimum electric field

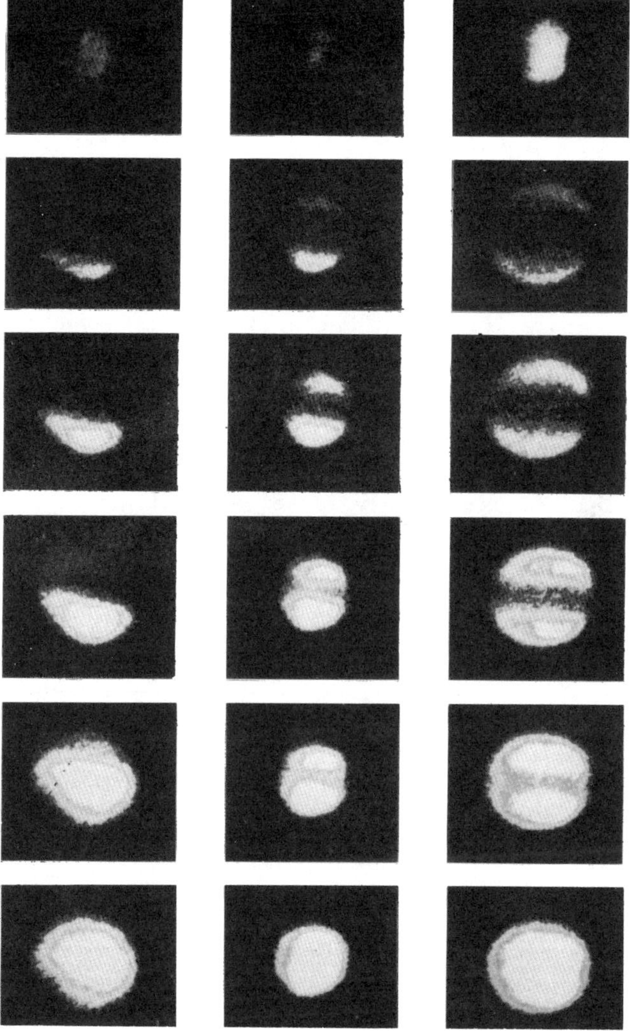

FIGURE III. Waveform-dependent electropermeabilization of NIH 3T3 cells monitored by enhanced fluorescence due to ethidium bromide binding to intracellular DNAs and RNAs.The positive electrode is at the bottom and the negative electrode at the top of each frame.The first row for each waveform shows the pre-pulse image. In all subsequent frames it has been subtracted. Time increases from the top to the bottom frame. Pulse duration was 0.400ms and frequency was 250kHz when applied. (*Left*) Unipolar square wave; field strength: 1.10 kV/cm. Time series of frames: 0.528, 1.508, 2.739, 6.303, 8.745 seconds. (*Middle*) Single square pulse; field strength: 4.95 kV/cm. Time series of frames: 0.297, 0.825, 1.419, 2.013, 8.943 seconds. (*Right*) Bipolar square wave; field strength: 2.25 kV/cm (peak-to-peak). Time series of frames: 0.132, 0.396, 0.891, 1.848, 3.267 seconds.

required to permeabilize each hemisphere of the cell membrane. This data in conjunction with eq.(2) was used to estimate the resting membrane potential and was found to range from -80 to -210 mV. Considering the heterogeneity in the cell population (e.g. cell age, membrane composition, etc.) the values obtained for the resting potential are reasonable. In terms of molecular transport, it should be pointed out that the short pulse durations used in all our experiments (msec time scale) exclude electroosmosis as a possible molecular transport mechanism for the events observed (which are in the msec to sec time scale). Furthermore, the rate of ethidium bromide uptake by the cells was found to strongly depend on the concentration of ethidium bromide, suggesting diffusion as a likely mechanism.When the electric field is excessively large, significant permeabilization is observed on both hemispheres facing the electrodes. The permeabilization, however, remains asymmetric as shown in the middle column of Figure III where a single square pulse of 4.95 kV/cm and 400µs duration was used. It is

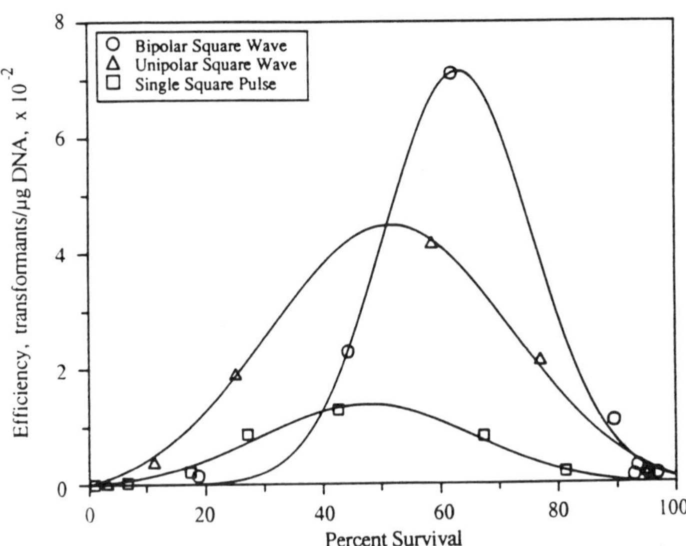

FIGURE IV. Relationship between transfection efficiency and percent survival. The data were obtained with 60 kHz (when applied), 0.400 msec pulse duration and field strength varying from 0.4 to 4 kV/cm. Initially the number of cells present are ≈1.1 x 10^5 with 10 mg/ml of plasmid added.

possible, under these circumstances, that polarization of the positively charged probe and hence its accumulation at the positive electrode side of the cell surface (and depletion at the negative side) may account for the observed asymmetry. On the right column of frames in Figure III, we show, unlike those observed with unipolar pulses, the bipolar waveform induces symmetrical permeabilization of the cell membrane. Because both sides of the membrane is permeabilized at less than lethal electric fields, we have found that bipolar waveforms offer a better efficiency in genetic transfection experiments (Tekle et al., 1991). Figure IV, shows that under optimal conditions for each waveform used, the bipolar waveform is at least 1.7- and 5.5-fold more efficient than unipolar square wave, and single square pulse methods, respectively.

ACKNOWLEDGEMENTS

We thank Drs Kenneth Spring, Blair Bowers, Sue Goo Rhee, and Ha Kun Kim for the help during the course of this work. E.Tekle is a recipient of the National Research Council (NRC) research associateship.

REFERENCES.

Chang DC (1989): Cell poration and cell fusion using an oscillating electric field. _Biophys. J._ 56: 641-652.

Chang DC and Reese TS (1990): Changes in membrane structure induced by electroporation as revealed by rapid-freezing electron microscopy. _Biophys. J._ 58: 1-12.

Chu G, Hayakawa H and Berg P (1987): Electroporation for the efficient transfection of mammalian cells with DNA. _Nucl. Acids Res._ 15: 1311-1326.

Dimitrov DS and Sowers AE (1990): Membrane electroporation-fast molecular exchange by electroosmosis. _Biochem. Biophys. Acta_ 1022: 381-392.

Mehrle W, Zimmermann U and Hampp R (1985): Evidence for asymmetrical uptake of fluorescent dyes through electropermeabilized membranes of Avena mesophyll protoplasts. _FEBS Lett._ 185: 89-94.

Mehrle W, Hampp R and Zimmermann U (1989): Electric field induced membrane permeabilization. Spatial orientation and kinetics of solute efflux in freely suspended and dielectrophoretically aligned plant mesophyll protoplasts. _Biochim. Biophys. Acta_ 978: 267-275.

Neumann E and Rosenheck K (1973): Potential difference across
    vesicular membranes. <u>J. Membr. Biol.</u> 14: 194-196.

Neumann E (1989): in <u>Electroporation and Electrofusion in
    Cell Biology.</u> Neumann E, Sowers AE and Jordan CA, eds.
    Plenum Press, New York. pp. 61-82.

Potter H, Weir L and Leder P (1984): Enhancer-dependent
    expression of human κ immunoglobulin genes introduced
    into mouse pre-B lymphocytes by electroporation. <u>Proc.
    Natl. Acad. Sci. USA</u> 81: 7161-7165.

Schwan HP (1983): in <u>Biological Effects and Dosimetry of
    Nonionizing Radiation.</u> Grandolfo M, Michaelson SM
    and Rindi A, eds. Plenum Press, New York. pp. 213-231.

Sowers AE (1988): Fusion events and nonfusion contents mixing
    events induced in erythrocyte ghosts by an electric
    pulse. <u>Biophys. J.</u> 54: 619-626.

Tatsuka M, Orita S, Yagi T and Kakunaga T (1988): An improved
    method of electroporation for introducing biologically
    active foreign genes into cultured mammalian cells. <u>Exp.
    Cell Res.</u> 178: 154-162.

Tekle E, Astumian RD and Chock PB (1990): Electro-
    permeabilization of cell membranes: effect of the
    resting membrane potential. <u>Biochem. Biophys. Res.
    Commun.</u> 172: 282-287.

Tekle E, Astumian RD and Chock PB (1991): Electroporation by
    using bipolar oscillating electric field: An improved
    method for DNA transfection of NIH 3T3 cells. <u>Proc.
    Natl. Acad. Sci. USA</u> 88: 4230-4234.

Yorifuji T, Tsuruta S and Mikawa H (1989): The effect of cell
    synchronization on the efficiency of stable gene
    transfer by electroporation. <u>FEBS Lett.</u> 245: 201-203.

MEMBRANE ELECTROCONFORMATIONAL CHANGES: PROGRESS IN THEORETICAL MODELLING OF ELECTROPORATION AND OF PROTEIN PROTRUSION ALTERATION

James C. Weaver
Harvard-MIT Division of Health Sciences and Technology
Massachusetts Institute of Technology
Cambridge, MA 02139

INTRODUCTION

Unambiguous, non-thermal effects of electric fields on biological cells have been reported by many investigators for moderate and strong fields (10 to 10^4 volt/cm; shorter exposure times for larger fields). This includes electroporation, which is believed due to transient and long lived pores in cell membranes. In order to develop theoretical models for such phenomena, we note that electrical interactions with cells are generally expected because of the heterogeneity of cells with respect to two basic parameters, electrical conductivity and electrical permittivity. This heterogeneity is particularly striking for the membrane/aqueous electrolyte interfaces, and leads to the general possibility of membrane forces. Possible responses to such forces include overall cell deformation, membrane indentations and membrane perforations (pores). Differential forces on the membrane and on membrane proteins are also possible, and include the possibility of altering protein protrusion from the membrane. To alter living cells, such changes must alter biochemical processes, e.g. transmembrane chemical fluxes. Here we describe progress in developing a quantitative theory of electroporation phenomena, and also a new possibility: changes in protrusion of membrane proteins. In the case of electroporation we emphasize predictions of measurable quantities, viz. the transmembrane voltage, U(t), and molecular transport, N_s, which is the total number of molecules of a given size and charge which cross a membrane due to a particular pulse. Modelling of both artificial planar bilayer membranes and cell membranes is possible. The artificial membranes are much simpler, and have therefore been the focus of initial theories. In the case of macromolecule protrusion, changes in protrusion could lead to alteration of binding site accessibility, and thereby provide a "signaling" possibility. Finally, by using "signal-to-(background + noise) ratio" (S/B+N) criteria, thresholds for electric field effects can be estimated. Electroporation has a large (S/B+N), while protein protrusion changes and other electroconformational change phenomena may involve much smaller values. By using the approximate criterion (S/B + N) $\approx$ 1, the smallest exposure for which an effect is expected can be estimated.

BACKGROUND

Many bioelectromagnetic effects have been reported, and clearly occur for strong and moderate fields, but there is considerable controversy relating to weak fields. Our interest is in the mechanisms of primary interaction, particularly for electric fields. Dramatic but non-thermal effects occur if strong electric fields are applied to cells or artificial planar bilayer membranes for short times (viz. electroporation), and clear effects occur at moderate fields (e.g. modulation of NaK-ATPase activity). In contrast, relatively few well-accepted experimental findings exist for weak fields. In all of these cases, however, changes in the geometry of membranes, membrane enzymes or membrane/protein complexes appear to provide a possible interaction mechanism, because changes in the transmembrane voltage change the net force on these structures.

METHODS

Biological systems are exceedingly complex and incompletely understood. In contrast, electrical fields are believed to be very well understood for the magnitudes and frequencies used in experiments. In order to combine the biological system (often an isolated cell) with physics, it is an accepted procedure to construct physical models of cellular subsystems, so that a problem becomes tractable. Here the cellular subsystem consists of the cell membrane and molecules associated with the membrane.

Table 1: Sequence of Events for Cell Alteration

EVENT	ATTRIBUTE
1st	Exposure of cell to field
2nd	Detection event; "signal" (S) exceeds "background" (B) + "noise" (N)
3rd	Alteration of a biochemical pathway
4th	Amplification via biochemistry
5th	Altered cell function achieved

An interaction with signal-to-(background + noise) ratio greater than one, i.e. $(S/B+N) > 1$, does not by itself imply that a cell is altered. Instead, as indicated above, the biochemistry must be altered. This clearly can occur by altered influxes and effluxes if pores occur, and plausibly occurs if membrane proteins alter their protrusion in an amount which changes binding site accessibility.

Table 2: Symbols Used

SYMBOL	MEANING
d	Thickness of lipid portion of a cell membrane
$\vec{E}_e(t)$	Electric field in the electrolyte
$\vec{E}_m(t)$	Transmembrane electric field ($E_m = U(t)/d$)
t	Time (differentiation denoted by a "dot", e.g. $\dot{N}_s(t)$)
ε_e	Electrical permittivity of the electrolyte
ε_m	Electrical permittivity of the membrane
σ_e	Electrical conductivity of electrolyte
σ_m	Electrical conductivity of a membrane
$U(t)$	Transmembrane voltage

General Basis for Electroconformational Changes

Changes in electrostatic free energy and in mechanical energy are general, and together determine the equilibrium configuration for a particular applied electric field. Cells are quite heterogeneous with respect to two basic electrical properties, conductivity (σ), and permeativity (ε). This heterogeneity and the overall membrane geometry lead to interfacial polarization, and an associated "amplification" of an applied electric field, $E_e(t)$. Thus, it is well known that that bigger electric field changes occur in $E_m(t)$, the electric field within the membrane. For this reason, membrane phenomena have long been regarded as important candidates for electric field interactions. Here we emphasize phenomena associated with deformation of the cell membrane. This includes deformation of the entire cell membrane, local distortions of the membrane (e.g. "dimples"), perforations of the membrane ("pores"), and relative protrusion changes of membrane proteins with respect to the membrane.

Table 3: Membrane Electroconformational Changes

TYPE OF CHANGE	SIGNIFICANCE
Overall membrane compression	Rupture and REB* incorrectly described[5]
Lipid-domain fluctuations	Alternative to transient pores[6]
Local distortions	Precursors to hydrophilic pores[7, 8]
Hydrophobic pore	Precursor to hydrophilic pore[9]
Hydrophilic pore	Present electroporation theory[9-15]
Composite pore	Candidate metastable pore
Membrane protein changes	Coupling to membrane proteins[16]
Protein protrusion changes	Signaling mechanism

*Reversible electrical breakdown.

Coupling to the cell in the form of modified biochemistry is essential if the external electric field is to actually alter the cell. Thus, possibilities such as overall cell deformation and localized membrane indentations are not compelling unless a deformation/biochemistry link is identified. In contrast, alteration of (1) membrane enzymes, (2) transmembrane fluxes through pores, or (3) protein binding site accessibility, can couple to one or more biochemical pathways.

Electroporation

Already important, electroporation has many potential applications in research, biotechnology and medicine because it is a physical method for for altering cell membranes. From the perspective of physics, dramatic electrical and mechanical phenomena occur within the membrane. These are initially driven by strong electric field interactions, but later probably by mechanical forces. From the perspective of biology, many different kinds of molecules experience greatly enhanced transport into or out of cells, and in many cases a large fraction of the cells survive. With both perspectives in mind, an understanding of electroporation should combine descriptions of membrane electrical behavior and of molecular transport.

Quantitative predictions are particularly important, as direct comparisons with experiments are then possible. Experimental measurements which are calibrated in terms of the transmembrane voltage and the numbers of transported ions and molecules are valuable for two reasons. First, a more fundamental and stringent test of theoretical models can be made, through direct comparisons. Second, knowledge of the amount of transport should aid in the development of applications of electroporation.

Table 4: Physical Ingredients of a Recent Model[15]

INGREDIENT	SIGNIFICANCE
$\dot{N}_c$	Pore creation rate (at radius r_{min})
$\dot{N}_d$	Pore destruction rate (at radius r_{min})
$n(r,t)$	Pore probability density function (pore sizes)
$\dot{n}(r,t)$	Dynamic behavior of pore population
σ_p	Born energy-modified conductivity within pores
$H(r,r_i)$	Hindered transport through pores
$U_p \leq U$	Local U reduced by spreading resistance
$R_E + R_N$	Membrane charging through external resistances
Circuit equation	Coupling to external environment
η_B	Transport of charged molecules[17]

One should be optimistic about achieving an understanding, because electroporation caused by short pulses universally occurs at a transmembrane voltage of about 1,000 mV for many different types of cell and artificial membranes. This independence of membrane composition suggests that a theory based on a small number of key parameters should succeed. Indeed, we find that a model based on transient pores can quantitatively describe several key aspects of the striking electrical behavior in both planar bilayer membranes and in cell membranes, can predict contribution of electrical drift to molecular transport, and should be capable of extension to include the contributions of convection and diffusion to molecular transport.

"Electroporation" refers to large, non-thermal cell membrane effects that occur for strong electric fields and very short exposure times.[2, 4] Although electroporation is incompletely understood, local membrane electroconformational changes (pore formation and pore evolution) are believed responsible for the dramatic electrical behavior and molecular transport across the cell membranes. Our approach involves the use of a "modular model" in which physical ingredients for uninterrupted membrane, pores, ionic and molecular transport through pores, and the external environment can be combined, while still enabling the ingredients to be changed (Table 4).

Changes in Protein Protrusion

A large number of important proteins reside in the cell membrane.[18] From a continuum electrostatics viewpoint, a protein residing in a cell membrane is characterized by significant spatial variations in local values of permittivity, conductivity and charge density. This gives rise to the general possibility of shifting a protein's equilibrium position normal to the membrane as U is changed. At least one experiment is consistent with this basic notion, which indicates the binding site accessibility changes as chemical means are used to alter U.[19]

Heterogeneity of Cell Membrane Electric Field Phenomena

There are several grounds for anticipating subpopulation responses of cells to electric fields. Unlike physical systems comprised of N_n nominally identical entities, cell populations are well known to exhibit a distribution of biological attributes. In addition to this "biological variability", one should expect heterogeneity in the response of cells to external electric fields for physical reasons.

Table 5: Sources of Heterogeneous Electric Field Responses

Cell size variation within a cell population
  ΔU (change in transmembrane voltage) caused by ΔE_e scales with size

Cell shape and orientation
  ΔU varies over cell membrane

Variation of local electric field experienced by cell
  Perturbation by nearby suspended cells (*in vitro*)
  Perturbation by surrounding tissue (*in vivo*)
  Caused by experimental apparatus[*]

Variation in cell membrane properties
  Membrane composition
  Membrane protein distribution
  Underlying structures (e.g. cytoskeleton)

Microscopic interaction processes fundamentally stochastic
  Local changes probabilistic (electrical energy and kT)

[*]Non-uniform electric fields, e.g. due to non-parallelplane electrodes, proximity of cell to an insulating microscope slide, or to the surface of culture flask.

The above factors are quite general, so that significant variability should be expected for any cellular response in which depends on U(t). As a consequence, experimental methods (e.g. image analysis) which are responsive to only one or a small number of individual cells may be misleading. Likewise, methods which respond to the combined effects of many cells (e.g. turbidity, radiolabeled substrate incorporation) will be responsive to the average behavior, and will not reveal this variability. As a general conclusion, the above potential sources of variability strongly suggest that subpopulations with different response should be expected.

Thresholds

How can the threshold for electric field effects be understood? The minimum exposure condition can be estimated by treating the membrane system as a physical detector of the field.[20-22] True detection is generally regarded as possible only if the detection process satisfies criteria involving signal size and the size of all of the confounding events. Here two classes of confounding effects are distinguished: (1) "background" electromagnetic fields of physical and biological origin, and (2) fundamental "noise" due to inescapable fluctuations associated with chemical and physical processes (Table 4). Physical background (e.g. mechanical vibration, ambient light variations) can in principle be lowered to arbitrarily small levels. Biological background, however, is fundamental to *in vivo* conditions[23, 24], and fundamental types of noise are inescapable.[25]

With this in mind, we assume that detection can only occur if the signal energy significantly exceeds the fluctuation energy associated with background and noise. This leads to a familiar, but somewhat arbitrary criterion

$$\frac{(S + B + N)}{(B + N)} = \frac{(Energy)_{S+B+N}}{(Energy)_{B+N}} = \frac{\int_0^{(\Delta t)} P(f,\Delta f)_{S+B+N}\, dt}{\int_0^{(\Delta t)} P(f,\Delta f)_{B+N}} \geq 2 \tag{1}$$

where the subscripts S, B and N denote "Signal", "Background" and "Noise". This condition is equivalent to requiring the signal-to-(background + noise) ratio to be one, i.e. $S/(B + N) \geq 1$ if the detection threshold is reached, and provides a basis for thresholds estimates. The signal-to-(background + noise) ratio is very large for electroporation conditions, and becomes progressively smaller as moderate and then weak electric field exposures are considered. In the case of electroporation, $S/(B + N) >> 1$, and the phenomena unambiguously occur. For other phenomena, $S/(B + N)$ may approach one. This may be particularly important for protein protrusion changes, and for other interaction mechanisms that are candidates for "weak" bioelectromagnetic field effects.

Table 6: Types of noise and background for electric fields.

TYPE OF NOISE	ATTRIBUTE
Noise	Intrinsic confounding fluctuations
Thermal noise	Fundamental; well established theory[26]
1/f noise	Universal; incomplete theory[27-29]
Shot noise	Fundamental; discrete, random entities[30]
	Molecules - governs cumulative flux
	Ions - governs total current flow
Background ("physical origin")	Confounding physical phenomena
	Electrical fields
	Magnetic fields
	Mechanical coupling (vibrations, etc.)
	Temperature variations
Background ("biological origin")	Physiological electrical environment[23, 24]
	Cardiac electric fields
	Neural electric fields
	Myoelectric electric fields
	Electrokinetic electric fields

RESULTS AND DISCUSSION

Most of our work to date involves an attempt to create quantitative models for electroporation. In order to illustrate types of predictions, results of a recent version of the electroporation theory are shown. The membrane protein protrusion problem has only recently been identified, and initial work begun.[31]

Electroporation

We have developed a theoretical model which successfully predicts several key aspects of electroporation. Examples of quantitative predictions are given in Figs. 1 - 4, for the case of an artificial planar bilayer membrane. Here the transmembrane voltage, U(t), is predicted. The underlying pore population (not directly measurable) is also predicted, via a pore probability density function, $n(r,t)$, which specifies the instantaneous number of pores with radii between r and $r + \Delta r$ as $n(r,t)\Delta r$. Using this type of result, the predicted electrical drift transport of a small charged molecule across a planar membrane is also given (Fig. 5).

This approach has also allowed estimates of the molecular transport through transient pores. Because of the high concentration of small, highly mobile ions (viz. Na^+ and Cl^-), the electrical behavior is assumed to be controlled entirely by ionic conduction through the pore population. The computation of molecular transport then uses the same, rapidly changing and heterogeneous pore population. An example, the estimate for the electrical drift-based transport of propidium iodide, is shown in Fig. 6.

Table 7: Measurable electroporation behavior quantitatively described.

BEHAVIOR	COMMENT
Planar Bilayer Membrane - transient pores only[15]	
$\quad$ U(t) during simple charging	Smallest pulses
$\quad$ U(t) during rupture	Moderate pulses
$\quad$ U(t) during incomplete REB	Large pulses
$\quad$ U(t) during REB	Still larger pulses
$\quad$ G(t) (conductance)	Above cases
$\quad$ $\dot{N}_s(t)$ (molecular flux)	Charged molecules[17, 32]
Cell Membrane - transient pores only[33, 34]	
$\quad$ U(t) during simple charging/discharging	Smallest pulses
$\quad$ U(t) during REB	Larger pulses
$\quad$ G(t) (conductance)	Above cases
$\quad$ $\dot{N}_s(t)$ (molecular flux)	Charged molecules

Pore Energetics

Starting with the work of Chizmadzhev's group[9], we and others have assumed that electroporation is due to hydrophilic pores created by transitions from hydrophobic pores, and that the hydrophilic pore free energy includes both a mechanical term[35, 36] and electrical term.[9-11] This is emphasized by writing the free energy in the form $\Delta E(r, U_p)$. Here U_p is the local transmembrane voltage at the site of the pore. We use $\Delta E = \Delta E_M + \Delta E_E + \Delta E_0$ where ΔE_M is the mechanical contribution, ΔE_E is the electrical contribution, and ΔE_0 is an arbitrary constant. The mechanical contribution is

$$\Delta E_M = 2\pi\gamma r - \pi\Gamma r^2 \tag{2}$$

Here Γ is the surface energy density of the membrane-water interface, and γ is the pore edge energy. For a planar membrane, rupture was originally attributed to appearance of one or more pores with radius $r > r_c$, such that $r_c = \gamma / \Gamma$.[35, 36] Subsequently Sugar and Neumann pointed out that although rupture could occur for planar membranes, it is not possible for the closed membranes of vesicles and cells.[37] These authors also emphasized the use of a minimum radius, $r_{min} = 1.0$ nm, based on headgroup packing constraints, which we have adopted.

484

The physical model for the pore energy is extremely simple. The pore is regarded as a cylindrical hole with electrical capacitance and with both an internal resistance and external "spreading resistance", such that the electrical contribution, ΔE_E, is[12, 13]

$$\Delta E_E = -\frac{\pi(\varepsilon_w - \varepsilon_l)U^2}{h^2} \int_{r_{min}}^{r} \alpha^2\, r\, dr\,, \quad \text{with} \quad \alpha(r) = \left[1 + \frac{\pi r \kappa_p(r)}{2h\kappa_e}\right]^{-1} \tag{3}$$

The quantity $\alpha(r)$ includes the "voltage divider" effect associated with the spreading resistance external to a pore and the internal pore resistance.

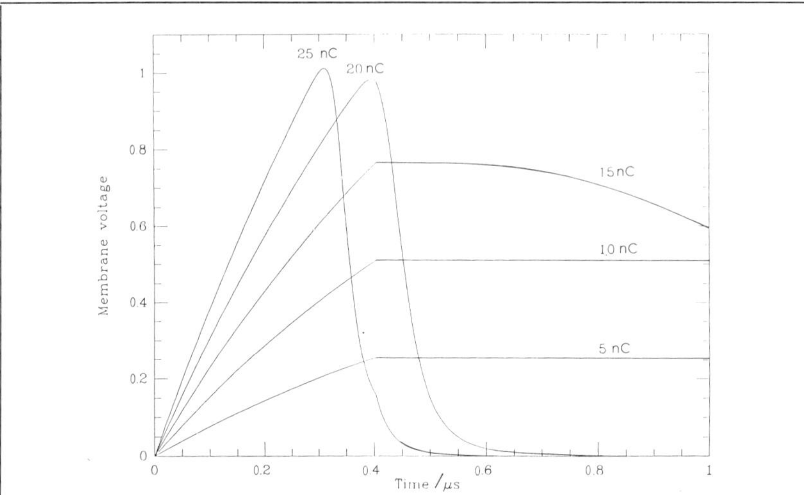

Figure 1: Prediction of U(t) for a short time scale for a planar bilayer membrane exposed to a single very short (0.4µsecond) pulse.[15] Some key features of reversible electrical breakdown (REB) are predicted by the model. Incomplete reversible electrical breakdown is also predicted, wherein the membrane discharge is incomplete, i.e. U(t) does not reach zero after the pulse. Each curve is labeled by the corresponding value of the injected charge Q. The curves for Q = 25 nanocoulombs and 20 nanocoulombs exhibit REB.

Pore Creation and Destruction

In order to form a pore, an energy barrier, Λ, must be overcome. We assume that Λ depends on U, and neglect any contribution of permanent dipoles.[38] Λ therefore depends on U^2, with an assumed form $\Lambda = \delta_c - a\,U^2$. ($\delta_c$ and a are constants). An absolute formation rate estimate[10] is

$$\dot{N}_c = \nu \exp\left[-\frac{\delta_c - aU^2}{kT}\right] \quad \text{where } \nu \text{ is an attempt rate} \tag{4}$$

Pore destruction is assumed to occur only at r_{min} (except for rupture events in planar membranes), and to be independent of U. This gives

$$\dot{N}_d = \chi\, n(r_{min})\, \exp\left[-\frac{\delta_d}{kT}\right] \qquad \text{where } \chi = \text{constant (dimensions of velocity)} \qquad (5)$$

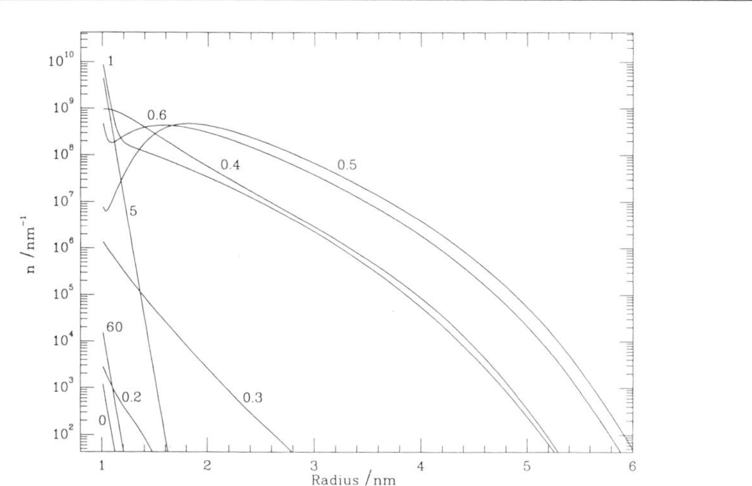

Figure 2: Prediction of n(r,t), the pore density function which describes the underlying heterogeneous population of transient pores.[15] Each curve is labeled by the corresponding value of the injected charge Q. For Q = 25 and 20 nanocoulombs, REB occurs, and the total number or pores, N, increases to about 10^8 in less than 0.5 μs and then decays exponentially with a time constant of ≈ 4.5 μs. For Q = 15 nanocoulombs, N increases rapidly to about 10^5 and remains almost constant for about 4 μs before the exponential decrease. For Q = 10 nanocoulombs, N increases to about 2 x 10^3 in about 5 μ and remains almost constant for about 30 μs before the decay phase. The membrane in this case ruptures. For Q = 5 nanocoulombs, N increases to about 40 in 80 μs. N will return to its initial value as the membrane discharges with a time constant of about 2 sec.

Pore Population Evolution Consistent with U(t)

The development of a heterogeneous pore population is described by a flux, J_p, in pore radius space

$$J_p = -D_p\left[\frac{\partial n}{\partial r} + \frac{n}{kT}\frac{\partial \Delta E}{\partial r}\right] \qquad (6)$$

D_p is the effective diffusion constant for the pore radius[39], which is independent of radius.[13] This approach leads to Smoluchowski's equation[13, 40]

$$\frac{\partial n}{\partial t} = D_p\left[\frac{\partial^2 n}{\partial r^2} + \frac{\partial}{\partial r}\left[\frac{n}{kT}\frac{\partial \Delta E}{\partial r}\right]\right], \qquad (7)$$

which contains both a diffusive term and a driving term related to the derivative of ΔE with respect to r. The effect of U(t) arises through ΔE, which is also a function of the local transmembrane voltage, U_p, and therefore U.

The assumption that pores are created and destroyed at r_{min} leads to the boundary condition for the flux of pores in radius space

$$J_p = \dot{N}_c - \dot{N}_d \qquad \text{at } r = r_{min} \tag{8}$$

Here $\dot{N}_c$ and $\dot{N}_d$ are the pore creation and destruction rates described previously.

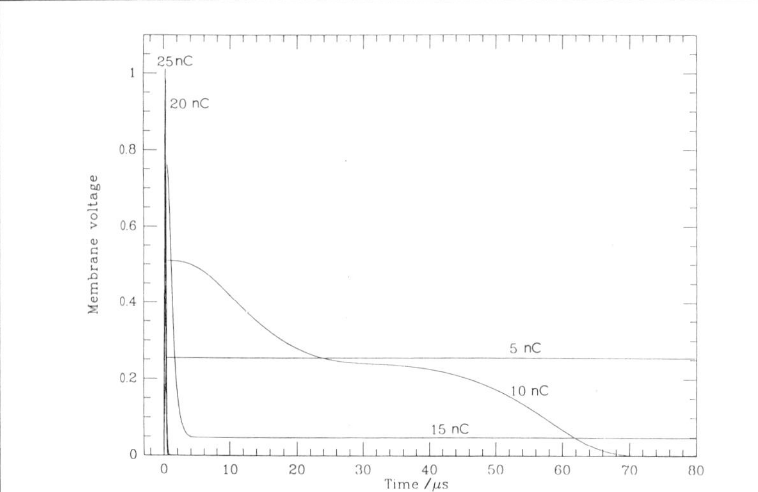

Figure 3: Prediction of U(t) over a longer time scale (0 to 80μseconds). Both rupture and simple charging of a planar bilayer membrane are shown.[15] Note that the characteristic sigmoidal behavior of U(t) is predicted by this version of the model, but that the time scale is somewhat shorter than found in experiments.[41] Each curve is labeled by the injected charge Q. The curves for Q = 25 and 20 nanocoulombs are the spikes at t = 0. The curve for Q = 15 nanocoulombs exhibits REB at t = 2 μs, but the membrane recovered before it could discharge completely. The curve for Q = 10 nanocoulombs exhibits rupture, but the curve for Q = 5 nanocoulombs shows that the membrane conductance did not increase enough to discharge the membrane.

Pore Conduction

Transmembrane movement of small hydrated ions (e.g. Na^+ and Cl^-) is assumed to occur by passage through pores with $r > r_i$. Entry of an ion into a small pore also requires consideration of the "Born energy"[42] and of hindered motion.[43] The conductivity, σ_e, of bulk electrolyte can be reasonably described by

$$\sigma_e = \sum_i (z_i e)^2 \eta_i c_i \,, \tag{9}$$

where c_i is an ion's concentration, η_i its mobility and $e = 1.6 \times 10^{-19}$ coulomb. z_i is the charge of the i-th type of ion, and the sum is over the different ion types of the electrolyte. However, the conductivity within a pore, σ_p, is reduced.

$$\sigma_p = \sum_i (z_i e)^2 \eta_i c_i H_i \exp\left[\frac{\mu^0_i}{kT}\right] \qquad (10)$$

μ^0_i is the standard chemical potential of an ion of type i inside the pore, and H_i, is a steric hindrance factor. We use the estimate[12]

$$\mu^0_i = \frac{(z_i e)^2}{4\pi\epsilon_l r} P\left[\frac{\epsilon_l}{\epsilon_w}\right] \qquad (11)$$

Here ϵ_l and ϵ_w are the dielectric constants of the lipid and the water. The function P has a maximum value of 0.25[42] The Born energy exclusion is particularly important for species with large $z^2_{s,effective}$.

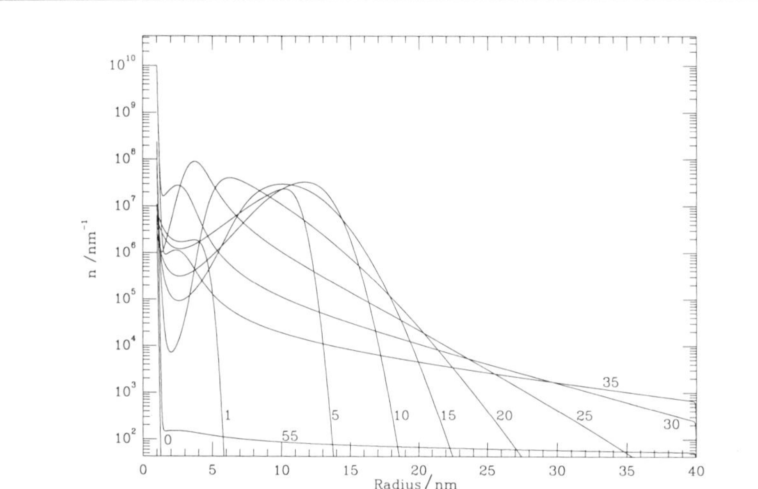

Figure 4: Prediction of the pore population distributions for the conditions of Fig. 3, as described by the pore density function, n(r,t).[15] Each curve is labeled by the corresponding value of the injected charge Q. For Q = 25 and 20 nanocoulombs, cases for which REB occurs, N increases to about 10^8 in less than 0.5 μs and then decays exponentially with a time constant of ≈ 4.5 μs. For Q = 15 nanocoulombs, N increases rapidly to about 10^5 and remains almost constant for about 4 μs before the exponential decrease. For Q = 10 nanocoulombs, N increases to about 2 x 10^3 in about 5 μ and remains almost constant for about 30 μs before the decay phase. The membrane in this case ruptures. For Q = 5 nanocoulombs, N increases to about 40 in 80 μs. N will return to its initial value as the membrane discharges with a time constant of about 2 s.

Hindered Transport within a Pore

The motion of a particle within a tube of almost the same size is sterically hindered. The reduction in transport can be approximated by the factor[43]

$$H(r,r_i) = \left[1 - \left[\frac{r_i}{r}\right]^2\right]\left[1 - 2.1\left[\frac{r_i}{r}\right] + 2.09\left[\frac{r_i}{r}\right]^3 - 0.95\left[\frac{r_i}{r}\right]^5\right] \qquad (12)$$

Note that $H \rightarrow 0$ as $r_i \rightarrow r$.

Reduction in Local Transmembrane Voltage at a Pore

An inhomogeneous electric field exists near the pore entrances, and the associated potential drop in the electrolyte is often termed the "spreading resistance", $R_s(r) \approx 1/2\sigma_e r$.[44] This results in a reduced transmembrane voltage, U_p, at the site of pore. The internal resistance of a pore is estimated by using the reduced conductivity, σ_p, so that $R_p = h/\pi r^2 \sigma_p$. These two resistances act in series, creating a "voltage divider" such that $U_p \leq U$. This reduces the electrical driving force. The current through a pore is $I_p = U/R_s(r) + R_p(r)$, and the total current I through the membrane is therefore

$$I = U \int_{r_{min}}^{\infty} \left[\frac{n(r, t)}{R_s(r) + R_p(r)} \right] dr \text{ and } G(t) = \int_{r_{min}}^{\infty} \left[\frac{n(r,t)}{R_s(r) + R_p(r)} \right] dr \tag{13}$$

as $r_{min} > r_i = r_+ = r_-$. Here $G(t)$ is the strongly time varying membrane conductance. which is used self-consistently with an equation for the external circuit (below) in order to describe $U(t)$.

Pore Population/External Environment Interaction

In order to obtain a complete model, it is important to include the current pathways within the membrane's environment, but external to the membrane. Thus, we have already explicitly included the "spreading resistance" in the external electrolyte near each pore. In addition, we include the essential features of the pulse generator, electrodes and bulk electrolyte. The membrane/experimental apparatus is represented by a parallel combination of membrane resistance, $R(t) = 1/G(t)$, and constant membrane capacitance C, and a single fixed resistance, R_N, which represents the charging pathway resistance and pulse generator output impedance. The capacitance changes negligibly during electroporation, because only a small fraction of the membrane is occupied by aqueous pores. However, $R(t)$, rapidly changes by orders of magnitude. In a charge pulse experiment a current pulse of amplitude I_i passes through R_N to create a voltage pulse, V_0. For $0 < t < t_{pulse}$ current flows into and/or across the membrane, and at $t = t_{pulse}$ the pulse is terminated by disconnecting the generator. The membrane can then discharge only through the membrane (planar membrane; not so for a cell). The appropriate circuit differential equations are

$$C \frac{d\,U}{dt} = \begin{cases} \dfrac{I_p R_N}{R_E + R_N} - U \left[G(t) + \dfrac{1}{R_E + R_N} \right] & \text{if } t < t_{pulse} \\[4mm] -\dfrac{U}{R} & \text{if } t > t_{pulse} \end{cases} \tag{14}$$

subject to the initial condition $U(0) = 0$. A self-consistent numerical solution to these equations is used in order to provide an electroporation simulation.

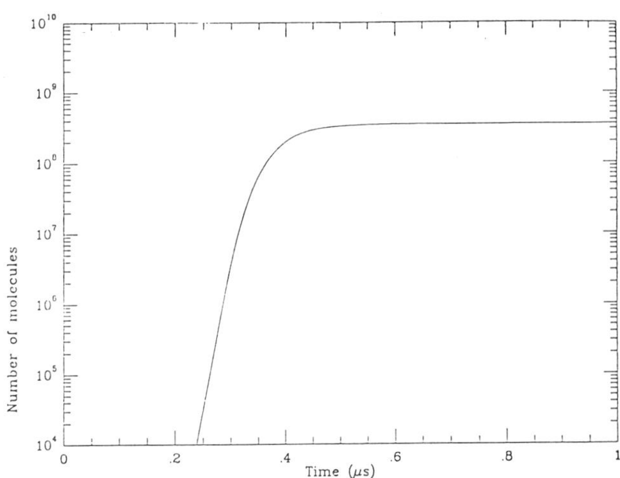

Figure 5: Predicted molecular transport across a planar membrane due to electrical drift, in the case of the membrane undergoing reversible electrical breakdown (REB). This estimate predicts the number of propidium iodide (PI; about 660 dalton) molecules which are transported by electrical drift through the transient pore population associated with electroporation. The PI molecule was treated as a sphere with radius r_s = 0.6nm and charge z_s = +2. As in Fig. 1, a 0.4μsecond square 20 ncoulomb pulse was used. Only transient aqueous pores are used in this version of the model. Future versions should include metastable pores, and also estimates of the contributions of diffusion and convection. The planar bilayer membrane (Figs. 3 and 4) is far larger than a cell membrane (Fig. 6), but can generally not tolerate a long pulse, because the planar membrane can rupture due to electroporation (Fig. 4), whereas as a cell membrane cannot.[37] As shown by a comparison with Fig. 6, the model predicts that the ability to use longer pulses with cells results in almost comparable molecular transport, in spite of the much larger area of the artificial planar membrane (Here $A_{cell}/A_{planar} \approx 10^{-10}$ m^2/1.45 x 10^{-6} m^2 = 7 x 10^{-5}. For the 20 ncoulomb pulse (voltage pulse of 2.5 volt), the transmembrane voltage is elevated only briefly (Fig. 3), so that in comparison to the cell, a combination of a significantly elevated transmembrane voltage and the evolution of a large pore population does not occur.

An initial approach to cell membrane electroporation has been explored by considering a "cubic cell", in which two opposing planar membranes represent the portion of cell membrane experiencing most of electroporation. This allows quantitative predictions of U(t), for all pulse sizes, and of molecular transport due to electrical drift.

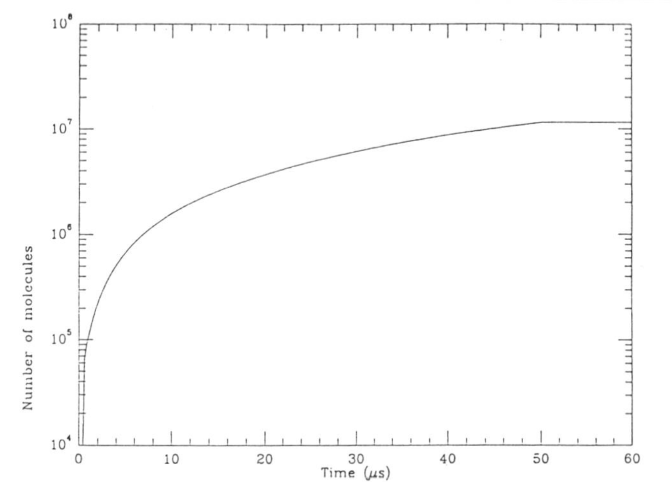

Time (μs)

Figure 6: Predicted molecular transport into a cubic cell due to electrical drift through pores, in the case of the membrane undergoing reversible electrical breakdown (REB). This estimate predicts the electrical drift contribution to the transport of propidium iodide (PI) molecules through the transient pore population in one of the two membranes of a cubic cell.[33] The curve's slope changes abruptly at $t = 50\mu$second, where N_s stops increasing because U(t) falls rapidly at the pulse end, which causes both the electrical drift driving force and the pore population to rapidly decrease. PI is highly charged, and therefore reasonably expected to have a significant electrical drift contribution. The extracellular bulk concentration is 80μM, the value used one experiment with intact yeast and propidium iodide.[45] For the purpose of estimating the Born energy (Table 4), the effective squared charge was taken to be $z_{s,eff}^2 = z_s^{1.5}$. Of order 10^7 molecules are predicted to be taken up. Other simulations (not shown) predict that far fewer molecules are transported as the size and charge of the molecule is increased.

Membrane Protein Protrusion Changes

An electrostatic continuum model approach is being pursued in which the protein, membrane and electrolyte are represented by assigning permittivities and charge densities to relatively simple geometries. For example, the protein is represented by a cylinder with hemispherical ends, with a portion of the cylinder assigned $\varepsilon = 2$ (a hydrophobic region) and the remainder $\varepsilon = 10$ (hydrophobic regions). Charge is separately specified by, for example, requiring the hemispherical end to contain a volume or surface charge density. Here we use embedded spherical charges. The membrane is initially represented by a flat sheet ($\varepsilon = 2$), and the electrolyte by $\varepsilon_e = 70$. The double layer at charged interfaces can be approximated by providing an excess charge density which decreases exponentially away from the interface. The free energy can then be computed under fixed charge, fixed potential or hybrid conditions.[46] Generally, as the transmembrane voltage, U(t), changes, the degree of protein protrusion will vary because of changes in the associated electrostatic forces.

In preliminary work we have used simple geometry (Fig. 7). As in the important, early calculations of Parsegian[42], our initial computation does not include the double layer.

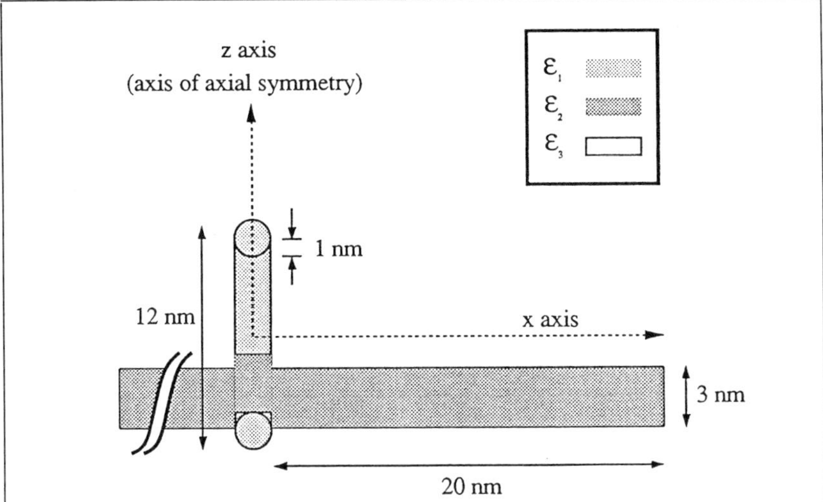

Figure 7: Typical configuration used for a protruding protein in a cell membrane. This simple case consists of a 10 nm long cylinder with 1 nm radius spherical ends (overall length 12 nm). Each spherical end is treated as a sphere with uniform volume charge density, such that the total charge on the top sphere is q_1 and on the bottom sphere is q_2. A 3 nm long hydrophobic region of the protein has $\varepsilon = 2$, and the remaining protein region has $\varepsilon = 10$. The membrane ($\varepsilon_m = 2$) has thickness 3 nm, and the surrounding electrolyte is assigned $\varepsilon_e = 70$.

The preliminary results of Fig. 8 are based on a membrane with a fixed transmembrane voltage of U = 100 mV.[47] By using several different positions, including some with the protein just outside the membrane, the depth of the potential well which confines the protein to the membrane was estimated to be $\Delta E_{binding} \approx 30$ (right peak) to 60kT (left peak). Additional computations are needed to refine the location and magnitude of these peaks. This preliminary result is consistent with a moderately long lifetime, i.e. the protein should not readily dissociate from the membrane because of thermal fluctuations. Addition of more charge groups should further increase $\Delta E_{binding}$. This preliminary estimate is made in the same spirit as Parsegian's original computation[42], in that double layers are not included. Other preliminary results[48] give a $\Delta E_{binding}$ of order 5kT for the protein/membrane association if $q_1 = 5e$ and $q_2 = 0$, which is consistent with the qualitative expectation that a macromolecule with charge on only one side of the membrane will not be well retained within the membrane. Overall, an electrostatic free energy approach should allow estimates of the well depth, $\Delta E_{binding}$, for different charge distributions within the simplified protein. By using a quadratic fit, the undamped frequency of oscillation can be estimated. Significant damping is expected, however, which can be estimated by treating the protein as a spheroid in a viscous medium.

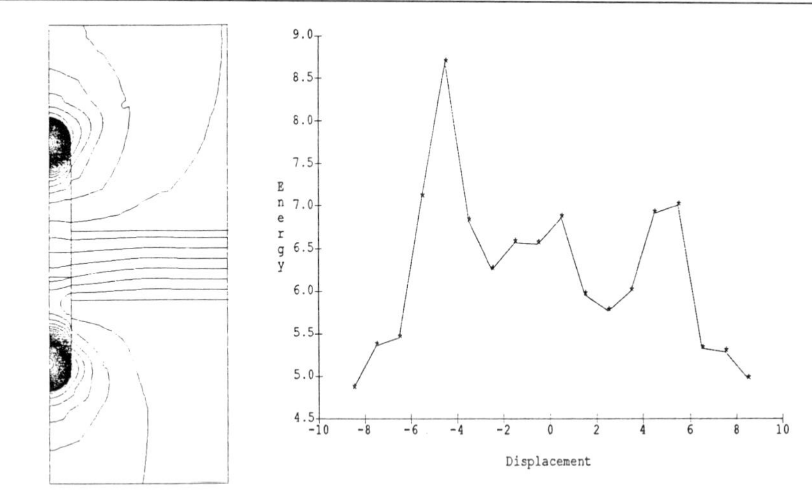

Figure 8: Preliminary results for the protein/membrane configuration of Fig. 7. Left: Equipotentials computed for the configuration of Fig. 7 with $q_1 = q_2 = 5e = 8.0 \times 10^{-19}$ coul. A fixed transmembrane voltage of magnitude 100mV exists, with a fixed surface charge density on the membrane such that the top surface has the same sign charge as the two charged spheres of the protein. Right: Electrostatic energy (in units of 10^{-19} joule) as a function of displacement (in units of nm) normal to the membrane.

It is emphasized that the above protein protrusion results were only recently obtained, and are therefore preliminary. Nevertheless, they allow a description of the approach to this aspect of membrane/protein electroconformation changes.

NOTATION AND VALUES OF PARAMETERS

The mathematical symbols used in the present version of the electroporation theory are given on the next page.

Table 8: Electroporation Theory Parameters and Values

PARAMETER		VALUE
a	coefficient of U^2 in Λ	1.9×10^{-20} farad
A_m	membrane area[*]	1.45×10^{-6} m^2
C	capacitance of membrane	9.61×10^{-9} farad
D_p	diffusion constant in pore radius space[*]	5×10^{-14} m^2 sec^{-1}
dt	time step size (in units of $D_p/(\Delta r)^2$)	0.5
h	membrane thickness[*]	2.8 nm
kT	Boltzmann's constant times temperature	4×10^{-21} joule
R_E	series resistance of electrolyte, electrodes, and wires	30 ohm
R_N	internal resistance of current source (pulse generator)	50 ohm
r_{max}	large pore initial size	40 nm
r_{min}	minimum pore radius[*]	1 nm
r_+	radius of positive ions	0.2 nm
r_-	radius of negative ions	0.2 nm
t_{pulse}	pulse length	0.4 μs
z_+	charge of positive ions (in units of proton charge)	+1
z_-	charge of negative ions (in units of proton charge)	-1
γ	pore edge energy density[*]	2×10^{-11} joule m^{-1}
Γ	membrane surface energy[*]	1×10^{-3} joule m^{-2}
δ_c	pore creation energy barrier[*]	2.04×10^{-19} joule
δ_d	pore destruction energy barrier[*]	2.04×10^{-19} joule
Δr	numerical grid spacing of simulation	0.0195 nm
ε_l	dielectric constant of lipid[*]	2.1 ε_0
ε_w	dielectric constant of water	80 ε_0
ν	pore creation rate prefactor[*]	10^{28} s^{-1}
κ_e	conductivity of bulk solution	0.98 ohm m^{-1}
χ	pore destruction rate prefactor[*]	5×10^{16} m sec^{-1}

[*]Parameters characterizing the membrane[15]

ACKNOWLEDGEMENTS

J. Zahn, M. Zahn, D. J. Mycue, R. S. Booker, 3rd, J. G. Bliss, A. Barnett and R. D. Astumian have contributed to many important discussions and/or the computations. This work supported partially by an equipment grant for a computer workstation from Stadtwerke Düsseldorf, AG, Düsseldorf, Germany, and has also benefited from the use of an electrostatics software package which was provided by Ansoft, Inc. (Philadelphia).

494

REFERENCES

1. Blank, M. and E. Findl (Eds), Mechanistic Approaches to Interactions of Electromagnetic Fields with Living Systems, Plenum, New York, 1987.

2. Neumann, E., A. Sowers, and C. Jordan (Eds), Electroporation and Electrofusion in Cell Biology, Plenum, New York, 1989.

3. Allen, J., S. F. Cleary, and F. M. (Eds) Hawkridge, Charge and Field Effects in Biosystems - 2, Plenum, 1990.

4. Chang, D. C., B. M. Chassy, J. A. Saunders, and A. E. Sowers (Eds), Handbook of Electroporation and Electrofusion, Academic Press, 1991. (in press)

5. Crowley, J. M., "Electrical Breakdown of Bimolecular Lipid Membranes as an Electromechanical Instability," Biophys. J., vol. 13, pp. 711 - 724, 1973.

6. Cruzeiro-Hansson, L. and O. G. Mouritsen, "Passive Ion Permeability of Lipid Membranes Modelled Via Lipid-Domain Interfacial Area," Biochim. Biophys. Acta, vol. 944, pp. 63 - 72, 1988.

7. Chizmadzhev, Yu. A., V. B. Arakelyan, and V. F. Pastushenko, "Electric Breakdown of Bilayer Membranes: III. Analysis of Possible Mechanisms of Defect Origin," Bioelectrochem. Bioenerget., vol. 6, pp. 63 - 70, 1979.

8. Bach, D. and I. R. Miller, "Glyceryl Monooleate Black Lipid Membranes Obtained from Squalene Solutions," Biophys. J., vol. 29, pp. 183 - 188, 1980.

9. Abidor, I. G., V. B. Arakelyan, L. V. Chernomordik, Yu. A. Chizmadzhev, V. F. Pastushenko, and M. R. Tarasevich, "Electric Breakdown of Bilayer Membranes: I. The Main Experimental Facts and Their Qualitative Discussion," Bioelectrochem. Bioenerget., vol. 6, pp. 37 - 52, 1979.

10. Weaver, J. C. and R. A. Mintzer, "Decreased Bilayer Stability Due to Transmembrane Potentials," Phys. Lett., vol. 86A, pp. 57 - 59, 1981.

11. Sugar, I. P., "The Effects of External Fields on the Structure of Lipid Bilayers," J. Physiol. Paris, vol. 77, pp. 1035 - 1042, 1981.

12. Pastushenko, V. F. and Yu. A. Chizmadzhev, "Stabilization of Conducting Pores in BLM by Electric Current," Gen. Physiol. Biophys., vol. 1, pp. 43 - 52, 1982.

13. Powell, K. T. and J. C. Weaver, "Transient Aqueous Pores in Bilayer Membranes: A Statistical Theory," Bioelectrochem. Bioelectroenerg., vol. 15, pp. 211 - 227, 1986.

14. Glaser, R. W., S. L. Leikin, L. V. Chernomordik, V. F. Pastushenko, and A. I. Sokirko, "Reversible Electrical Breakdown of Lipid Bilayers: Formation and Evolution of Pores," Biochim. Biophys. Acta, vol. 940, pp. 275 - 287, 1988.

15. Barnett, A. and J. C. Weaver, "Electroporation: A Unified, Quantitative Theory of Reversible Electrical Breakdown and Rupture," Bioelectrochem. and Bioenerg., vol. 25, pp. 163 - 182, 1991.

16. Astumian, R. D., P. B. Chock, T. Y. Tsong, and H. V. Westerhoff, "Effects of Oscillations and Energy-Driven Fluctuations on the Dynamics of Enzyme Catalysis and Free-Energy Transduction," Phys. Rev. A, vol. 39, pp. 6416 - 6435, 1989.

17. Barnett, A. and J. C. Weaver, Molecular Transport Across a Planar Bilayer Membrane Due to Electroporation Caused by a Short Pulse: Prediction of the Electrical Drift Contribution. (in preparation)

18. Gennis, R. B., Biomembranes: Molecular Structure and Function, Springer-Verlag, New York, 1989.

19. Balázs, M., J. Matkó, J. Szöllösi, L. Mátyus, M. J. Fulwyler, and S. Damjanovich, "Accessibility of Cell Surface Thiols in Human Lymphocytes is Altered by Ionophores or OKT-3 Antibody," Biochem. Biophys. Res. Comm., vol. 140, pp. 999 - 1006, 1986.

20. Weaver, J. C., "Transient Aqueous Pores: A Mechanism for Coupling Electric Fields to Bilayer and Cell Membranes," in Mechanistic Approaches to Interactions of Electromagnetic Fields with Living Systems, ed. M. Blank and E. Findl, pp. 249 - 270, Plenum, New York, 1987.

21. Weaver, J. C. and R. D. Astumian, "The Response of Cells to Very Weak Electric Fields: The Thermal Noise Limit," Science, vol. 247, pp. 459 - 462, 1990.

22. Astumian, R. D., B. Robertson, and J. C. Weaver, The Thermal Noise Limit for the Detection of Weak Electric Fields by Living Biological Cells: Basic Issues. (in preparation)

23. Nuccitelli, R., "Endogenous Ionic Currents and Electric Fields in Living Systems," Bioelectromagnetics, 1992. (submitted)

24. Wachtel, H., "Endogenous Bioelectric Fields Derived from Nerve, Muscle and Bone Activity," Bioelectromagnetics, 1992. (submitted)

25. Fishman, H. M. and H. R. Leuchtag, "Electrical Noise in Physics and Biology," Current Topics in Membranes and Transport, vol. 37, pp. 3 - 35, 1990.

26. Nyquist, H., "Thermal Agitation of Electric Charge in Conductors," Phys. Rev., vol. 29, p. 614, 1927.

27. Keshner, M. S., "1/f Noise," Proc. IEEE, vol. 70, pp. 212 - 218, 1982.

28. Barnes, F. S. and M. Seyed-Madani, "Some Possible Limits on the Minimum Electrical Signals of Biological Significance," in Mechanistic Approaches to Interactions of Electromagnetic Fields with Living Systems, ed. M. Blank and E. Findl (Eds), pp. 339 - 347, Plenum, New York, 1987.

29. Weissman, M. B., "1/f Noise and Other Slow, Non-Exponential Kinetics in Condensed Matter," Rev. Mod. Phys., vol. 60, pp. 537 - 571, 1988.

30. Villars, F. M. H. and G. B. Benedek, *Physics with Illustrations from Medicine and Biology, Volume 2: Statistical Physics,* pp. 3-1, Addison-Wesley, 1974.

31. Weaver, J. C., "Theory of Electroporation," in *Biomembrane Electrochemistry,* ed. M. Blank and I. Vodyanoy, American Chemical Society. (submitted)

32. Weaver, J. C. and A. Barnett, "Progress Towards A Theoretical Model of Electroporation Mechanism: Membrane Electrical Behavior and Molecular Transport," in *Handbook of Electroporation and Electrofusion,* Academic Press, 1991. D. C. Chang, B. M. Chassy, J. A. Saunders and A. E. Sowers (Eds) (in press)

33. Weaver, J. C., A. Barnett, and J. G. Bliss, An Approximate Theoretical Model for Isolated Cell Electroporation: Quantitative Estimates of Electrical Behavior. (in preparation)

34. Barnett, A. and J. C. Weaver, Electrical Behavior and Molecular Transport During Cell Membrane Electroporation: Quantitative Estimates. (in preparation)

35. Litster, J. D., "Stability of Lipid Bilayers and Red Blood Cell Membranes," Phys. Lett., vol. 53A, pp. 193 - 194, 1975.

36. Taupin, C., M. Dvolaitzky, and C. Sauterey, "Osmotic Pressure Induced Pores in Phospholipid Vesicles," Biochmistry, vol. 14, pp. 4771 - 4775, 1975.

37. Sugar, I. P. and E. Neumann, "Stochastic Model for Electric Field-Induced Membrane Pores: Electroporation," Biophys. Chemistry, vol. 19, pp. 211 - 225, 1984.

38. Neumann, E., M. Schaefer-Ridder, Y. Wang, and P. H. Hofschneider, "Gene Transfer into Mouse Lyoma Cells by Electroporation in High Electric Fields," EMBO J., vol. 1, pp. 841 - 845, 1982.

39. Deryagin, B. V. and Yu. V. Gutop, "Theory of the Breakdown (Rupture) of Free Films," Kolloidn. Zh., vol. 24, pp. 370 - 374, 1962.

40. Pastushenko, V. F., Yu. A. Chizmadzhev, and V. B. Arakelyan, "Electric Breakdown of Bilayer Membranes: II. Calculation of the Membrane Lifetime in the Steady-State Diffusion Approximation," Bioelectrochem. Bioenerget., vol. 6, pp. 53 - 62, 1979.

41. Benz, R. and U. Zimmermann, "Pulse-Length Dependence of the Electrical Breakdown in Lipid Bilayer Membranes," Biochim. Biophys. Acta, vol. 597, pp. 637 - 642, 1980.

42. Parsegian, V. A., "Energy of an Ion Crossing a Low Dielectric Membrane: Solutions to Four Relevant Electrostatic Problems," Nature, vol. 221, pp. 844 - 846, 1969.

43. Renkin, E. M., "Filtration, Diffusion and Molecular Sieving Through Porous Cellulose Membranes," J. Gen. Physiol., vol. 38, pp. 225 - 243 , 1954.

44. Newman, J., "Resistance for Flow of Current to a Disk," J. Electrochem. Soc., vol. 113, pp. 501 - 502, 1966.

45. Bartoletti, D. C., G. I. Harrison, and J. C. Weaver, "The Number of Molecules Taken Up by Electroporated Cells: Quantitative Determination," FEBS Letters, vol. 256, pp. 4 - 10, 1989.

46. Zahn, M., Electromagnetic Field Theory: A Problem Solving Approach, Wiley & Sons, New York, 1979.

47. Mycue, D. Q., J. D. Zahn, M. Zahn, and J. C. Weaver, 1991. (preliminary results)

48. Booker, R. S., Modeling Change in Macromolecule Protrusion from a Membrane, 1991. S. B. Thesis, Dept. of Electrical Engineering and Computer Science, MIT

ELECTROFUSION YIELD MODIFIED BY MEMBRANE-ACTIVE SUBSTANCES

Lei Zhang, Hermann Berg

ZIMET, Laboratory of Bioelectrochemistry, Jena, F.R.Germany

INTRODUCTION

Nowadays electrofusion of membranes of animal cells or
protoplasts of plants and microorganisms turns out to be
an effective tool in cell biology. Usually cells are
brought together tightly before or after a strong single
uni- or bipolar field pulse yielding electroporation and
subsequently fusion. For this close membrane contact several techniques are possible :
- in cell suspension of low conductivity "pearl chain"
  formation occurs by dielectrophoresis,
- in cell suspension of physiological salt concentration
  cell clusters are formed by centrifugation or in presence of high concentrations of polyethylene glycol
  $(PEG \geqslant 20\%)$

The electrofusion yield depends not only on the electric
parameters, the membrane composition and the electrolyte but
additionally also on membrane-active substances (Zhang
et al., 1991), (Chapel et al., 1986), namely in positive
or negative directions. A systematic study of these modifications is in the very beginning and explanations have
to take into account not only the interaction forces between lipids, channel proteins and other biopolymers
(Sowers, 1990) on the one hand and anionic, cationic
amphoteric and neutral substances on the other but also
the disturbances in the ion distribution (e.g. double
layer structure) at the outer membrane surface. Using
barley protoplasts as a membrane model we tested four
types of compounds - representing a broad range of molecular masses - dissolved in a medium of low conductivity.
Our aim is to find some rules for regulating electrofusion
by membrane-active substances.

EXPERIMENTAL AND METHODS

Protoplasts from barley (Hordeum vulgare) were produced in the usual way as described previously (Zhang et al., 1991) and resuspended in 0.55 M mannitol immediately before the experiments.

Electrofusion equipment (Zhang et al., 1991) consists of
- a pulse generator for rectangular pulses (0-100 V, 0-500 µs) in combination with a sine wave generator (1 MHz). Both are connected with :
- a meander chamber : 60 parallel flat electrodes separated by a distance of 200 µm from each other. Electrode material : sputtered NiCr/Al of 1.2 µm thickness, protected by a 0.035 µm SiO_2 layer,
- the electric field is inhomogeneous because of the difference in protoplast diameter (~20 µm) and thickness of the electrodes,
- microscopic observation device,

Substances: 4 groups of membrane-active substances are compiled in the table and in the legends of figures.

Anions	Cations	Ampholytes	Neutral Sub.

M_m	M_m	M_m	M_m
EDTA 372	Daunomycin 542	Serumal-bumin 70000	Dextran 9600
Dodecyl-sulfate (Na^+) 288	Netropsin 539	Trypto-phan 204	PVP 10000
Blue dextran 10900	Tetracaine 248	Pronase 28000	Tween-80 1300
	CTA 365		Triton-100 631
			Cyclodex-trin hy-drate 1300
			PEG 1000 4000 6000 40000

Table. Electrofusion yield responsed different groups of substances can be distinguished. All substances were solved in 0.5 M mannitol solution. (PVP: polyvinylpyrrolidone; CTA:N-cetyl-N,N,N-trimethylammonium bromide)

MEASUREMENTS (Zhang et al., 1991)

All measurements were performed at room temperature (19°C). The day-to-day variation of protoplast stability could not be entirely eliminated so data are relative to a control. The fusion yield was determined by counting the number of dielectrophoretic membrane contacts between protoplasts before and after pulsation, in absence or in presence of increasing concentrations of these agents. In the diagrams is indicated the relative fusion yield F_r:

$$F_r = \frac{F}{F_{contr.}} \quad \text{with} \quad F = \frac{N_b - N_a}{N_b}$$

where N_b = the number of contacts of protoplasts before the pulse, N_a = the number of contacts of protoplasts after the pulse.
Additionally the relative conductivity ($\mathscr{K}_r$) is shown: that means the concentration dependence of the conductivity (determined in $mScm^{-1}$) of substance divided by the conductivity of 0.01 M KCl standard : 1.231 $mScm^{-1}$ at 19°C.

RESULTS AND DISCUSSION

The negatively charged protoplasts react sensitively to osmotic differences in solution and to pulsation conditions by morphological responses, e.g. deformation, shrinking, flattening of attached membranes, even leakage and loss of chloroplasts as shown in (Zhang et al., 1991). Moreover a strong influence is the age of protoplast on fusion yield. Therefore in relation to the control data is inevitable for a systematic study. In total only 20 substances which can be divided into 4 groups, namely cations, anions, ampholytes and neutral substances were tested in our measurements.
Characteristic results of the concentration dependence of F_r are shown in figures 1, 2, 3 (concentration always as mg/ml for comparison with low and high molecular compounds). Depending on conductivity, detergent activity, molecular mass, three basic types of concentration dependence from fusion yield were found (compare also table):

A) convex shapes for organic cations,

B) concave shapes for organic anions,

C) isothermic shapes (exponential increase) for ampholytes (only at low concentration) and neutral substances.

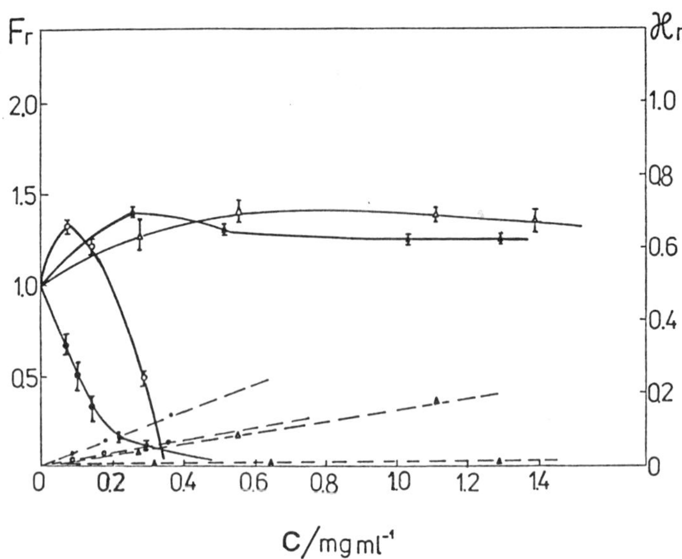

FIGURE 1. Influence of EDTA (●), CTA (o), Tween (▲) and
Serum albumin (Δ) concentration respectively on F_r and $\mathcal{H}_r$.
Pulse parameters :
1.4 kV/cm, 45 /us, 3 pulses (for EDTA and CTA)
1.5 kV/cm, 28 /us, 1 puls   (for Tween and Serum albumin).
CTA is in 10 times lower concentration and Serum albumin is
in 100 times higher concentration respectively shown in the
curves. F_r is shown as continued line; $\mathcal{H}_r$ is in dotted line.

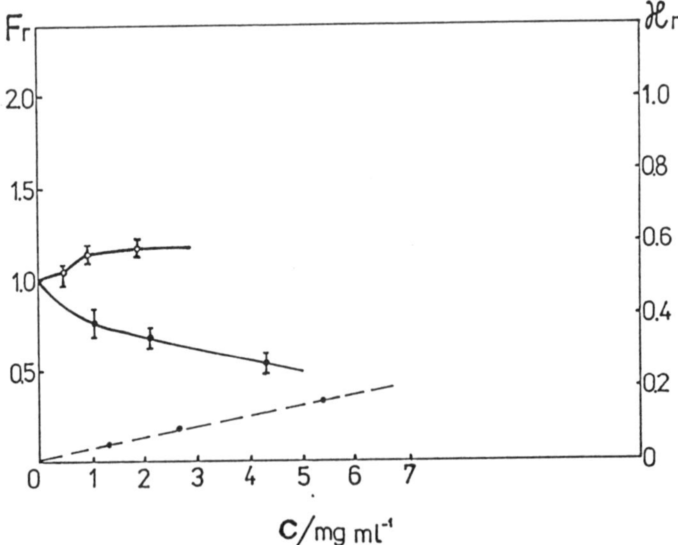

FIGURE 2. Influence of Dextran (o) and Blue dextran (●)
concentration respectively on F_r and $\mathcal{H}_r$.
Pulse parameters : 1.4 kV/cm, 45 /us, 3 pulses.

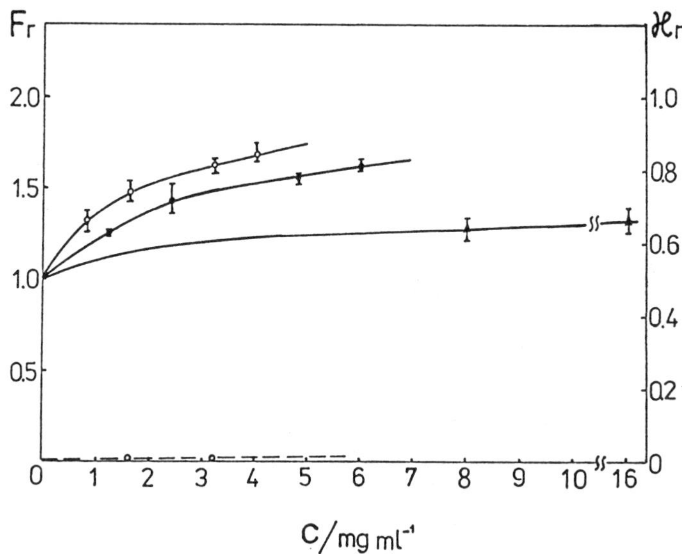

FIGURE 3. Influence of different molecular mass of PEG and its concentration respectively on F_r. PEG 4000 (o), PEG 6000 (●) and PEG 40000 (▲).
Pulse parameters: 1.5 kV/cm, 28 µs, 1 pulse.

The main reason for these 3 basic types is the strong influence of their conductivity on their membrane activity in such a way :
  - high conductivity by strong electrolytes dissipates the pulse energy causing always a decrease of fusion yield at higher concentration for cases A) and B).
  - Decreasing conductivity of organic cations (type A) may change convex curve optimally to such experimental of type C).
For the differences between type A) and B) at lower concentration are responsible for
  - the shielding of the negative surface charges of membranes by organic cations on the one hand
  - the increase of repulsion by organic anions on the other.
Only for type C) the effective pulse energy is the same in the whole concentration range of neutral substances.

Figure 1 shows typical examples, namely
  A) the convex curve of the cation CTA with a steep increase at low concentrations,
  B) the concave shape for the anion EDTA,
  C) the exponential increase for the ampholyte serum albumin and for the neutral compound Tween-80.

There are slight differences in case C) between ampholytes
and neutral compounds. Whereas the curve of ampholytes
decreases at high concentration with rising conductivity
of the solution; the dependence of neutral substances
reach a constant plateau (compare figures 2 and 3). This
strong influence of conductivity and repulsion can be
seen from the curves of dextran (fig. 2) contrary to its
anion blue dextran, with nearly the same molecular mass.
The influence of molecular mass is demonstrated for PEG
exhibiting isotherm like curves fig. 3. Here PEG (M_m=4000)
has the highest effectivity. An explanation for this M_m-
dependence is different on surface coverage, because
a short chain molecule have relatively higher percentage
of adsorption on the surface of the membrane than a long
chain, which tends to extend their loops into the solution. Such PEG effects occur at 100 times lower concentrations as in the case of the well known fusion experiments in classical genetics.

Nevertheless at lower concentration of PEG in our measurement can get higher fusion yield by given impulse energy;
on the other hand it is possible to decrease the field
strength in comparison with PEG-free solution in order
to get the same fusion yield. In such a way harmful effects on the viability of cells can be avoided by serum
albumin or dextran, too.
Moreover, neutral substances and - with reservations -
for ampholytes, too, their isothermic dependences turn
out to be novel indication for intensity of their interactions (adsorption, insertion) with membrane constit
uents.

Extension to other compounds, especially drugs for
animal cells, neurotransmitters and anaesthetics etc.
will prove our A, B, C rules.

REFERENCES

Zhang L, Fiedler U and Berg H (1991): Modification of
        Electrofusion of Barley Protoplasts by Membrane-
        active Substances. In: Bioelectrochem. Bioenerg.,
        26 (in press).
Chapel M, Montane M, Ranty B, Teissie J and Alibert G
        (1986): Viable Somatic Hybrids are Obtained by
        Direct Current Electrofusion of Chemically Aggre-
        gated Plant Protoplasts. In: FEBS Lett. 196: 79-85.
Sowers A (1990): Low Concentrations of Macromolecular
        Solutes Significantly Affect Electrofusion Yield
        in Erythrocyte Ghosts. In: Biochim. Biophys. Acta,
        1025: 247-281.